# Plant Systematics

# McGraw-Hill Series in Organismic Biology

**CONSULTING EDITORS**

Professor Melvin S. Fuller, *Department of Botany, University of Georgia, Athens*

Professor Paul Licht, *Department of Zoology, University of California, Berkeley*

**Gunderson:** *Mammalogy*
**Jones and Luchsinger:** *Plant Systematics*
**Kramer:** *Plant and Soil Water Relationships: A Modern Synthesis*
**Leopold and Kriedemann:** *Plant Growth and Development*
**Patten and Carlson:** *Foundations of Embryology*
**Price:** *Molecular Approaches to Plant Physiology*
**Ralph:** *Introductory Animal Physiology*
**Ross:** *Biology of the Fungi*
**Weichert and Presch:** *Elements of Chordate Anatomy*

# Plant Systematics

Samuel B. Jones, Jr., Ph.D.
*Professor of Botany*
*University of Georgia*

Arlene E. Luchsinger, M.S., A.M.L.S.
*Biological Sciences Bibliographer*
*University Library*
*University of Georgia*

**McGraw-Hill Book Company**

New York   St. Louis   San Francisco   Auckland   Bogotá   Düsseldorf
Johannesburg   London   Madrid   Mexico   Montreal   New Delhi
Panama   Paris   São Paulo   Singapore   Sydney   Tokyo   Toronto

**PLANT SYSTEMATICS**

This book was set in Times Roman by A Graphic Method Inc.
The editors were James E. Vastyan and Susan Gamer;
the cover was designed by Carla Bauer;
the production supervisor was Dominick Petrellese.
The drawings were done by Fine Line Illustrations, Inc.
Fairfield Graphics was printer and binder.

**Library of Congress Cataloging in Publication Data**

Library of Congress Cataloging in Publication Data

Jones, Samuel B date
    Plant systematics.

    (McGraw-Hill series in organismic biology)
    Bibliography: p.
    Includes index.
    1. Botany—Classification. I. Luchsinger,
Arlene E., joint author. II. Title.
QK95.J63     581'.01'41     78-14961
ISBN 0-07-032795-5

# Contents

# Preface

Like other fields of biology, systematics—or taxonomy—has a long history. However, modern taxonomy serves as a unifying force in biology because data are utilized from a multitude of disciplines, including biochemistry, electron microscopy, and paleontology, to develop classifications.

Recent developments in the field of plant systematics, coupled with progress in the disciplines which contribute information to classification, have created a need for a new textbook in systematic botany. The field of taxonomy requires thoughtful reevaluation and the communication of a comprehensive balanced approach designed for today's students. This book provides an up-to-date synthesis of an exciting and active field of botany.

*Plant Systematics* is designed to meet the needs of undergraduate students in a one-semester or, preferably, a two-semester or two-quarter course in plant taxonomy. The level of presentation is based on the assumption that the students will have had an introductory biology course. We hope that it will prove stimulating to students who may have had additional training.

The documentation of information by citations of the literature also makes this text suitable for beginning graduate students. The selected references at the end of each chapter provide help for the student who desires further information on a topic. The references are not a complete bibliography for the subject matter of each chapter; rather, they are intended to guide the student in pursuit of a subject of interest. Over half of the references provided are for relevant literature published during the 1970s. All bibliographic citations have been verified and are given in a form which will facilitate their location in academic libraries.

The subject matter is divided into fourteen chapters and three appendixes. Many of the chapters are independent, and their sequence of presentation may be varied.

Chapters 1 through 7 place primary emphasis upon the modern and dynamic application of academic and theoretical considerations to systematics.

Chapters 8 and 9 deal with curatorial routines and the practical procedures of collecting, preparing, and identifying specimens.

Reflecting its descriptive nature, the terminology of systematic botany is extensive. Chapter 10 furnishes ready references for definitions and illustrations of technical terms.

Chapter 11 gives information concerning the bibliographic aids necessary for taxonomic research.

Chapters 12, 13, and 14 provide descriptions and information for over 100 families of pteridophytes, gymnosperms, and angiosperms occuring in North America north of Mexico. The flowering plant families are arranged according to Dr. Arthur Cronquist's latest classification. Diagnostic illustrations of many of the families of flowering plants are included. Technical terms unique to the pteridophytes and gymnosperms are defined, and some are illustrated.

Appendix 1 is a glossary of important Greek and Latin terms; Appendix 2 provides annotated listings of over 250 references to aid in plant identification; Appendix 3 gives Cronquist's listing of classes, subclasses, orders and families of the angiosperms; and Appendix 4 gives the classification of the angiosperms according to Thorne's system.

## ACKNOWLEDGMENTS

In undertaking the task of preparing a textbook, it is necessary to use information from the publications of many writers. The authors are grateful to all these persons. In addition, the gratitude of the authors is expressed to the many persons who have offered encouragement, assistance, and cooperation in the preparation of the manuscript including: W. W. Anderson, J. G. Bruce, W. C. Burnett, R. H. Chapman, M. T. Clegg, D. J. Crawford, T. Croat, A. Cronquist, W. H. Duncan, A. M. Evans, D. Isely, M. S. Fuller, F. C. Galle, A. M. Harvill, F. W. Gould, J. W. Hardin, L. K. Kirkman, A. Q. Howard, S. C. Keeley, D. B. Lellinger, R. McVaugh, J. R. Massey, R. Ornduff, M. Reines, T. Stuessy, A. Takhtajan, R. D. Thomas, B. L. Turner, L. Urbatsch, D. B. Walker, and J. J. White. Many others kindly provided illustrations, and they are cited individually in each caption.

Dr. T. M. Barkley, Kansas State University, criticized the entire manuscript. Without his invaluable editorial comments and encouragement, our task would have been much more difficult. Valerie A. Jones prepared illustrations for Chapters 10 and 14. Katie E. Bishop, Carla J. Ingram, and Hazel K. Henderson typed the manuscript. Nancy C. Coile made numerous editorial suggestions and proofread the manuscript; Alathea J. Fortner helped with the proofreading. The authors received assistance from the Department of Botany of the University of Georgia and the University of Georgia Libraries. Gratitude is extended to our spouses, Carleen A. Jones and Dale Luchsinger, and our children for their moral support and patient understanding throughout this project.

Comments and suggestions for improvements of *Plant Systematics* are welcomed from both students and teachers of plant systematics.

*Samuel B. Jones, Jr.*
*Arlene E. Luchsinger*

# Chapter 1

# Introduction to Systematic Botany

Systematic botany is an old and important division of the study of plants. Primeval humans were interested in the food and medicinal plants which grew in their environment. They recognized hundreds of different plants. This recognition of plants for food, medicine, fiber, and so on established the first taxonomic groups of plants. As humans developed a tribal life and language, written languages made it possible for observations to be collected and recorded. This accumulated knowledge could then be passed from one generation to the next.

The relatively simple activity of recognition of useful and harmful plants marked the beginning of systematic botany. It has developed into a highly complex science concerned with arranging plants into natural groupings and the naming of these groups. Systematic botany acquaints us with the fascinating differences among the species of plants. A plant's name is the key that unlocks the door to its total biology. Scientists who study the biochemistry, ecology, physiology, and practical uses of plants need to use plant names. However, systematic botany is not meant only for scientists; it may be used by other people with varied interests and training.

Systematic botany deals with the identification and naming of plants and with their arrangement into groups of closely related organisms, such as genera

1

or families. It includes all activities that are part of the effort to organize and record the diversity of plants. In this book, we are concerned only with the systematics of the spore-producing vascular plants, which include the ferns, horsetails, and the like, and the seed-bearing plants, i.e., gymnosperms and flowering plants. However, many of the same principles may be applied to the algae, fungi, and mosses.

## DEFINITIONS

*Systematic botany* is the broad field concerned with the study of the diversity of plants and their identification, naming, classification, and evolution. Since there is no agreement or etymological basis for the distinctions between *systematics* and *taxonomy* (Radford et al., 1974; Ross, 1974), these two terms are used interchangeably in this text. *Classification* is the arrangement of plants into groups having common characteristics. Then these groups are arranged into a system. Similar species of flowering plants are placed into a *genus*; similar genera are grouped into *families*; families with common features are arranged into *orders*, orders into *subclasses*, subclasses into a *class*, and classes into *divisions*. Classification results in the placing of plants into a hierarchy of ranks or categories such as species, genera, families, and so on. In addition to expressing relationships based upon common features, classification serves as a filing and information retrieval system.

   *Identification* is the recognition of certain characters of flower, fruit, leaf, or stem and the application of a name to a plant with those particular characters. Recognition occurs when the specimen under consideration is similar to a previously known plant. If comparison of the specimen with similar species reveals that it differs from them, it may be named as a new species.

   A *taxon* (plural, *taxa*) is a convenient term which is applied to any taxonomic group at any rank: e.g., species, genus, or family. *Nomenclature* is the orderly application of names to taxa in accordance with the International Code of Botanical Nomenclature. These rules provide procedures for selection of the correct name or formulation of a new name. *Description* is the listing of features or characteristics of a plant. Each plant name is accompanied by a description. The term *flora* refers either to the plants growing in a particular geographic area or to a systematic listing or description of those plants.

## OBJECTIVES

Plant taxonomy has four objectives: (1) to inventory the world's flora; (2) to provide a method for identification and communication; (3) to produce a coherent and universal system of classification; and (4) to demonstrate the evolutionary implications of plant diversity. Although the inventory of the earth's flora is largely complete in the northern temperate zone, much remains to be done in the tropics. As documented knowledge of the earth's plant life accumulates, the data must be systematized and made available.

Once names are provided, there must be a method to identify the taxon as being similar to another known entity. This is accomplished by descriptions, keys, catalogs, illustrations, manuals, and other publications that aid in the identification of specimens. Technology is presently available for the use of computers to aid in identification (Morse, 1971; Shetler, 1974). In the future, computer programs may be routinely used in plant identification. Computers are now being used to search the current literature for information published about plants. For example, searches may be made in the computer data bases of several indexing and abstracting services using the generic name *Vernonia*. If *Vernonia* appears in the title or as one of the key words, the computer will make available the bibliographic information needed to locate the article it has indexed. Much time is saved using computer searches instead of searching the literature indexes manually.

In the century since Darwin, biologists have been able to demonstrate that evolutionary lineages occur among taxa. The evolutionary development or lineage of a taxon is its *phylogeny*. Phylogenetic relationships exist which show that diverse species and the lineages they represent did not arise spontaneously, but may have had a common ancestral form. Modern classification attempts to use all available information about plants to develop a phylogeny. Since the fossil record is often fragmentary, especially for flowering plants, information must be gathered from a wide variety of sources to formulate a hypothesis regarding phylogeny.

## PHASES OF PLANT SYSTEMATICS

Since one of the primary concerns of plant taxonomy is to provide an inventory of plants of the world, one phase is *exploration and discovery*. This is an activity of even preliterate peoples. However, the most active recorded period began with the voyages of discovery by Western Europeans in the 1400s. The peak of botanical exploration, in terms of plants described, was the late 1800s, but exploration and discovery continue today, especially in the tropics. Plant material collected on these early expeditions to the far corners of the earth was sent to botanists in Europe for naming and description. In the late 1700s and early 1800s, botanists were overwhelmed with the amount of plant specimens collected for study. Collections of pressed and dried specimens called *herbaria* (singular, *herbarium*) were developed. Botanists frequently became interested in particular taxa and exchanged plant specimens with other botanists for extensive study. Some of these collections later formed the nuclei of world-renowned herbaria.

By the late 1800s, many botanical centers were established in Europe and North America. Their herbaria grew rapidly as access to remote regions of the earth increased. In this phase, many species were described, named, and classified into genera and families for the first time. The *flora*, or the plants that inhabit a given area, became known through herbarium specimens.

After adequate herbarium material is accumulated for a given geographic or political region, another phase called *synthesis* begins. Classification based on *morphology*, the form and structure of plants, is developed. This presents the means of identifying the plants of that area. Botanists who have the goal of developing improved classification investigate the distribution of characters in taxa. This synthesis phase reached its plateau in the late 1800s and continues with considerable study today.

The third or *experimental* phase is the combining of data for interpretation in evolutionary or phylogenetic terms. The emphasis is on understanding the course of evolution and its cause at the specific or generic level. Studies are currently ongoing in Western Europe, North America, Russia, Australia, New Zealand, and Japan, for example, where workers are concerned with morphological and chemical variation and cytological features such as chromosome number, behavior, and morphology. Controlled experiments are conducted in genetics. Hybridization, which is common in plants, is studied. Such research has led to revised concepts of evolutionary theory. Most information obtained from experimental studies is used in improving our knowledge of species and genera and has therefore had an impact on the classification of higher families and orders. With experimental methods, variation within species has become well-documented, and the cause of variation in plants may often be deduced.

The foundation of angiosperm classification was largely developed before the general acceptance of Darwin's evolutionary ideas of the late 1800s and was based on *typology*. The doctrine should be well understood, because its effects are evident in much of systematics today. The early adherents of typology reasoned that each natural taxon had an idealized pattern or "type." In its extreme form, such reasoning led to the denial of variability within taxa. Today, however, a loosely defined typology is a useful and necessary kind of thought in taxonomy. The endless array of plant groups encountered by taxonomists must be organized mentally. Thus, taxonomists seek repetitive patterns as aids in recognizing mosses, ferns, gymnosperms, dicots, grasses, sunflowers, oaks, and so on. Typology was most influential during the early years of the exploration and discovery phase. It also continues into the synthesis phase.

## SYSTEMATIC BOTANY—A RELEVANT FIELD IN TODAY'S WORLD

Today the science of identifying, naming, and classifying all plants is a challenging field of study. The potential economic uses of plants may not be immediately evident, but we must know which plants are related to one another. Wild relatives of our cultivated plants often contain genes that can provide the desirable qualities, such as disease resistance, needed by plant breeders for crop improvement. Students are interested in learning about plants as part of a greater environmental awareness. Nonscientists can more readily understand evolution and variation by observing the relationships of form and ecology that are so conspicuous among flowers. Probably no other science has so captured the interest and enthusiasm of the amateur. Just examine the number of picture books available in good bookstores for identifying plants.

## Duties of a Taxonomist

The activities of a taxonomist are basic to all other biological sciences, since systematics provides an inventory of plants, schemes for identification, methods of naming, and a system of classification of plants. Taxonomists have a serious responsibility to both society and science to provide correct names and a natural classification. Not only is systematics basic to other scientific fields, but it also depends on them for information useful in constructing classifications. A sound classification suggests problems worthy of study by ecologists, chemists, plant breeders, pharmacologists, horticulturists, and foresters. Most important, plant taxonomists must have a spirit and a willingness to serve others in all fields that relate to plants.

## Need for Names

People communicating with one another about plants and their relationships use the names of plants. Other fields of endeavor depend on taxonomists to supply names. Scientific names are of little interest in themselves, but the act of naming, or changing names, is important and not merely an academic sport of systematic botanists. Names provide a designation for each plant. Nameless items, whether new or previously known, cannot be utilized or understood until named or identified.

Through research and study there will always be more to learn about plants. In recognizing genera, families, and higher ranks, students will acquire an acquaintance with a system of classification, the diversity of the organisms, and the inherent patterns of variation.

## Critical Problems and Opportunities

There are opportunities for students to make a contribution to society through plant systematics (Stuessy, 1975; Turner, 1971). Many floras or classifications were done over 100 years ago and are out-of-date. Classifications change as additional knowledge accumulates, so that some names used in the older floras are now incorrect. Since the 1800s many new species have been named, but they have never been critically examined with reference to overall classifications.

Because the taxonomy of cultivated plants is difficult, the classification of these groups is inadequate. Worldwide taxonomic revisions and monographs of potentially useful plants are badly needed. Information about the wild relatives of cultivated plants must be assembled if we are to meet with the increasing demands for food.

The information necessary for nature interpretation and appreciation sessions offered in local, state, and national parks can be provided by taxonomists. Enrollments in local flora and plant taxonomy classes are increasing, especially when taxonomists realize their responsibilities to society and are interested in sharing their knowledge and enthusiasm about plants.

Inventories of tropical areas should be completed before their ecosystems are destroyed by agricultural practices and encroaching civilization. Most tropical rain forests have never been inventoried, yet the few that remain are rapidly being cleared (Gómez-Pompa, Vázques-Yanes, and Guevara, 1972). The many species that inhabit these areas will be lost forever. Also lost will be the bits and pieces of information that are necessary in developing phylogenetic classifications. Decisions as to whether natural ecosystems should be modified or left alone will depend upon our knowledge of their plant life.

## SELECTED REFERENCES

Constance, L.: "Systematic Botany—An Unending Synthesis," *Taxon,* **13**:257–273, 1964.

Davis, P. H., and V. H. Heywood: *Principles of Angiosperm Taxonomy,* Van Nostrand, Princeton, N.J., 1965.

Gómez-Pompa, A., C. Vázques-Yanes, and S. Guevara: "The Tropical Rain Forest: A Non-renewable Resource," *Science (Wash. D.C.),* **177**:762–765, 1972.

Lawrence, G. H. M.: *Taxonomy of Vascular Plants,* Macmillan, New York, 1951.

Morse, L. E.: "Specimen Identification and Key-Construction with Time-sharing Computers," *Taxon,* **20**:269–282, 1971.

Radford, A. E., W. C. Dickison, J. R. Massey, and C. R. Bell: *Vascular Plant Systematics,* Harper and Row, New York, 1974.

Raven, P. H.: "Trends, Priorities, and Needs in Systematic and Evolutionary Biology," *Brittonia,* **26**:421–444, 1974.

Rollins, R. C.: "Taxonomy of Higher Plants," *Amer. J. Bot.,* **44**:188–196, 1957.

Ross, H. H.: *Biological Systematics,* Addison-Wesley, Reading, Mass., 1974.

Shetler, S. G.: "Demythologizing Biological Data Banking," *Taxon,* **23**:71–100, 1974.

Simpson, G. G.: *The Principles of Animal Taxonomy,* Columbia, New York, 1961.

Sokal, R. R.: "Classification: Purposes, Principles, Progress, Prospects," *Science (Wash. D.C.),* **185**:1115–1123, 1974.

Solbrig, O. T.: *Principles and Methods of Plant Biosystematics,* Macmillan, London, 1970.

Stuessy, T. F.: "The Importance of Revisionary Studies in Plant Systematics," *Sida,* **6**:104–113, 1975.

Turner, B. L.: "Training of Systematists for the Seventies," *Taxon,* **20**:123–130, 1971.

# Historical Background
# of Classification

The history of classification is an exciting facet of plant systematics. Preliterate people arranged plants by their usefulness, whether edible, poisonous, or medicinal. As the purposes of classification expanded, the criteria for classification changed. The groupings of plants of the early hunter-gatherers eventually gave way to a classification reflecting plant affinities. Today we group plants by their presumed natural relationships at the specific, generic, familial, and higher levels.

## BEGINNINGS BY PRELITERATE HUMANS

The discoveries of the use of plants for food and later as medicine began at a very early stage in human evolution. Prehistoric people knew and used nearly all the important crop plants that we cultivate today. Food gatherers modified wild species by selecting plants with such features as tastiness in vegetables or higher yield in grains. Contemporary studies indicate that primitive peoples in remote areas today recognize and have precise names for large numbers of plants in their local environment. Some of these peoples regularly use plants for fish or arrow poisons, others for drugs to treat wounds or sickness, and still others for narcotic or hallucinatory purposes. Classification by preliterate people is at least partly based on the useful and harmful properties of plants.

## EARLY WESTERN CIVILIZATIONS

Early Western civilizations developed in areas such as Babylonia and Egypt, where cultivation of crops was feasible. Since agriculture supported these civilizations, botanical lore was of great importance. But it was not until the development of writing and convenient writing materials such as papyrus, made from a Nile River sedge (*Cyperus papyrus*), that experience and knowledge of plants could be easily recorded.

### Theophrastus (about 370–285 B.C.)

The Greek philosopher Theophrastus was a disciple of Aristotle and is often called the "father of botany." After Aristotle's death in 323 B.C., he inherited Aristotle's library and garden. Theophrastus is credited with several hundred manuscripts, but only two botanical discourses survive, *Enquiry into Plants* and *The Causes of Plants*. These are available in translation (Theophrastus, 1916, 1927). His writings summarize what was known of plants at the time and are similar to notes for lectures. Theophrastus classified plants into herbs, undershrubs, shrubs, and trees. He described approximately 500 different species of plants. Theophrastus noted many differences in plants, such as corolla types, ovary positions, and inflorescences. He distinguished between flowering and nonflowering plants. He was aware of structural features, such as the pericarp of fruits, and the distinctions among plant tissues. The botanical information contained in the writings of Theophrastus was sound and sophisticated. The original contributions of his work were not improved upon until after the Middle Ages. Many names for plants today are derived from those used by Theophrastus.

### Caius Plinius Secundus, "Pliny the Elder" (A.D. 23–79)

Pliny, a Roman naturalist and writer, held important military and governmental posts. He attempted to compile everything known in the world into an extensive 37-volume encyclopedia entitled *Historia naturalis* ("Natural History"). Nine of the volumes were devoted to medicinal plants. Although it contained many fables, *Historia naturalis* had a profound influence on botany in Europe until after the Middle Ages.

### Pedanios Dioscorides (first century A.D.)

The Roman military surgeon Dioscorides was the most important botanist after Theophrastus. Traveling extensively with the Roman armies, he had firsthand knowledge of plants used as remedies. Apparently interested in improving medical service in the Roman empire, Dioscorides prepared a truly outstanding book, *Materia medica*, which included descriptions of some 600 species of medicinal plants. Excellent illustrations were later added. For 1500 years this book was attentively studied. No drug was recognized as genuine unless named in *Materia medica*. The most beautiful and famous copy was prepared about A.D. 500 for Emperor Flavius Olybrius Anicius as a gift to his daughter, Princess Juliana. It is available in facsimile, and the illustrations can be easily associated with familiar plants. The original manuscript is now in Vienna (Dioscorides, 1959). Many of the names used by Dioscorides appear today as familiar generic names. *Materia medica* was not a deliberate attempt at classification; however, some plants such as legumes, which today are regarded as closely related, were grouped together. *Materia medica* contained less botany than the works of Theophrastus, but its usefulness in medicine caused it to be considered the definitive work of plant knowledge until the end of the Middle Ages.

## THE MIDDLE AGES

During the European Middle Ages, little progress was made in original scientific study of plants. Wars and the decay of the Roman Empire caused the destruction of much literature. Manuscripts were lost at a faster rate than they could be laboriously copied in the newly founded monasteries. Botanical knowledge was largely confined to the previously known works of Theophrastus, Pliny, and Dioscorides.

### Islamic Botany

From around A.D. 610 to about A.D. 1100, some classical botanical works were preserved by the Moslem society because the Islamic scholars had a great admiration for Aristotle and other Greek scholars. Since their scientific interests were of a practical nature, pharmacy and medicine of the Islamic people were highly developed. Islamic botanists produced practical lists of drug plants, but developed no original schemes of classification.

### Albertus Magnus, "Doctor Universalis" (about 1193–1280)

Albertus Magnus wrote on aspects of natural history and medicine during the Middle Ages. His botanical work *De vegetabilis* not only dealt with medicinal plants, as had the previous works by the Greeks and Romans, but it also provided descriptions of plants. These excellent descriptions were based on firsthand observations of the plants. Albertus Magnus attempted a classification of plants and is believed to have been the first to recognize, on the basis of stem structure, the differences between monocots and dicots.

## HERBALISTS

With the coming of the Renaissance, there was a revival of scientific spirit, and interest in botany increased. The invention of printing with movable type in about 1440 allowed botanical books to be produced which were available to a wider audience than the former hand-copied manuscripts. Botanical books were produced with descriptions and illustrations made from woodblocks or metal plate engravings. They were intended to be used for identifying medicinal plants.

These books, or *herbals*, were in turn sought and used by the gatherers and diggers of medicinals, who were called *physicians* or *herbalists*. The herbals were written for utilitarian purposes; they were often cheaply done and of poor intellectual quality, sometimes containing independent observations but generally little that might be considered real scientific advancements. Often published locally, the herbals were entitled *Gart der Gesundheit* or *Hortus sanitatis*.

### German Herbalists: Brunfels, Bock, Cordus, Fuchs

Germany in the sixteenth century was a center of botanical activity. The most outstanding contributions of this period were in the form of herbals by Otto Brunfels (1464–1534), Jerome Bock (1489–1554), Valerius Cordus (1515–1544), and Leonhard Fuchs (1501–1566). Brunfels, Bock, and Fuchs are sometimes called the "German fathers of botany" and are better known than Cordus.

Brunfels's *Herbarium vivae eicones* contains excellent illustrations. Brunfels was the earliest German Renaissance writer of note on botany. Bock's *Neu Kreuterbuck* contained excellent descriptions and some beginnings toward a system of classification. *Historia plantarum* was completed by Cordus in 1540 but was not published until 1561, some 17 years after his death. Cordus's herbal contains descriptions of 446 species in flower and fruit. He prepared botanical descriptions in a systematic format based on studies of living plants. Fuchs's *De historia stiprium*, elaborate with illustrations and descriptions, was better-documented and was the most noteworthy of the herbals of that period. Much knowledge of plants was summarized by the German herbalists, but no coherent classification scheme was produced, perhaps because relatively few plants were concerned and there was no real need for an elaborate taxonomic scheme at this point in time.

### Herbals of Other Countries or Civilizations

English botanical activity of the sixteenth century was represented by William Turner (1510–1568) and John Gerard (1542–1612). The Dutch were represented by Charles de L'Ecluse (Latinized as Carolus Clusius) (1526–1609), Rembert Dodoens (Dodonaeus) (1517–1585), and Mathias de l'Obel (Lobelius) (1538–1616). Pierandrea Mattiolia (Matthiolus) (1500–1577), an Italian botanist, published in numerous editions a popular annotated and illustrated work derived from the writings of Dioscorides. Botanical activity was flourishing throughout Europe.

Although we have focused on Western Europe, the botanical activity in other civilizations should not be overlooked. The Aztecs of Mexico developed botanical gardens, cultivated plants for both food and as ornamentals, and used medicinal herbs. The *Badianus Manuscript*, an Aztec herbal, was published in 1552 by two Aztecs (*Badianus Manuscript*, 1940).

Chinese civilization was older and much more advanced during the Middle Ages than civilization in Western Europe. The Chinese were printing on paper with movable block type before A.D. 1000. They were very interested in plants and introduced many species into cultivation. Botanical works were apparently produced in China around 3600 B.C.; however, the oldest manuscript still in existence dates back to about 200 B.C.

Agriculture was developed in India nearly 2000 B.C., and many different crops were cultivated. One interesting Indian botanical work, thought to have been written around the first century, indicates that methods of cultivation were well known.

## TRANSITION OF THE 1600s

The aggressive exploration of the New World in the 1600s was responsible for many new plant discoveries. In Europe the herbalists studied these plants and added them to their herbals. Because of the large number of new species, botanists needed to develop a more precise system for naming and arranging plants.

### Andrea Caesalpino (1519–1603)

Caesalpino, an Italian botanist, followed Aristotelian reasoning and logic. In *De plantis libri*, which appeared in 1583, he sought a classification using a philosophical rather than a purely utilitarian approach by basing his classification on the features of plants. Caesalpino profoundly influenced botanists of a later date including Tournefort, Ray, and Linnaeus.

### Caspar Bauhin (1560–1624)

In 1623, Bauhin, a Swiss botanist, published *Pinax theatri botanici,* which was a list of 6000 plants. *Pinax* provided a much-needed synonymy of plant names by listing for each plant all the names given to it by different botanists. This proved to be a valuable botanical reference. Bauhin used some binomial nomenclature (names consisting of two words) and appeared to have an understanding of a concept of grouping species into genera. The genera were not described but were defined by the characters of the included species.

## John Ray (1627–1705)

The English biologist Ray published numerous works. However, his two most significant botanical publications were *Methodus plantarum nova* (Ray, 1682) and *Historia plantarum* (Ray, 1686–1704), in three volumes. The last edition of *Methodus*, published in 1703, treated 18,000 species, many of which came from areas other than Europe. This was a great increase over the 500 or so that were treated by the herbalists. Ray developed a system of classification based upon form relationships by grouping together plants which resembled one another. He attempted to define species in the terms of morphology and reproduction. His classification was the greatest advance in theoretical botany of the seventeenth century. Ray's system later influenced the thinking of the de Jussieu and de Candolle families.

In Ray and Bauhin can be seen the beginnings of *natural classification*, i.e., the grouping together of those plants that resemble one another. Today it is said that plants that resemble one another are usually closely related. However, it was not until the time of Darwin that the idea of lineages was firmly established.

## Joseph Pitton de Tournefort (1656–1708)

The French botanist Tournefort is best known for his *Institutiones rei herbariae*, published in 1700. *Institutiones* was very popular because of the ease in identifying its 9000 species arranged into 700 genera. Compared with Ray's *Historia*, his system of classification was inferior because it was artificial. Its purpose was not to group closely related species but to aid in identification. Tournefort placed great emphasis on the genus. Although the roots of this concept are much older, dating to Aristotle, he is sometimes known as the "father of the genus concept." Tournefort used the concept of genus consistently and provided descriptions of the genera. From that point on, the concept of genus was well established in classification. There is evidence in his writing that he began to develop a system of groups at a higher level than that of the genus—i.e., aggregates of genera.

## CARL LINNAEUS (1707–1778) AND THE LINNAEAN PERIOD

Linnaeus, born in Råshult, Sweden, became the renowned botanist of the eighteenth century (Figure 2-1). He is sometimes called the "father of taxonomy." Today Linnaeus is best remembered for his consistent use of a referable system of nomenclature, i.e., the binomial system of nomenclature.

Linnaeus entered the University of Lund in 1727 to study medicine. Dissatisfied with Lund, in 1729 he moved to the University of Uppsala, where he came under the influence of the Dean, Olaf Celsius, who introduced him to Professor Rudbeck, a professor of botany. On an expedition to Lapland in

**Figure 2-1** Linnaeus (1707–1778).
*(From Swingle, 1946; with permission.)*

1732, Linnaeus greatly increased his knowledge of natural history. In 1735 Linnaeus went to the Netherlands and quickly finished his medical degree at the University of Harderwijk. While in the Netherlands, Linnaeus became the personal physician to the wealthy banker George Clifford, who had extensive botanical and horticultural interests. Clifford became the patron of Linnaeus. The three years spent in the Netherlands and traveling in Europe were important and creative years in Linnaeus's life. Before returning to Sweden in 1738, he published several books on natural history. He was also able to meet many of the prominent naturalists of the time, including John Frederik Gronovius and Hermann Boerhaave from the Netherlands, Professor J. J. Dillen and Sir Hans Sloane in England, and the de Jussieu brothers in France. After establishing a medical practice upon his return to Sweden, he became professor of medicine and botany at the University of Uppsala in 1741, a position he held until his death in 1778.

His three best-known works are the following: *Systema naturae*, 1735, et seq., presenting his system of classification in outline form; *Genera plantarum*, 1737, providing descriptions of many genera; and *Species plantarum*, 1753, a two-volume catalog used for plant identification. These have been reprinted many times and are available in most scientific libraries.

Linnaeus was hailed by many of his contemporaries as starting a new epoch in botany. We now view his work as the culmination of an attempt to create a workable system of identifying and classifying plants. Even though the system was useful for identifying plants, natural relationships were not stressed and unlike plants were often grouped together. Linnaeus divided plants into 24 classes based in large part on the number, union, and length of stamens. Plants with one stamen were placed in class Monandria, with two stamens class Diandria, and then Triandria, Tetrandria, Pentandria, and so on. Classes were divided into orders based on the number of styles in each flower. The strength of Linnaeus's artificial "sexual system" is its simplicity, since botanists could easily use it to apply names to plants.

Today we consider the greatest contribution of Linnaeus to be his consistent use of a precise and referable system of binomial nomenclature. "Referable" means that the name of the species immediately indicates the genus into which it is classified. In *Species plantarum*, each plant had a generic name, a polynomial descriptive phrase intended to serve as a definition of the species analogous to our dichotomous keys, and a trivial name (i.e., specific epithet) printed in the margin. An example is shown below (see also Figure 2-2):

> glauca 8.  *SERRATULA foliis ovato-oblongis accuminatis ferratis, floribus corym-*
> *bofis, calycibus fubrotundis.*

The generic name is *Serratula*; the trivial name is *glauca*, and the polynomial descriptive phrase is *foliis ovato-oblongis accuminatis ferratis, floribus corym-bofis, calycibus fubrotundis*. *Serratula glauca* forms the binomial. Information was also included about previous publications and illustrations, and about specimens in his herbarium and where the specimens were collected. The convenience of using the trivial name in combination with the generic name was immediately obvious, and the binomial system of nomenclature was widely accepted. Linnaeus was not the first to use two names for each species; Bauhin and others had done so, but not consistently. Linnaeus's nomenclatural scheme survives intact to this day.

Linnaeus recognized the limitations which are to be expected when only one set of characteristics is used (e.g., stamen number), but the convenience for identifying plants was great. This convenience may have even impeded botanical progress toward the development of a natural classification. His taxonomic scheme was purely artificial and was basically devoid of informational content. It was the influence of the nomenclature that caused the taxonomic system to impede the progress of botanical classification.

In any analysis of the work of Linnaeus, the conditions of the eighteenth century and the confused state of botanical nomenclature must be considered. Linnaeus made use of the works of Bauhin, Caesalpino, Ray, Tournefort, and others and put together a synthesis of their ideas. Linnaeus and his followers are best associated with a botanical era, that of the artificial sexual system.

Linnaeus's teaching abilities resulted in his added importance as a naturalist. He attracted hundreds of enthusiastic students. Among his better-known students were Peter Kalm, Fredrick Hasselquist, Peter Forskål, Peter Thunberg, and Daniel Solander.

His collections were sold by his widow in 1783 to J. E. Smith, an English botanist and one of the founders of the Linnaean Society of London. The herbarium of Linnaeus is now housed in London at the Linnaean Society. The herbarium sheets have been photographed and are available in microfiche in many research libraries. Much has been written on the life and works of Linnaeus.

1146 SYNGENESIA: POLYGAMIA ÆQUALIS.

Cirſium inerme, caulibus adſcendentibus, foliis linearibus infra cinereis. *Gmel. ſib.* 2. p. 71. t. 28.? *ſed flo-
res majores.*
*Habitat in* Sibiria. *D. Gmelin.*
Caulis *angulatus, eorymboſus, ramis itidem corymboſis,
ut terminetur denſiſſima ſylva florum, fere infinitorum.*
Folia *ſaligna, ſubtus albo villo veſtita.* Calyces *cy-
lindrici ſquamis glabris, acutis, purpuraſcentibus. Si-
milis præcedenti, ſed folia baſi parum decurrentia, ſubtus villoſa,* & Calyces *copioſiores, argutiores, glabri
magis* & *lætius colorati.*

novebora-   6. SERRATULA foliis lanceolato-oblongis ſerratis pen-
cenſis.          dulis. *Hort. cliff.* 392. *Roy. lugdb.* 143.
             Serratula noveboracenſis maxima, foliis longis ſerratis,
                 *Dill. elth.* 255. t. 263. f. 342.
             Serratula noveboracenſis altiſſima, foliis doriæ molli-
                 bus ſubincanis. *Moriſ. hiſt.* 3. p. 133. *Raj. ſuppl.* 208.
             Centaurium medium noveboracenſe luteum, ſolidagi-
                 nis folio integro tenuiter crenato. *Pluk. alm.* 93. t.
                 109. f. 3.
             *Habitat in* Noveboraco, Virginia, Carolina, Canada,
                 Kamtſchatca. ♃

præalta     7. SERRATULA foliis lanceolato oblongis ſerratis pa-
                 tentibus ſubtus hirſutis. *Mill. dict.* t. 234.
             Serratula virginiana, perſiæ folio ſubtus incano. *Dill.
                 elth.* 356. t. 264. f. 343.
             Serratula præalta, anguſto plantaginis aut perſicæ folio.
                 *Bocc. muſ.* 2. p. 45. t. 32.
             Eupatoria virginiana, ſerratulæ noveboracenſis latiori-
                 bus foliis. *Pluk. alm.* 141. t. 280. f. 6.
             *Habitat in* Carolina, Virginia, Penſylvania.
             *Receptaculum nudum, nec villoſum.* Tozzet. app. 166.

glauca.     8. SERRATULA foliis ovato oblongis acuminatis ſer-
                 ratis, floribus corymboſis, calycibus ſubrotundis.
                 *Gron. virg.* 116.
             Serratula marilandica, foliis glaucis cirſii inſtar denti-
                 culatis. *Dill. elth.* 354. t. 262. f. 341.
             Centaurium medium marianum, folio integro cirſii no-
                 ſtratis more ſpinulis fimbriato. *Pluk. mant.* 40.
             *Habitat in* Marilandia, Virginia, Carolina. ♃

ſquarroſa.  9. SERRATULA foliis linearibus, calycibus ſquarro-
                                                                      ſis

**Figure 2-2** A photograph of a page from C. Linnaeus, *Species plantarum*, 2 vols., Vienna, 1764, with the trivial (i.e., specific) name printed in the margin. Number 8, *Serratula glauca*, was based upon a specimen collected in Virginia by John Clayton and published in J. F. Gronovius, *Flora virginica*, 2 vols., Lugduni Batavorum, 1739–1743. The same plant had earlier been called *Centaurium medium marianum* in L. Plukenet, *Almagesti botanici mantissa*, London, 1700, and named *Serratula marilandica* in J. J. Dillenius, *Hortus elthamensis*, 2 vols., Lugduni Batavorum, 1732. This species is now known as *Vernonia glauca* (L.) Willd. *(Photograph, University of Georgia Libraries.)*

Additional information may be found in references cited at the end of this chapter (Daniels et al., 1976; Fries, 1923; Stafleu, 1971; Stearn, 1957).

## NATURAL SYSTEMS

By the end of the 1700s, most botanists realized that there were "natural affinities" among plants. Opposition developed to the artificial sexual classification scheme of Linnaeus, because his system often placed unlike plants together (e.g., cacti and cherries). This opposition was especially strong in France, where the sexual system had never been fully accepted. There was gradual abandonment of (1) the use of single characters to classify plants, and (2) the selection of characters based on theory rather than experience or experimentation. The development of a classification system reflecting natural relationships, which could also be used to aid in identification, became a major focus of botanical activity.

A "natural system" of classification implies that plants presumed to be related are cataloged together. In its original context, the natural system was designed to reflect God's plan of creation and not one of lineages.

### Michel Adanson (1727 – 1806)

The French naturalist Adanson traveled in tropical Africa, where the flora and fauna made him aware of the failings of one-character taxonomy. His classification system was based on the idea of giving all observable characters equal weight. Adanson rejected the idea that certain characters are more valuable than others. He described groupings similar to our modern orders and families (a category suggested earlier by Ray) as set forth in *Familles des plantes* (1763). The science of taxonomy, however, has shown that some characters are in fact more valuable than others, but modern taxonomists try to use as many characters as possible. Adansonian methods have found advocates in recent years in numerical taxonomists who use computers to develop classifications from all measurable features of plants.

### J. B. P. de Lamarck (1744 – 1829)

The French biologist Lamarck is generally best known for his unsuccessful attempts to provide an explanation for the idea of evolution, but he is well known to taxonomists for his treatment of the plants of France. In *Flore française,* published in 1778, he set forth his procedures for determining which plant precedes another in a natural series, and he stated rules for the grouping of species and the treatment of orders and families.

## de Jussieu Family

There were four botanists in the de Jussieu family: the brothers Antoine (1688–1758), director of the Jardin des Plantes in Paris; Joseph (1704–1779), explorer and collector in South America; and Bernard (1699–1777), founder of the Royal Botanical Garden at Versailles; and their nephew Antoine-Laurent (1748–1836), founder in 1793 of the Musée d'Histoire Naturelle de Paris.

Sometime before the French Revolution, a decision was made to arrange the plants in the garden at Versailles according to the Linnaean system. The garden was never arranged in this manner, because Bernard de Jussieu felt that plants that "looked alike" (i.e., were presumed to be related) should be grouped together. This was not always true with Linnaeus's artificial sexual system, where unlike things were grouped together because they had the same number of stamens. It should, however, be remembered that the idea of evolution and lineages was not established until Darwin's time. In 1763 Antoine-Laurent de Jussieu joined his uncle and began to work on natural groupings of plants in the garden.

In 1789, during the French Revolution, Antoine-Laurent de Jussieu published *Genera plantarum secundum ordines naturales disposita*, a work resulting from his experiences in arranging the garden. The publication of *Genera plantarum* caused the idea of natural systems to be widely accepted. Antoine-Laurent de Jussieu recognized 100 carefully characterized groups of plants that are now called *families* (he called them *orders*). That this treatment was remarkable is indicated by the acceptance of most of the groupings today, along with the features by which they are recognized. The de Jussieu system was superior to the artificial sexual system of Linnaeus and was fundamental to further progress toward a natural system of classification. The philosophy of the natural system was firmly established in the scientific community by Antoine-Laurent de Jussieu. It would continue until the idea of organic evolution became firmly entrenched in the late 1800s.

## de Candolle Family

The reputation of the Swiss-French de Candolle family was established by Augustin Pyramus de Candolle (1778–1841). Although born in Geneva, de Candolle (Figure 2-3) received his botanical training in Paris and was greatly influenced by the French botanists of that day. In 1813 he published *Théorie élémentaire de la botanique*, which set forth the principles of plant taxonomy. His classification scheme differed somewhat from the de Jussieu system, but it was a natural classification scheme. A major effort of A. P. de Candolle from 1816 to his death in 1841 was *Prodromus systematis naturalis regni vegetabilis*, which was an attempt to classify and describe every known species of vascular plant. Although never finished, it remains the only worldwide treatment of certain groups. The first seven volumes were written by A. P. de Can-

**Figure 2-3** A. P. de Candolle (1778–1841). *(From Swingle, 1946, with permission; original from A. Gundersen, Brooklyn Botanic Garden.)*

dolle himself, and the next ten were written by specialists and until 1873 were edited by his son, Alphonse (1806–1893).

### George Bentham (1800–1884) and Sir Joseph Dalton Hooker (1817–1911)

Attention of many botanists during the first half of the nineteenth century was focused on the development and improvement of the natural systems of classification. This era was climaxed by the publication of Bentham and Hooker's *Genera plantarum* in 1862–1883. Bentham was a self-trained English botanist, and Hooker was director of the Royal Botanic Gardens at Kew, near London. *Genera plantarum* is a three-volume work in Latin, giving names and descriptions of all genera of seed plants. The classification system by which it is arranged is clearly derived from the systems of de Jussieu and de Candolle. *Genera plantarum* was one of the last great works where the species concepts were based on the idea that species are fixed entities, unchanged through time and placed on earth by the Creator. Although it appeared after Darwin, it is pre-Darwinian in concept, despite the fact that Hooker and Bentham were both supporters of Darwin. Nevertheless, their delimitation of families and genera is based on natural relationships, reflecting the heritage of de Jussieu. The reason that *Genera plantarum* is so useful, even today, is that Bentham and Hooker prepared the generic descriptions from observations made of the plants themselves rather than by copying descriptions from the literature. Their generic descriptions are accurate, complete, and precise; the large genera are divided into sections and subsections; and the geographical range is given. Their system of classification was widely accepted and used in the British colonial floras and elsewhere. British herbaria, such as Kew and the British Museum, are still arranged according to it.

## THE INFLUENCE OF DARWIN'S THEORY OF EVOLUTION ON SYSTEMATICS

By the middle of the nineteenth century, the stage was set for the idea of evolution. The observations and collections of world travelers and the endless flow of specimens back into Europe made it impossible to ignore knowledge regarding the fossil record and variation in species of plants and animals. The history of the theory of evolution is as fascinating as that of the history of plant systematics; however, we shall consider only its impact on plant systematics.

Charles Darwin (1809–1882) and A. R. Wallace (1823–1913) wrote a paper entitled "The Tendency of Species to Form Varieties and Species by Natural Means of Selection," which was read before the Linnaean Society of London on 1 July 1858. On 24 November 1859 Darwin's *Origin of Species* was published, and every copy of the first edition was sold that very day. Biology was irrevocably changed. By the twentieth century, the concept that species are not unchanging creations was widely accepted. Rather than endless generations of species created all alike, species are dynamic, variable population systems which change with time and form lineages of closely related organisms. This idea provided a conceptual framework which had great impact on systematics.

## TRANSITIONAL PHYLOGENETIC SYSTEMS

As the impact of Darwinism became apparent, taxonomists began to integrate evolutionary concepts into their classifications. There were no abrupt departures from previous systems, since the post-Darwinian systems were firmly based on plant morphology. Taxonomists consciously tried to arrange the natural groups of plants in an evolutionary sequence proceeding from the simplest to the most complex. Botanists struggled with the problem of structures, which seemed simple but in reality resulted from the reduction or fusion of more complex ancestral features. Attempts were made to develop classification systems which were evolutionary, but which may seem naive to present-day taxonomists.

### August Wilhelm Eichler (1839–1887)

Germany was the leading center for the study of plant morphology during the second half of the nineteenth century. Classification at that time was exclusively morphological, and German plant morphologists began to consider their data in light of evolutionary theory. A number of workers made contributions. Among the most important was Eichler, who in 1883 developed what was to become a widely accepted classification of the plant kingdom. He divided the plant kingdom into the nonseed plants (the Cryptogamae) and the seed plants (the Phanerogamae). The former group contains algae, fungi, bryophytes, and seedless vascular plants, whereas the latter is divided into gymnosperms and angiosperms; angiosperms are further separated into monocotyledons and dicotyledons. Eichler did not accept, or at least did not understand, the idea of secondary reduction (i.e., from complex back to simple flowers), which affected the placement of certain groups in his classification scheme.

### Adolf Engler (1844–1930) and Karl Prantl (1849–1893)

Engler (Figure 2-4) was a professor of botany at the University of Berlin and director of the Berlin Botanical Garden from 1889 to 1921. He proposed a system of classification based on that of Eichler, differing only in detail. Along with his associate Prantl, he published *Die natürlichen Pflanzenfamilien* in many volumes from 1887 to 1915. This monumental work includes keys and descriptions for all the plant families. *Die natürlichen Pflanzenfamilien* is abundantly illustrated, and the morphological, anatomical, and geographical literature are summarized. Engler and Prantl prepared many of the treatments, but a number were contributed by specialists. The size and scope of *Die natürlichen Pflanzenfamilien* gave it practical value far greater than its value as a classification device. Engler's taxonomic scheme was widely accepted because of the detail and elaborate form of its presentation. Most non-British herbaria and many floras still follow his sequence in arranging plants. Engler did not intend for the linear sequence of his system to be understood as evolutionary. He believed that simple, unisexual flowers were primitive, and in his earlier work he rejected the idea of secondary reduction. His system of classification has undergone continual revision by his followers and has been published many times

**Figure 2-4** Adolf Engler (1844–1930). *(From Swingle, 1946, with permission; original from A. Gundersen, Brooklyn Botanic Garden.)*

in successive editions of *Syllabus der Pflanzenfamilien*. Volume 2 of the twelfth and latest edition, edited by Melchior and treating the angiosperms, appeared in 1964. Through the years the general outline has changed little, although some families have been shifted within the classification. The present Englerian system is generally regarded as obsolete, since most botanists do not agree with its assumptions regarding the primitive flower.

## INTENTIONAL PHYLOGENETIC SYSTEMS

Taxonomic schemes which try to reflect evolution are said to be *phylogenetic*. By the turn of the twentieth century, the impact of Darwinism and genetics was fully felt in plant systematics, and systems of classification were proposed that were intentionally phylogenetic. One of the advantages of phylogenetic systems is that they are rich in informational content, since the identity of a plant implies knowledge of its affinities and evolutionary relationships. Actually, the natural systems and the phylogenetic systems are similar in many respects, and the families and genera differ little in content. The difference is in the placement of families and orders, as well as what is considered to be the primitive flower.

The Englerian school considered simple, unisexual flowers to be primitive. Examples of an Englerian primitive flower would be something like the unisexual flowers found in willow (*Salix*) or cottonwood (*Populus*). The phylogenetic systems view primitive flowers in the ancestral sense. Because it has numerous free sepals, petals, and stamens and spirally arranged carpels, the Magnoliaceae-Ranunculaceae flower is primitive in the latter system. Certain families of dicots possess characters which are considered primitive by most phylogenists. These characters include monosulcate pollen, unsealed carpels, free spirally arranged floral parts of indefinite number, and wood without vessels (Walker, 1976). These primitive families are sometimes referred to by a loose informal term, *ranalian complex*. This term is derived from the name of the order Ranales, which is centered around the families Magnoliaceae and Ranunculaceae and their allies.

"Primitive = simple" in the Englerian context is a concept derived from reason, while in phylogenetic context, "primitive" means those things that were here first. According to the paleobotanists, the ranalian complex of characters were here first; therefore, these characters are primitive.

### Charles E. Bessey (1845—1915)

A professor of the University of Nebraska, Bessey (Figure 2-5) received part of his training under Asa Gray, a noted botanist at Harvard. Bessey was the first American to make a major contribution to the theory of plant classification. In 1894 Bessey, greatly influenced by the evolutionary ideas of Darwin, published an intentionally phylogenetic taxonomic system based upon the prin-

**Figure 2-5** Charles E. Bessey (1845–1915). *(With permission of the School of Life Sciences, University of Nebraska.)*

ciples of organic evolution. He considered the flowering plants to be monophyletic, i.e., derived from one evolutionary line and representing a continuum of lineages. His system was based on a set of *dicta* or concepts of primitive features found in ancient plants, versus advanced characters of more recently evolved plants. His system is basically a modified Bentham and Hooker classification with emphasis on primitive ranalian stock giving rise to the other dicotyledons and to the monocotyledons. His final classification was published in outline form shortly after his death in 1915.

### John Hutchinson (1884 – 1972)

A British botanist, Hutchinson was associated with the Royal Botanic Gardens at Kew. In his *Families of Flowering Plants* (1973) and *Genera of Flowering Plants* (1964–1967), he proposed a system of classification resembling Bessey's but differing in several fundamental points. Essentially, Hutchinson derived the flowering plants from hypothetical proangiosperms (i.e., the transitional plants from gymnosperms to angiosperms) and divided them into three lines: the Monocotyledones, Herbaceae Dicotyledones, and Lignosae Dicotyledones. The primitive monocotyledons were derived from the primitive herbaceous dicotyledons, their point of origin being in the Ranales. Hutchinson considered the woody versus herbaceous habit of fundamental importance among the dicotyledons. The woody line was derived from the woody Magnoliales and the herbaceous line from the herbaceous Ranales. Hutchinson's treatment of individual families and genera, and especially that of the monocotyledons, is excellent. The division of the dicotyledons into woody and herbaceous lines is considered unfortunate and unnatural because it often places closely related families some distance apart. This has detracted from the overall value of his work.

## CONTEMPORARY SYSTEMS OF CLASSIFICATION

At present, the classification of the plant kingdom continues to be modified as new information becomes available. In recent years, the classification of flowering plants has benefited from data from paleobotany, biochemical systematics, and ultrastructure using scanning and transmission electron microscopes. These kinds of data, used in conjunction with information derived from traditional sources such as comparative anatomy and gross morphology, further refine the classification.

The American plant taxonomist Robert Thorne proposed a synopsis of his classification in 1968 and a somewhat more detailed exposition in 1976. Armen Takhtajan (1969) of the Soviet Union and Arthur Cronquist (1968) of the New York Botanical Garden presented detailed outlines of their classifications. The systems of Thorne, Takhtajan, and Cronquist are Besseyan in origin and tradition. These recent classifications continue to improve, and no doubt the future will see more revisions as botanists attempt to develop phylogenetic classifications based on evolutionary principles using newly available information. Table 2-1 (page 24) summarizes the historical development of classification.

Selected references to botanical history which provide a detailed yet readable synthesis are provided at the end of the chapter (Barkley, 1974; Becker, 1974; Core, 1955; Gardner, 1972; Greene, 1909; Hawks and Boulger, 1928; Reed, 1942; Sachs, 1906; and Sirks and Zirkle, 1964). Stafleu (1967, 1976) provides an excellent selective guide to botanical publications.

## SELECTED REFERENCES

Adanson, M.: *Familles des plantes,* 2 vols., Paris, 1763. (Facsimile edition, Lehre, 1966.)

Anderson, F. J.: *An Illustrated History of the Herbals,* Columbia University Press, New York, 1977.

Arber, A.: *Herbals, Their Origin and Evolution,* The University Press, Cambridge, England, 1938. (The classic reference on herbals.)

*Badianus Manuscript* (translated and edited by E. W. Emmart), Johns Hopkins, Baltimore, 1940. (An Aztec herbal.)

Barkley, T. M.: "History of Taxonomy," in A. E. Radford et al. (eds.), *Vascular Plant Systematics,* Harper and Row, New York, 1974, pp. 13–34.

Bauhin, C.: *Pinax theatri botanici,* Basel, 1623.

Becker, K. M.: "Systems of Classification," in A. E. Radford et al. (eds.), *Vascular Plant Systematics,* Harper and Row, New York, 1974, pp. 583–644. (A detailed comparison of the various systems of classification.)

**Table 2-1   Historical Development of Classification Systems**

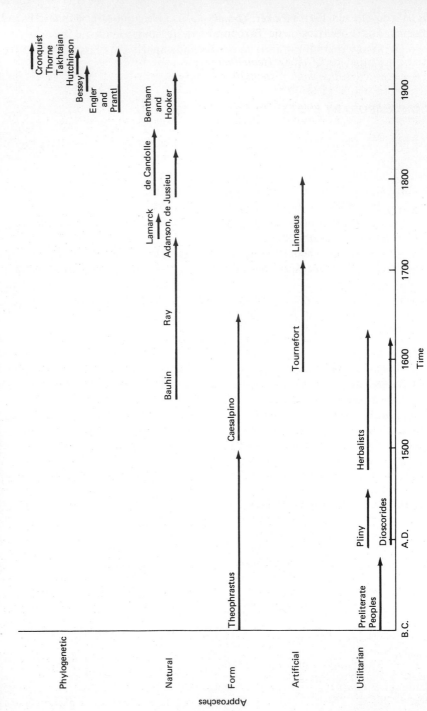

Bentham, G., and J. D. Hooker: *Genera plantarum,* 3 vols., London, 1862–1883.

Bessey, C. E.: "Phylogenetic Taxonomy of Flowering Plants," *Ann. Mo. Bot. Gard.,* **2:**109–164, 1915.

Candolle, A. P. de: *Théorie élémentaire de la botanique,* Paris, 1813.

—— et al.: *Prodromus systematis naturalis regni vegetabilis,* 17 vols. and 4 index vols., Paris, 1824–1873.

Caesalpino, A.: *De plantis libri,* Florence, 1583.

Core, E. L.: *Plant Taxonomy,* Prentice-Hall, Englewood Cliffs, N.J., 1955, pp. 9–61. (An excellent history of plant taxonomy.)

Cronquist, A.: *The Evolution and Classification of Flowering Plants,* Houghton Mifflin, Boston, 1968.

Daniels, G. S., et al.: "Linnaean Symposium at the Hunt Institute for Botanical Documentation," *Taxon,* **25:**3–74, 1976. (A series of eight special papers.)

Darwin, C.: *Origin of Species,* London, 1859.

——: *Origin of Species,* 6th ed., J. Murray, London, 1873. (The 6th edition is the edition most frequently cited.)

Dengler, R. E. See Theophrastus.

Dioscorides: *The Greek Herbal of Dioscorides,* R. T. Gunther (ed.), Hafner, New York, 1959.

Eichler, A. W.: *Blüthendiagramme construirt und erläutert,* 2 vols., Leipzig, 1875–1878. (Facsimile edition: Eppenhain, 1954.)

Emmart, E. W. See *Badianus Manuscript.*

Engler, A.: *Syllabus der Pflanzenfamilien,* 12th ed., 2 vols., H. Melchior and E. Werdermann (eds.), Gebrüder Borntraeger, Berlin, 1954–1964.

—— and K. Prantl: *Die natürlichen Pflanzenfamilien,* 23 vols., Leipzig, 1887–1915.

—— and ——: *Syllabus der Pflanzenfamilien,* 2d ed., Berlin, 1898. (In many editions, some of which were edited by other people, the 1st edition is cited as *Syllabus der Vorlesungen...,* 1892.)

—— and ——: *Die natürlichen Pflanzenfamilien,* 2d ed., W. Engelmann, Leipzig, 1924– . (The 2d edition is not yet complete. Twenty-one volumes have thus far been published.)

Fries, T.M.: *Linnaeus, the Story of His Life,* adapted by B. D. Jackson, Witherby, London, 1923.

Gardner, E. J.: *History of Biology,* 3d ed., Burgess, Minneapolis, 1972. (A good general reference on the history of biology.)

Garnsey, H. E. F. See Sachs, J.

Green, J. R.: *A History of Botany 1860–1900,* Clarendon, Oxford, 1909.

Greene, E. L.: "Landmarks of Botanical History," *Smithsonian Misc. Coll.,* **54:**1–329, 1909. (A basic reference on botanical history.)

Gunther, R. T. See Dioscorides.

Hawks, E., and G. S. Boulger: *Pioneers of Plant Study,* The Sheldon Press, London, 1928.

Hutchinson, J.: *The Genera of Flowering Plants,* 2 vols., Clarendon, Oxford, 1964–1967.

——: *The Families of Flowering Plants,* 3d ed., Clarendon, Oxford, 1973.

Hort, A. See Theophrastus.

Jackson, B. D. See Fries, T. M.

Jussieu, A. L. de: *Genera plantarum secundum ordines naturales disposita,* Paris, 1789.

Lamarck, J.: *Flore française,* 3 vols., Imprimerie Royale, Paris, 1778.

Linnaeus, C.: *Systema naturae,* Lugduni Batavorum, 1735. (Has been issued in many reprints, revisions, and re-editions.)

————: *Genera plantarum,* Lugduni Batavorum, 1737.

————: *Species plantarum,* 2 vols., Stockholm, 1753. (Facsimile edition: London, 1957–1959; there are numerous later editions of this work.)

Melchior, H. See Engler, A.

Ray, J.: *Methodus plantarum nova,* London, 1682. (Facsimile edition: Weinheim, 1962.)

————: *Historia plantarum,* 3 vols., London, 1686–1704.

Reed, H. S.: *A Short History of the Plant Sciences,* Chronica Botanica Company, Waltham, Mass., 1942. (An excellent, readable source of more detailed information on botanical history.)

Sachs, J.: *History of Botany (1530–1860),* translated by H. E. F. Garnsey and revised by I. B. Balfour, Clarendon, Oxford, 1906.

Sirks, M. J., and C. Zirkle: *The Evolution of Biology,* Ronald, New York, 1964. (Excellent literature citations.)

Stafleu, F. A.: "Taxonomic Literature," *Regnum Veg.,* **52,** 1967. (A selective guide to botanical publications and authors.)

————: "Linnaeus and the Linnaeans," *Regnum Veg.,* **79,** 1971.

———— and R. Cowan: "Taxonomic Literature," 2d ed., *Regnum Veg.,* **94,** 1976- . (This will be a multivolume work when complete.)

Stearn, W. T.: "An Introduction to the *Species plantarum* and Cognate Botanical Works of Carl Linnaeus," in C. Linnaeus, *Species plantarum,* facsimile of the 1st ed., 1753, Ray Society, London, 1957.

Swingle, D. B.: *A Textbook of Systematic Botany,* McGraw-Hill, New York, 1946.

Takhtajan, A.: *Flowering Plants—Origin and Dispersal,* Smithsonian Institution Press, Washington, 1969.

Theophrastus: *Enquiry into Plants,* 2 vols., translated by A. Hort, W. Heinemann, London, 1916.

————: *De causis plantarum,* translated by R. E. Dengler, Philadelphia, 1927.

Thorne, R. F.: "Synopsis of a Putatively Phylogenetic Classification of the Flowering Plants," *Aliso,* **6**(4):57–66, 1968.

————: "A Phylogenetic Classification of the Angiospermae," *Evol. Biol.,* **9:**35–106, 1976.

Tournefort, J. P. de: *Institutiones rei herbariae,* 3 vols., Paris, 1700.

Walker, J. W.: "Comparative Pollen Morphology and Phylogeny of the Ranalean Complex," in C. B. Beck (ed.), *Origin and Early Evolution of Angiosperms,* Columbia, New York, 1976, pp. 241–299.

# Chapter 3

# Plant Nomenclature

The assignment of names to plants is called *nomenclature*.* It involves principles governed by rules developed and adopted by the International Botanical Congresses. The rules are formally listed in the International Code of Botanical Nomenclature (Stafleu, 1972) and are often referred to as the "Code." The ultimate goal of this precise system, as embodied in the Code, is to provide one correct name for each taxon. The rules of nomenclature are subdivided into *articles*, which must be adhered to, and *recommendations*, which are optional.

Although classification schemes may change with time, the scientific names of plants are relatively stable. The plant retains its name although the family or higher taxonomic categories are changed. Much effort has been devoted to establishing procedures for naming taxa and for changing names that were incorrectly assigned.

Nomenclature and classification are inseparable. The placement of a plant or group of plants in the classification scheme may be determined by knowing its name. When the generic name of a plant is known, it is possible with the proper bibliographic aids to determine the family to which that genus is usually

---

*Nomenclature* is pronounced nō-́men-clā-́chur and its adjective form is *nomenclatural*. The word *nomenclatorial* is the adjective form of *nomenclator*, a person or book that deals with names.

assigned. Such a bibliographic tool is *A Dictionary of the Flowering Plants and Ferns* (Willis, 1973).

## BASIS OF SCIENTIFIC NAMES

The present system of nomenclature is the result of a historic series of changes which gradually became formalized. The oldest plant names we now use are the common names used in ancient Greece and Rome. Today all plant names have a Latinized spelling or are treated as Latin regardless of their origin. This custom originates from medieval scholarship and the use of Latin in most botanical publications until the middle of the nineteenth century. The assignment of names was relatively unstructured until the seventeenth century when the number of plants known to botanists began to increase greatly. This resulted in a need for a more precise naming system for plants. During several centuries before 1753, names were often composed of three or more words. These names are called *polynomials*. For example, in the herbal of Clusius (1583), the name *Salix pumila angustifolia altera* is used for a species of willow.

This complex name-description system was not workable because it was not readily expandable. In 1753, with Linnaeus's *Species plantarum*, the *binomial* format was substituted for the polynomial. This two-word format made naming more convenient and provided a readily expandable system. Our present formal nomenclature began with the publication of Linnaeus's *Species plantarum*. Since 1753 nomenclatural procedures have become standardized through periodic legalistic revision so that plants are not named haphazardly.

### Scientific Names versus Common Names

Latinized scientific names often appear formidable. There is a natural inclination to avoid words with unfamiliar and difficult pronunciations. Although scientific names may be difficult to pronounce, guides to pronunciation do exist (Johnson, 1971). To help in remembering names, Appendix 1 provides the meanings of some Latinized names.

Why do botanists use Latinized scientific names instead of common names? Common names present a number of problems. First, common or vernacular names are not universal and may be applied only in a single language. (Scientific names, on the other hand, are universal and are recognized throughout the world.) Second, common names usually do not provide information indicating the generic and family relationships. Third, if a plant is well known, it may have a dozen or more common names. For example, *Chrysanthemum leucanthemum* is called daisy, white daisy, ox-eye daisy, shasta daisy, or white weed; *Centaurea cyanus* is variously known as cornflower, bluebottle, bachelor's button, or ragged robin. Fourth, sometimes two or more plants may have the same common name. In Georgia, *Sida* in the family Malvaceae is called ironweed; but in the Midwest, *Vernonia* in the family Compositae is called ironweed. Fifth, another problem is that many species, particularly rare ones, do not have common names.

### Composition of Scientific Names

The genus name and specific epithet together form a binomial called the *species name*. The term *species name* is often erroneously used to refer to the specific epithet alone, but the species name consists of both the generic name and the specific epithet. A complete scientific name must be followed by the third element, the name of the person or persons who formally described the plant. For example, the complete scientific name for white oak is *Quercus alba* Linnaeus; the genus is *Quercus*, the specific epithet *alba*, and the author or authority, Linnaeus. The author element of a name is often abbreviated, and "L." is normally used for the authority in place of "Linnaeus." To be absolutely correct, the species name of white oak is not *"alba,"* but is *Quercus alba* L. Therefore, a complete scientific name of a species consists of three elements: (1) genus (plural, genera), (2) specific epithet, and (3) the author or authority.

**Generic Names**   The generic name is a singular Latinized noun or a word treated as a noun. It is always written with an initial capital letter. After a generic name has been spelled out at least once, it may be abbreviated by using the initial capital letter; for example, *"Q."* for *Quercus*. Generic names may not consist of two words unless they are joined with a hyphen. Latin inflectional endings are used for both generic names and specific epithets. Section 3 of the International Code of Botanical Nomenclature deals with what makes a generic name. Stearn's *Botanical Latin* (1973) is an excellent reference for the mechanics and grammar of botanical Latin.

The name may be taken from any source, and it may commemorate some person of distinction. Genera such as *Linnaea* for Linnaeus or *Jeffersonia* for Thomas Jefferson are commemorative. Many ancient common names, such as *Asparagus* and *Narcissus*, were converted into generic names directly from Greek. Features of plants, such as the liver-like leaves of *Hepatica*, gave generic names to still others, the word *Hepatica* being derived from the Latin word for liver. Information about a plant is sometimes expressed in a generic name because it indicates in a general way the kind of plant under consideration. With familiar genera we can recognize the plants by their generic names, e.g., *Rosa* as a rose and *Pinus* as a pine, both of which are ancient colloquial names.

**Specific Epithets**   Specific epithets may be derived from any source and may honor a person, or they may be derived from an old common name, a geographic location, or some characteristic of the plant, or they may even be composed arbitrarily (see Article 23 of the Code). The specific epithet is often an adjective illustrating a distinguishing feature of the species. Specific epithets consisting of two words must be hyphenated, as in the case of *Capsella bursa-pastoris* (L.) Medic.

The specific epithet usually agrees with the gender of the generic name if the specific epithet is an adjective. If the specific name is an adjective placed in a genus which has the masculine ending *-us*, a species might be spelled *albus*, but if it is a genus with a feminine spelling, it would be spelled *alba*. In spite of its *-us* ending, *Quercus* is feminine for the purposes of botanical Latin: thus, *Quercus alba*. It is customary to treat all trees in botanical Latin as feminine, as was usually the situation in classical Latin. A specific epithet may also be a noun in apposition carrying its own gender. When the noun is in apposition, it is normally in the nominative case—e.g., *Pyrus malus* for the common apple. When a specific epithet is named after a person and ends in a vowel or *-er*, the letter *-i* is added (e.g., *glazioui*), but if it ends in a consonant, the letters *-ii* are added (e.g., *ramondii*) (Recommendation 73C of the Code). When named for a female, it ends in *-iae* or *-ae*; e.g., *luciliae*. Specific epithets derived from geographical names usually are terminated by *-ensis, -(a)nus, -inus, -ianus*, or *-icus*; examples are *quebecensis*, *philadelphicus*, and *carolinianus* (Recommendation 73D).

The Code recommends that all specific epithets be written with a small initial letter, but capital letters may be used when epithets are derived from a person's name, from former generic names, or from common names. Both the generic name and the specific epithet are customarily underlined when written or typed; when printed, they are in italics or boldface. The authority is never underlined.

**Authority**   The name of the person or persons following the genus and specific epithet indicates the authority or author. It is a source of historical information regarding the name of the plant (Clausen, 1938). By giving the author's name, one may discriminate among names. The authority may be abbreviated; for example, "L." for Linnaeus or "Michx." for André Michaux. Frequently a name will have two authorities, with the first in parentheses. For example, with *Vernonia acaulis* (Walter) Gleason, the positioning of these two authorities shows that this species was first described by Walter, who supplied the specific epithet *acaulis*. Walter put it in a genus other than *Vernonia*, and at some later point Gleason transferred this species to *Vernonia*. When the rank of a taxon is changed or when a species is transferred from one genus to another, the name of the describing author is placed in parentheses and is followed by the name of the person who made the change. Transfers are sometimes necessary in taxonomic studies when new information suggests that taxonomic boundaries be realigned. Name changes should be made only after careful consideration of taxonomic relationships and must follow the requirements of the International Code of Botanical Nomenclature (Weatherby, 1946).

## RULES OF NOMENCLATURE

The increased number of plants known to European botanists in the eighteenth century required the development of order and stability in plant nomenclature. The first elemental rules of naming plants were proposed by Linnaeus in 1737 and again in 1751. In the latter part of the eighteenth century, *priority*, or the use of the oldest name, was recognized as the cornerstone of nomenclature, ensuring that each plant had a unique name. Botanists who did not adhere to this principle created confusion in the naming of plants. A. P. de Candolle in his *Théorie élementaire de la botanique* (1813), set forth a detailed set of rules regarding the process of assigning names. Later, the rules of A. P. and A. de Candolle evolved into our present International Code of Botanical Nomenclature. Numerous plants were inescapably named two or more times by accident, so in the years following Linnaeus a complex synonymy developed. Steudel in 1821 and 1840-1841 published an index of plant names, *Nomenclator botanicus*, which listed all names known to have been assigned to plants. This was useful for checking names and synonyms. It was the forerunner of *Index kewensis* (discussed in Chapter 11).

In 1867 the First International Botanical Congress was convened in Paris by Alphonse de Candolle, the son of A. P. de Candolle. Botanists from many countries met and adopted a set of rules for nomenclature. Most of the rules had been proposed by Alphonse de Candolle. The rules were an excellent beginning, but practical applications revealed some inherent deficiencies. The need to modify the rules became evident in the late 1800s when botanists at Kew Gardens in England, at the Botanical Gardens at Berlin, at the New York Botanical Garden, and elsewhere began to stray from the rules adopted in Paris.

Botanical Congresses were held in 1892, 1905, 1907, and 1910 in an attempt to resolve nomenclatural problems and establish internationally acceptable rules. Nomenclatural procedures were standardized on a worldwide basis with general agreement reached by the International Botanical Congress of 1930. Subsequent Congresses have been held on a regular basis and have offered only minor modifications in the rules. A detailed history of the development of the Code is discussed in Lawrence (1951) and Smith (1957). Many of the terms used in the Code are elaborated upon and easy to find in McVaugh, Ross, and Stafleu (1968).

### Principles

Today botanists throughout the world use the International Code of Botanical Nomenclature, which is written in English, French, and German (Stafleu, 1972). A set of nomenclatural principles forms the philosophical basis of the Code:

  1   "Botanical nomenclature is independent of zoological nomenclature."

The Code provides solely for the nomenclature of plants. The same name that has been assigned to a plant may also be used by zoologists for naming an animal. For example, *Cecropia* refers both to a moth and to a large genus of tropical trees in the family Moraceae.

2   "The application of names of taxonomic groups is determined by means of nomenclatural types." The "type" principle provides that each species name must be associated with a particular specimen, the nomenclatural type. The type for a genus is a species, for a family it is a genus, and so on.

3   "The nomenclature of a taxonomic group is based upon the priority of publication." This very important principle provides that the correct name is the earliest properly published name that conforms to the rules. Earliest published names take precedence over names of the same rank published later. Priority for plant nomenclature begins 1 May 1753 for vascular plants and some other groups, but not for all plants. This was the date of publication of Linnaeus's *Species plantarum*.

4   "Each taxonomic group with a particular circumscription, position, and rank can bear only one correct name, the earliest that is in accordance with the rules, except in specified cases."

5   "Scientific names of taxonomic groups are treated as Latin regardless of their derivation." This rule requires that generic and specific epithets, as well as other names, be Latin or treated as if they were Latin. The Code must be consulted for details on selecting the proper grammatical endings for names of all taxa.

6   "The rules of nomenclature are retroactive unless expressly limited."

Rules adopted by the Congresses operate to affect nomenclatural matters carried out before the passing of the rules. Botanists should consult the latest International Code of Botanical Nomenclature when confronted with solving current nomenclatural problems.

## Procedures

Detailed procedures based upon these principles are divided into *Rules* and *Recommendations*. The Code states, "The objective of the Rules is to put the nomenclature of the past into order and to provide for that of the future; names contrary to a rule cannot be maintained." The Recommendations dealing with minor points provide guidance and uniformity in naming plants. However, names which are contrary to the Recommendations cannot be rejected for that reason.

## Ranks of Taxa

The formal taxonomic hierarchy is a system of categorical ranks with associated names (Scott, 1973). Generally, the species is the basic unit of classification (Article 2). Each species belongs to a series of taxa of consecutively higher rank. The International Code of Botanical Nomenclature provides the series of ranks with names which are the hierarchical categories (Articles 3 and 4). The ranks, in descending sequence, provided by the Code are shown in Table 3-1 (page 34), along with an example of each.

The Code, in effect, defines the categories only by listing their sequence. It may not be necessary to use all the categories provided by the Code for a small order, family, or genus, but the sequence of categories must not change (Article 5). However, certain categories (i.e., species, genus, family) are essential if nomenclature is to function. The categories commonly used in the flowering plants are the class, subclass, order, family, genus, species and sometimes either subspecies or variety or even sometimes both. Categories such as subfamily, tribe, subgenus, section, and so on may be used and are frequently necessary in large and complex groups. In actual practice, species are grouped into genera and genera into families and so on through the sequence of categories. Each rank in turn is more inclusive than the lower categories. This categorization gives order and accessibility to the classification of plants and provides a meaningful system of information input or retrieval.

The Code requires standardized grammatical endings for the categories from division down to subtribe. However, an exception is the use of certain family names which have been sanctioned by the Code because of old, traditional usage. These names do not end in the usual family ending of -aceae (Article 18). The names of these families, along with their alternative names, are Palmae (Arecaceae), Gramineae (Poaceae), Cruciferae (Brassicaceae), Leguminosae (Fabaceae), Guttiferae (Clusiaceae), Umbelliferae (Apiaceae), Labiatae (Lamiaceae), and Compositae (Asteraceae). Botanists are authorized by the Code to use either of these alternatives. Some manuals use the older names and others use the -aceae names. Family names ending with -aceae are based upon generic names; for example, Brassica is the base of Brassicaceae.

## The Type Method

Names are established by reference to a nomenclatural type. Taxonomists use the type method as a legal device to provide the correct name for a taxon. The nomenclatural type of a species, a *type specimen*, is a single specimen or the plants on a single herbarium sheet. The type specimen for the species *Vernonia alamanii* DC. is located in the de Candolle Herbarium in Geneva, Switzerland. The type of genus is a species; e.g., the type of the genus *Vernonia* is *V. noveboracensis* (L.) Michx. The type of family is a genus; e.g., the genus *Aster* is the type genus for the family Compositae (Asteraceae).

The nomenclatural type is not necessarily the most representative of a taxon; it is the specimen or specimens with which the name of that taxon is permanently associated. The type specimen in no way reflects the typological concept of an idealized specimen (as discussed in Chapter 1). The type specimen has nothing at all to do with variation, but only indicates the attachment of a name to a particular specimen.

**Table 3-1  Series of Ranks Provided by the International Code of Botanical Nomenclature**

| Ranks of taxa | Example | Endings of ranks above genus |
|---|---|---|
| Division | Magnoliophyta | -phyta |
| Class | Magnoliopsida | -opsida |
| Subclass | Asteridae | -idae |
| Order | Asterales | -ales |
| Suborder | | -inales |
| Family | Asteraceae (or Compositae) | -aceae |
| Subfamily | | -oideae |
| Tribe | Vernonieae | -eae |
| Subtribe | Vernonineae | -ineae |
| Genus | Vernonia | |
| Subgenus | | |
| Section | Lepidoploa | |
| Subsection | Paniculatae | |
| Series | Verae | |
| Subseries | | |
| Species | Vernonia angustifolia Michx. | |
| Subspecies | V. angustifolia ssp. angustifolia | |
| Variety | | |
| Subvariety | | |
| Form | | |
| Subform | | |

When a species new to science is collected, several things must be done: (1) it is given a name; (2) a Latin diagnosis or description is prepared; (3) a type is designated; and (4) the name and description are published. All these must be done in accordance with the Code. An example of a publication of the name and description of a new species is shown in Table 3-2. In this description, a type was designated and deposited in the New York Botanical Garden Herbarium and a Latin description was provided (see Article 7 and the Guide for the Determination of Types).

The Code designates several kinds of types (Article 7). The *holotype* is the one specimen used or designated by the author in the original publication as the nomenclatural type. If a holotype was designated by the author, it may not be rejected; and any type chosen after the original publication cannot be regarded as the holotype. Today it is essential that a holotype be designated for a newly described species and deposited in an established public herbarium.

An *isotype* is a duplicate specimen of a holotype collected at the same place and time as the holotype. A *lectotype* is a specimen chosen by a later worker from original material studied by the author of the species, when no holotype was designated or when the holotype has been lost or destroyed. A *syntype* is one of two or more specimens cited by an author of a species when no holotype was designated, or it is any one of two or more specimens originally designated as types. A *paratype* is cited in the original publication. It is a

---

**Table 3-2   Description of a Species New to Science**

---

**Vernonia cronquistii** S. B. Jones, sp. nov. TYPE: México: Guerrero: semi-open slopes in pine-oak forest in the mountains along the highway ca. 62 rd miles N of Acapulco, and 20 mi S of Chilpancingo, *Cronquist* 9705 (Holotype: NY! Isotypes: GH! MEX! MICH! MO! NY!).

Herba perennis, erecta, 1.5-metralis; caules purpurei necnon glabri. Folia caulina (6.5) 8-12 (15) cm longa, 1.9-4.5 cm lata (ratione longitudinis cum latitudine ca. 3-4), ad medium dilatata, ovato-lanceolata, supra scabridiuscula, infra glabrescentia, apicibus acuminatis, basibus anguste cuneatis, marginibus serratis; petioli 0.5-1.2 cm longi glabrescentes. Inflorescentiae paniculatae-umbellatae. Capitula 10-14(18)-flora, cum pedunculis 0.5-1.3 cm longis. Involucra anguste campanulata 5.5-8.5 mm longa, 3-7.5 mm lata; phyllaria ciliata, laxe imbricata, purpurea, eis interioribus lineari-lanceolatis, 4.2-7.5 mm longis, 0.9-1.5 mm latis, apicibus acutis vel cuspidatis, eis exterioribus lanceolatis, 1-2 mm longis, 0.6-0.9 mm latis. Pappi setae albae, eis interioribus 5-6.1 mm longis, eis exterioribus 0.6-1.1 mm longis. Corollae (7.3) 9-11 (12.6) mm longae, Vernoniapurpureae, glabrae. Antherae 2.7-3.3 mm longae. Achaenia 2.2-3.1 mm longa, piloso-hispida, ca. 9-11 nervata. Chromosome number $n = 17$. Flowering and fruiting occur from October to December. This species is distributed from Guerrero to Oaxaca along the Sierra Madre del Sur. . . . It occurs on semi-open slopes in pine-oak or pine forests at elevations of 700-950 m. It is named in honor of Dr. Arthur Cronquist who made the type collection and has provided encouragement to me with my studies of *Vernonia*.

Additional specimens examined include: México: Guerrero: Rincón de la Via, *Kruse* 739 (ENCB); Plan de Carrizo, Galeana, *Hinton* 11035 (GH, K, MICH, NY, US); Oaxaca: 5-6 km NE Putla rd to Tlaxiaco, *McVaugh* 22273 (ENCB, MICH).*

---

*The specimens named in the last paragraph are paratypes. *Source: Rhodora,* **78**:194, 1976.

specimen other than the holotype or isotype. If the author cited two or more specimens as types, the remaining cited specimens are paratypes. A *neotype* is selected when all the original specimens and their duplicates have been lost or destroyed. If there is no holotype, a lectotype must be selected from among the isotypes or syntypes. If none are known to exist, a neotype may be selected. A lectotype has precedence over a neotype because a lectotype was studied by the original author.

The early botanists did not designate types as is done today. To these early botanists, species were based upon all specimens, illustrations, and descriptions within the limits of the species. These elements or everything associated with the name at first publication are known collectively as the *protologue*. Recommendation 7B of the Code suggests that when the elements of the protologue are heterogeneous, the lectotype should be selected to preserve current usage.

### Priority of Names

*Priority* is concerned with the precedence of the date of valid publication and determines the acceptance of one of two or more names that are otherwise acceptable. A name is said to be *legitimate* if it is in accordance with the rules and *illegitimate* if it is contrary to the rules (Article 6). The rule of priority states, "For any taxon from family to genus inclusive, the correct name is the earliest legitimate one with the same rank, except in cases of limitation of priority by conservation" (Article 11). The Code contains several limitations on the principle of priority. "The principle of priority does not apply to names of taxa above the rank of family" (Article 11).

To avoid disadvantageous changes caused by strict application of priority, some generic and family names are conserved by action of the International Botanical Congresses (Article 14). Conserved names are referred to as *nomina conservanda*. This means that some names, even though they are not the oldest legitimate names, are used in preference to the older names. Occasionally, family or generic names, perhaps published in obscure publications or otherwise not used, will have priority over well-known names despite not having been in regular use. Adoption of such generic names usually requires formally transferring specific epithets to the resurrected genus. To avoid the confusion this would cause, names are conserved by decisions of an International Botanical Congress. A list of the conserved family and generic names may be found in Appendixes 2 and 3 of the Code. Names of species and of ranks below species may not be conserved. "For any taxon below the rank of genus, the correct name is the combination of the earliest available legitimate epithet in the same rank with the correct name of the genus or species to which it is assigned, except..." in some special cases (Article 11). Priority of nomenclature for vascular plants (except fossils) begins with the publication of Linnaeus's *Species plantarum* in May, 1753 (Article 13).

## Effective and Valid Publication of Names

To become a part of the legal botanical nomenclature, names of taxa must meet certain requirements when published. These requirements are explicitly stated by the Code. "Publication is effected, under this Code, only by distribution of printed matter (through sale, exchange, or gift) to the general public or at least to botanical institutions with libraries accessible to botanists generally. It is not effected by communication of new names at a public meeting, by the placing of names in collections or gardens open to the public, or by the issue of microfilm made from manuscripts, typescripts or other unpublished material. Offer for sale of printed matter that does not exist does not constitute publication" (Article 29).

Currently, publication of handwritten descriptions or descriptions printed in nursery catalogs or seed exchange lists is not considered to be effective publication (Article 29). A plant name is not effectively published if printed on a label attached to herbarium specimens even if the specimens are widely distributed (Article 31). Effective publication refers to the place and form of publication of the names of plants. The botanical community must communicate plant names in widely distributed scientific literature.

For valid publication, a name must be *effectively published* in the form specified by the Code. It must be accompanied by a description or a reference to a previously published description for that taxon (Article 32). Since 1935 all diagnoses of new taxa must be written in Latin to be validly published. The *diagnosis* is a statement by the author giving the distinguishing features of the taxon. The description itself need not be in Latin, although it is recommended. (See Table 3-2.)

## Citation of Author's Name

To be accurate and complete, the name of a taxon should include a citation of the author or authors who originally described that taxon (Article 46): e.g., *Vernonia arkansana* DC., for A. P. de Candolle; *Vernonia* Schreb., for J. D. C. von Schreber; and the tribe Vernonieae Cass., for Henri Cassini. There are many sources of explanations of abbreviated names of authors, including the *Manual of the Vascular Plants of Texas* (Correll and Johnston, 1970) and Gray's *Manual of Botany* (Gray, 1950), both of which are excellent.

The author citation expedites locating the original plant description, which helps determine the type and date of publication for the taxon (Clausen, 1938). Sources providing references to original descriptions are *Index kewensis* on an international basis and the *Gray Herbarium Index* for New World plants. Either source provides references to the original descriptions. Another function of the author citation is to identify the name. Through unfortunate error an author may publish a name that is preoccupied, i.e., the specific epithet may have been used for another taxon in the same genus. The author citation permits one

to distinguish between the two names. Of course, only the earlier name is legitimate. Author citations can aid botanists in tracing the transference of species
from one genus to another. For example, the author citation for *Vernonia
noveboracensis* (L.) Michx. reveals that Andre Michaux transferred to the
genus *Vernonia* a species originally described in another genus by Linnaeus.
The original species name used by Linnaeus was *Serratula noveboracensis* L.
Since *Serratula* L. was published in 1753 and *Vernonia* Schreb. in 1791, it appears as though *Vernonia* violates the rule of priority. Reference to Appendix 3
of the Code indicates that *Vernonia* Schreb., (1791) nom. cons., non L. 1753,
has been conserved by international agreement over *Serratula* L. 1753, so
Michaux's combination is legitimate.

When names are published by two authors, the author citations are linked
by either & or *et* (Latin, "and")—e.g., *Opuntia pollardii* Britt. et Rose, for
N. L. Britton and J. N. Rose. The author citation *Carex stipata* Muhl. ex Willd.
indicates that the species was described by G. H. E. Muhlenberg but was
published by K. L. Willdenow, who attributed the name to Muhlenberg.

### Retention, Choice, and Rejection of Names

The Code has rules outlining the proper procedures for selecting the correct
name when taxa are divided, transferred, or rejected. That is, a genus might be
divided into two genera or a species transferred from one genus to another
genus; or if a name is illegitimate, it is rejected. A brief synopsis of the major
points of the most important rules concerning retention, choice, and rejection
of names is presented here. Chapter 5 of the Code should be consulted for a
complete account of the topic.

A change in the diagnostic limit separating the taxon from its nearest
relatives is not justifiable cause for a change in the name of the taxon (Article
51). For example, change in the concept of diagnostic characters of a genus or
species is not a reason to change a name. If a genus is divided into two or more
genera, the original generic name is retained for the genus that includes the
nomenclatural type species for the genus (Article 52). Likewise, when a species
is divided into two or more species, the original specific epithet must be retained for the species that includes the type specimen (Article 53). This same
rule applies to infraspecific taxa, i.e., subspecies and varieties.

When a species is described in one genus and later transferred to another
genus, the specific epithet, if legitimate, must be retained (Article 55).
*Chrysocoma acaulis* Walt., 1788, is now treated as *Vernonia acaulis* (Walt.)
Gleason, 1906. *Chrysocoma acaulis* Walt. is the basionym of *Vernonia acaulis*
or the name-bringing synonym associated with Walter's type specimen, which
is located in the Museum of National History in Paris. For many years this
species was called *Vernonia oligophylla* Michx., 1803. In 1906 Gleason properly recognized that there was an older specific epithet and basionym for this
taxon, *C. acaulis* Walt., 1788. If a taxonomist considers two previously distinct

species to be the same species, the earlier epithet must be selected for the newly combined species. The latter name is then considered to be a taxonomic synonym. In such cases the identity of the basionym is important in determining the correct name for a species. When a subspecies or variety is transferred to another genus or species, it is no different than transferring specific epithets, for epithets hold their priority within rank (cf. Article 56 and Recommendation 60A). If it is transferred without change of rank, the original epithet must be retained, unless there is some nomenclatural barrier.

When two or more taxa of the same rank are united, the oldest legitimate name or epithet is selected (Article 57). For example, if the genera *Sloanea* L., 1753, *Echinocarpus* Blume, 1825, and *Phoenicosperma* Miq., 1865, are united, *Sloanea* L. is the oldest name and would be correct (Article 57). The other two names become taxonomic synonyms.

When identifying plants, one may notice that many manuals will cite one or more synonyms for certain species treated in that flora. This practice is helpful because familiar names may become synonyms. A fine example of this is Radford et al., *Manual of the Carolina Flora* (1968). Following the description of *Solidago graminifolia* (L.) Salisbury, they cite as synonyms: *Euthamia graminifolia* (L.) Nutt. -S (for Small, 1933); *S. graminifolia* var. *nuttallii* (Greene) Fernald -F, G (for Fernald, 1950, and Gleason and Cronquist, 1963). From this information you can compare treatments of the taxa in different manuals.

During plant identification work, it may be necessary to refer to taxonomic revisions. In a revision, synonyms will be given in a complex and formal listing. The following is an example:

*Vernonia leiocarpa* DC. Prodr. 5: 34. 1836. TYPE: MEXICO: *Karwinski* s.n. (HOLOTYPE: G-DC, as IDC microfiche G-DC!).

*Cacalia leiocarpa* (DC.) Kuntze, Rev. Gen. Pl. 2: 970. 1891.

*Eremosis leiocarpa* (DC.) Gleason, Bull. New York Bot. Gard. 4: 232. 1906.

*Eremosis melanocarpa* Gleason, Bull. New York Bot. Gard. 4: 232. 1906. TYPE: GUATEMALA: Santa Rosa: Chupadero, *Heyde & Lux 3416* (HOLOTYPE: NY!; ISOTYPES: F! GH! MO! US!).

*Vernonia melanocarpa* (Gleason) Blake, Contr. Gray Herb. 52: 18. 1917.

*Vernonia leiocarpa* was described by A. P. de Candolle in *Prodromus*, Volume 5, page 34, in 1836. The type was collected in Mexico by Karwinski s.n. (Latin for *sine numero*, meaning "without [collection] number"). G-DC indicates that the type is located in the de Candolle Prodromus herbarium at the Conservatoire Botanique de Genéve, Switzerland. Each herbarium is assigned an abbreviation by the International Association for Plant Taxonomy. These abbreviations are found in Holmgren and Keuken (1974). The type specimen was viewed on IDC microfiche of the de Candolle Prodromus herbarium.

The exclamation point (!) is an abbreviation for *vidi* (Latin, "I have seen it"), and it indicates that the author of the revision has seen the specimen cited.

Synonyms in a revision are listed in chronological order. For example, in 1891 Kuntze transferred the specific epithet to the genus *Cacalia*, making the combination *Cacalia leiocarpa* (DC.) Kuntze. In 1906 Gleason made the combination *Eremosis leiocarpa* (DC.) Gleason, by transferring *V. leiocarpa* DC. to *Eremosis*. Gleason also described *Eremosis melanocarpa* Gleason, based on the type collected in Guatemala by Heyde and Lux. *3416* is the collection number of the particular specimen of Heyde and Lux. This holotype is located in NY (New York Botanical Garden) and was examined by the author of the revision, as indicated by "!." Isotypes examined are located in F (Field Museum), GH (Gray Herbarium), MO (Missouri Botanical Garden), and US (United States National Museum). Blake did not recognize *Eremosis* and transferred the specific epithet *melanocarpa* to *Vernonia*. The result was the combination *Vernonia melanocarpa* (Gleason) Blake. The author of this taxonomic revision believed *V. melanocarpa* to be a taxonomic synonym of *V. leiocarpa*. A review of the formal taxonomic treatment provides a taxonomic history of the entity. There is a growing and useful practice to give all the names that are based on the same type specimen in one paragraph, thus using a paragraph for each basionym, its nomenclatural type, and its taxonomic synonyms. The *basionym* is the epithet with which the type is associated.

Another important rule of the Code is the following: "A legitimate name or epithet must not be rejected merely because it is inappropriate or disagreeable, or because another is preferable or better known, or because it has lost its original meaning" (Article 62). Therefore, the name *Scilla peruviana* cannot be rejected because it grows in the Mediterranean area rather than in Peru. *Vernonia crinita* Raf. is better known than the older *V. arkansana* DC., but *V. crinita* Raf. cannot be retained just because it is better known.

A name is a *later homonym* if it is spelled like a name previously and validly published for a taxon of the same rank based on a different type (Article 64). Different genera or different species within a genus cannot have the same name. If they do, the earlier name is legitimate and the later name is a later homonym. *Tapeinanthus* Boiss. ex Benth, 1848, is a later homonym of *Tapeinanthus* Herb., 1837. *Astragalus rhizanthus* Boiss., 1843, is illegitimate because it is a later homonym of *Astragalus rhizanthus* Royle, 1835.

The Code deals with other matters, including spelling of names and epithets (Nicolson, 1974; Nicolson and Brooks, 1974). The names of plants must be spelled as they were originally published unless there was a spelling or typographic error (Article 73).

A frequent source of confusion is the naming of infraspecific taxa (Clausen, 1941; Fosberg, 1942; Weatherby, 1942). If a subspecies or variety is described in a species not previously divided into infraspecific taxa, there is automatically a "type" subspecies or variety. This bears the same epithet as the species, but is not followed by an authority. This means that a species with infraspecific taxa must have at least two subspecies or varieties. For example, the two subspecies of *Vernonia obtusa* (Blake) Gleason are *Vernonia obtusa* ssp. *obtusa* and *Vernonia obtusa* (Blake) Gleason ssp. *parkeri* S. B. Jones. (Ssp. *obtusa* is not followed by an author citation.) The Code requires that the epithet be repeated and the original type specimen be the type of ssp. *obtusa*. The logic for this is simple. Creation of a subspecies (or variety) automatically creates two subspecies: (1) the entity that the author has in mind when erecting the new subspecies, and (2) the remaining materials within the species. The former receives the new name (e.g., *Vernonia obtusa* ssp. *parkeri*, S. B. Jones), while the latter is signified by repetition of the specific epithet (e.g., *Vernonia obtusa* ssp. *obtusa*).

Hybrids between different species in the same genus or between closely related genera are sometimes described and named. In order to be validly published, the names of hybrids follow the same rules as those that relate to names of nonhybrids (Article 40). Additional rules and recommendations needed for the naming of hybrids are found in the Code in Appendix 1, Names of Hybrids. Some of the more important points include the following: Hybrids between two species of the same genus are shown either by a formula—e.g., *Salix aurita* L. x *S. caprea* L.; or if desired, by a formal name—e.g., *Quercus* x *beadlei* Trel., a hybrid of *Q. alba* x *Q. michauxii* Nutt. It is permissible under the Code to give them either a name or a formula. The opinion of plant taxonomists differs as to whether hybrids should or should not be named (Wagner, 1969, 1975).

Article 28 of the Code deals with nomenclature for cultivated plants. Plants that are brought in from the wild and cultivated retain the names applied to the same taxa in their native habitat (Article 28). Horticultural plants which are produced in cultivation through hybridization, selection, or other processes and that are worthy of being named receive *cultivar* names. Cultivar names are written with a capital initial letter. They are either preceded by the abbreviation *cv.* (meaning "cultivar") or placed in single inverted commas. Cultivar names may be used after generic, specific, or common names. Examples of cultivars are *Camelia japonica* cv. Purple Dawn and *Citrullus* cv. Crimson Sweet (or watermelon cv. Crimson Sweet, or *Citrullus lanatus* cv. Crimson Sweet). The use of the term *variety* to refer to cultivars is improper, but the usage was traditional for a long time and is only now going out of style (Stuart, 1974). Detailed information on specialized nomenclatural situations dealing with cultivated plants may be found in the *International Code of Nomenclature for Cultivated Plants* (Gilmour, 1969). The nomenclature for cultivated plants must follow this Code.

Some of the more important general rules governing cultivar names are the following:

1 New cultivar names must now be in modern languages and not be Latin names.
2 If the botanical name of a species to which they belong is changed, cultivar names remain unchanged.

**3** Two or more cultivars in the same genus are not permitted to bear the same name.

**4** Since 1 January 1959, new cultivar names must not be the same as a botanical or common name of a genus or a species.

**5** New cultivar names published after 1 January 1959 require a description which may be given in any language.

**6** It is recommended that cultivar names be registered with a registration authority to prevent duplication or misuse of cultivar names.

The International Code of Botanical Nomenclature is a response to the fact that science requires a precise system of naming plants. The Code deals with the terms used to denote the ranks of taxa as well as with the scientific names applied to plants. There are valid reasons for the occasional but necessary changes of familiar plant names. Examples and problems for practice with the application of the Code may be found in St. John (1958) and Benson (1962). The use of the case method with these problems is an excellent way for a potential taxonomist to develop a working knowledge of the Code.

Davis and Heywood (1963) observe, "To most systematists, however, nomenclature is a time-consuming necessity that comes between them and the plant. Nevertheless, it is one of the tools of the taxonomists' trade and for that reason its principles must be mastered." It should be emphasized that for detailed knowledge of the rules of nomenclature, the Code itself must be consulted. Here we have considered only general principles and some of the more important points.

## SELECTED REFERENCES

American Joint Committee on Horticultural Nomenclature: *Standardized Plant Names*, 2d ed., H. B. Kelsey and W. A. Dayton (eds.), J. Horace McFarland Co., Harrisburg, Pa., 1942.

Benson, L.: *Plant Taxonomy, Methods and Principles*, Ronald, New York, 1962.

Candolle, A. P. de: *Théorie élémentaire de la botanique*, Paris, 1813.

Clausen, R. T.: "On the Citation of Authorities for Botanical Names," *Science (Wash. D.C.)*, **88**:299–300, 1938.

———: "On the Use of the Terms 'Subspecies' and 'Variety'," *Rhodora*, **43**:157–167, 1941.

Clusius, C. [l'Écluse, C. de]: *Rariorum aliquot stirpium*, 1583. (Facsimile, Akademische Druck-U. Verlagsanstalt, Graz, Austria, 1965.)

"Composite List of Weeds," *Weed Sci.*, **19**:437–476, 1971.

Correll, D. S., and M. C. Johnston: *Manual of the Vascular Plants of Texas*, Texas Research Foundation, Renner, Tex., 1970.

Davis, P. H., and V. H. Heywood: *Principles of Angiosperm Taxonomy*, Van Nostrand, Princeton, N.J., 1963.

Fernald, M. L. See Gray, A.

Fosberg, F. R.: "Subspecies and Variety," *Rhodora*, **44**:153–157, 1942.

Gilmour, J. S. L. (ed.): "International Code of Nomenclature of Cultivated Plants," *Regnum Veg.,* **64,** 1969.

Gray, A.: *Manual of Botany,* 8th ed., M. L. Fernald (ed.), American Book, New York, 1950.

Harvard University, Gray Herbarium: *Gray Herbarium Index,* 10 vols., G. K. Hall and Co., Boston, Mass., 1968.

Heller, J. C.: "The Early History of Binomial Nomenclature," *Huntia,* 1:33–70, 1964.

Holmgren, P. K., and W. Keuken: "Index Herbariorum," *Regnum Veg.* 92, 1974.

*Index kewensis plantarum phanerogamarum,* 2 vols., 15 suppl., Oxford, 1893–1970.

Jeffrey, C.: *Biological Nomenclature,* 2d ed., E. Arnold, London, 1977.

Johnson, A. T.: *Plant Names Simplified,* 2d ed., W. H. & L. Collingridge Ltd., London, 1971.

Kelsey, H. P. See American Joint Committee on Horticultural Nomenclature.

Lawrence, G. H. M.: *Taxonomy of Vascular Plants,* Macmillan, New York, 1951.

Linnaeus, C.: *Critica botanica,* Leyden, 1737.

———: *Philosophia botanica,* Stockholm, 1751. (Translated into English by H. Rose, 1775.)

———: *Species plantarum,* 2 vols., Stockholm, 1753. (Facsimile edition: London, 1957–1959.)

McVaugh, R., R. Ross, and F. A. Stafleu: "An Annotated Glossary of Botanical Nomenclature," *Regnum Veg.,* **56,** 1968.

Nicolson, D. H.: "Orthography of Names and Epithets: Latinization of Personal Names," *Taxon,* 23:549–561, 1974.

——— and R. A. Brooks: "Orthography of Names and Epithets: Stems and Compound Words," *Taxon,* 23:163–177, 1974.

Radford, A. E., H. E. Ahles, and C. R. Bell: *Manual of the Vascular Flora of the Carolinas,* The University of North Carolina Press, Chapel Hill, 1968.

Rollins, R. C.: "The Need for Care in Choosing Lectotypes," *Taxon,* 21:635–637, 1972.

St. John, H.: *Nomenclature of Plants,* Ronald, New York, 1958.

Scott, P. J.: "A Consideration of the Category in Classification," *Taxon,* 22:405–406, 1973.

Smith, A. C.: "Fifty Years of Botanical Nomenclature," *Brittonia,* 9:2–8, 1957.

Stafleu, F. A. (ed.): "International Code of Botanical Nomenclature," *Regnum Veg.,* **82,** 1972. (Appreciation is extended to Dr. Stafleu for permission to quote passages from the Code.)

Stearn, W. T.: *Botanical Latin,* 2d ed., David & Charles, Newton Abbot, England, 1973.

Steudel, E. G.: *Nomenclator botanicus,* 2 vols., Stuttgart, 1821–1824.

———: *Nomenclator botanicus,* 2d ed., 2 vols., Stuttgart, 1841.

Stuart, D. C.: "Some Problems at the Cultivar Level," *Taxon,* 23:179–184, 1974.

Wagner, W. H.: "The Role and Taxonomic Treatment of Hybrids," *BioScience,* **19:**785–789, 1969.

———: "The Spoken 'X' in Hybrid Binomials," *Taxon,* 24:296, 1975.

Weatherby, C. A.: "Subspecies," *Rhodora,* 44:157–167, 1942.

———: "Changes in Botanical Names," *Amer. Midl. Nat.,* 35:795–796, 1946.

Willis, J. C.: *A Dictionary of the Flowering Plants and Ferns,* 8th ed., revised by H. K. A. Shaw, Cambridge, London, 1973.

# Chapter 4

# Principles of Plant Taxonomy

Taxonomy is based upon the similarities and dissimilarities between organisms. Historically, taxonomy has been a descriptive science based on the variation and form of morphological characters. The classification schemes of the taxonomists of the 1700s and 1800s placed similar-appearing organisms together in species, comparable species into genera, and genera with resemblances into families. Taxonomists sought the "ground plan" of the Creator.

With the advent of Darwinism in the late 1800s, the concepts of species relatedness and evolution were incorporated into classification. Darwin's evolutionary theory provided the concepts that species represent lineages and that species within a genus have evolutionary affinities with branching relationships. Genera and families represent progressively older divergences in the lineages. Before Darwin, it was believed that each species was created individually by God and remained unchanged through time. This is the idea of *special creation,* a hypothesis now rejected by the scientific community.

Because similar appearance often reflects approximation of relationships, the Darwinian concept of lineages did not radically alter previous classifications. It did lead to the development of deliberate phylogenetic classification schemes. *Phylogenetic classifications* presume to express genetic relationships. The concept that species have lineages gave rise to the idea of *monophyletic* classification. In both theory and practice, genera, families, and so on should reflect the notion that the members of each group share a common origin. This concept had little effect on the species, genus, or family levels of classification, but the arrangement of families changed drastically from a linear sequence to a branching format designed to reflect evolutionary relationships.

Today most taxonomic treatments are implicitly *phylogenetic*. They attempt to recognize and bring together related groups of plants. New research methods have been developed, and taxonomists now use not only gross morphology, but chemistry, anatomy, ultrastructure, and a variety of sophisticated techniques to determine relationships that result in the classification of taxa. Changes in classification occur because of the discovery of new information.

## CATEGORIES

To communicate with others and to sort the vast numbers of plants, systematists name plants and arrange them into a hierarchy of categories. A *hierarchy* is an orderly array composed of a series of inclusive levels called *categories*. The categories are the ranks to which taxa are assigned e.g., species, genus, family, and so on. A diagram of a taxonomic hierarchy is shown in Figure 4-1. The higher categories, such as family or order, are more inclusive than the lower categories of genus or species. Table 3-1 lists the series of hierarchical categories provided by the International Code of Botanical Nomenclature. Although this contains many categories, normally only the order, family, genus, and species are used. The other categories are regularly used to reflect evolutionary relationships in very large and complex groups.

The categories are defined by the Code according to their position relative to other categories. The Code does not define or explain what is meant, in a biological sense, by a species, a genus, or a family. The classifications used in botany are based on the fundamental category of the species. Species are grouped into the higher category of genus, and then genera are grouped into families. Species can be subdivided into lesser categories such as subspecies or varieties.

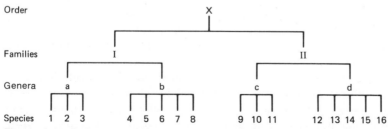

**Figure 4-1** A diagram of the common hierarchical categories indicating the increasing inclusive nature of the higher categories. *(Modified from Ross, 1974; with permission.)*

## Monophyletic Requirement of Categories

A *monophyletic* taxon is one derived from one ancestral population system, whereas a *polyphyletic* taxon has been derived from two or more ancestral population systems or taxa. It has become a common principle of systematics that taxa should be monophyletic. If assemblages of plants seem to be polyphyletic, they should be divided into monophyletic groups. Consequently, monophyletic categories at all ranks in the hierarchy express evolutionary lineages and relationships. Because of the sparse fossil record, it is often difficult to determine the origin of the higher ranks in the angiosperms. In actual practice, determination of the monophyletic line represents a hypothesis inferred from the study of present-day members of the group.

## Species

In taxonomic practice, a group of individual plants which are fundamentally alike are generally treated as a species. Ideally, a species should be separated by distinct morphological differences from other closely related species. This is necessary in order to have a practical classification that can be used by others. However, it is sometimes difficult to delimit a species precisely. If each minor variation in plant populations were to be made the basis of species distinction, there would be no end to the number of species. For systematic purposes, differences in features should not be allowed to obscure resemblances.

A species is a concept that cannot be defined in exact terms and is not absolute and inelastic. Rigorous definitions of species are not possible because the criteria may change with the characteristics of each group. Much of the disagreement about what is actually a species is vague or minor and tends to obscure the large amount of agreement on what constitutes a species.

In developing concepts of species, specimens should be regarded as samples of living, reproducing populations of genetically related individuals. Many different kinds of species have developed by diverse evolutionary and genetic mechanisms. Different species may have various strategies of reproductive isolation that reduce or prevent interbreeding. For example, many species are sexual, a few are asexual; some have arisen by polyploidy, changes in chromosome numbers, or other mechanisms. Since species represent lineages, systems of living populations may be found at various stages or levels of morphological divergence from one another in reproductive isolation. The importance of a concept of divergence was stressed by Fosberg (1942), who observed that many kinds of evolutionary processes produced numerous sorts of species at various stages of divergence from one another.

The *biological species concept* envisions the species as a distinct population system of central importance in nature and in evolutionary biology. According to this concept, biological species are kept separate from closely related species inhabiting the same general region by reproductive isolating mechanisms that prevent or greatly reduce gene exchange between them. Two such reproductively isolated species can coexist in the same general area and not lose their identity. In actual taxonomic practice, the biological species concept is difficult to apply to plant species. A diploid and a tetraploid cannot effectively exchange genes; but for practical purposes, different ploidy levels within a morphological species are traditionally regarded as representing the same species. When many plant species come into contact, they are able to exchange genes and may produce fertile hybrids. First-generation hybrids between plant species range in fertility from completely fertile to partially or completely sterile. For this and other reasons, the ability or lack of ability to produce hybrids and exchange genes cannot be used as a general criterion for the erection of species boundaries in most plants.

**Infraspecific Taxa.** A species embraces the variation within its populations. To manage some of the recurring variation, three infraspecific categories are used by plant systematists to provide formal taxonomic recognition of variation within species. These are *subspecies*, *variety* (Latin, *varietas*), and *form* (Latin, *forma*).

The categories *subspecies* and *variety* are applied to populations of species in various stages of differentiation. A common process of evolution and speciation in many plants is the gradual divergence of a former homogeneous species or population system into two or more diverse population systems (Figure 4-2). Their divergence is usually related to adaptation to differing geographical areas or climates, or to local but sharply distinct ecological habitats. During

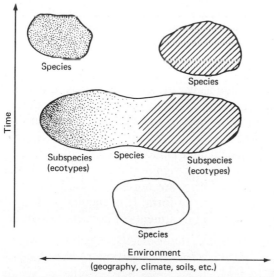

**Figure 4-2** Differentiation of a single species in a relatively homogeneous environment into subspecies in differing environments, and eventually into two species following further differentiation of the populations.

the process of becoming adapted to different habitats, the local populations become genetically distinct. This genetic differentiation is reflected in both morphology and physiology. Often these differentiated populations, or *ecotypes,* occupy adjacent ranges where they interbreed and intergrade at the points of contact. Consequently, there can be gradual or sharp discontinuities in the variation patterns between the divergent populations, depending upon the environmental gradient. These ecotypes most often form the basis of a subspecies or variety. Fosberg (1942) suggests that the stages of divergence can be indicated by the infraspecific category in which the populations are placed. This gives the taxonomist a great deal of flexibility by providing several infraspecific ranks that may be used to fit various evolutionary situations.

Variety was the first infraspecific category to be used for plants. To Linnaeus, it was primarily an environmentally induced variation. The term *subspecies* was introduced early in the 1800s by Persoon (1805-1807) and was considered a major variation in the species. Gray (1887) regarded the variety as a considerable change in the ordinary appearance of a species. In 1926 Hall argued that the term *variety* had so many uses and had been applied to so many different things, from races to cultivars, that its use should be discontinued. Others, such as Fernald, regarded the variety as a geographical subdivision of a species and the subspecies as parts of a species complex. The California school founded by Hall has used the term *subspecies* and the eastern Fernald school has favored use of *variety* for subdivisions of species.

In usual taxonomic practice, subspecies and varieties are recognizable morphological variations within species. Their populations have their own patterns of variation correlated with geographical distributions or ecological requirements. Whether the category "subspecies" or "variety" is employed often depends upon the custom or philosophy of the systematist. Sometimes this is largely based upon where or under whom the taxonomist was trained. Also, epithets have priority within rank, and historically some groups have numerous varieties and others, subspecies. If one category is used to the exclusion of the other, it would necessitate a swarm of nomenclatural transfers.

The category *form* is generally used to recognize and describe sporadic variations, such as occasional white-flowered plants in a normally purple-flowered species. If these white flowers are not correlated with any other features of the plants, including geographical and ecological distributions, they might be described as a form. Little taxonomic significance is attached to the minor and random variation upon which forms are normally based. For this reason, few modern systematists use the rank of form. It is encountered most frequently in older taxonomic treatments or in treatments of horticultural interest.

### Genera

Like species, the genus represents a concept. From a practical standpoint, the *genus* is an inclusive category whose species have more characteristics in common with each other than with species of other genera within the same family. Genera, therefore, are aggregates of closely related species. Lawrence (1951) writes that from an evolutionary viewpoint, the genus is an important concept because the sum of all characteristics is used to group closely related species, allowing the genus to be viewed as a unit. The function of the genus concept is to bring together species in a phylogenetic manner by placing the closest related species together within the generic classification (Rollins, 1953).

The genera in some families, e.g., Magnoliaceae, show striking differences in features, whereas genera in other families, e.g., Compositae, are separated on less obvious characters. A possible reason for this difference is that extinction of species and genera has been greater in the older families (Magnoliaceae), resulting in well-marked distinctions in features among genera.

There is no size requirement for a genus. A genus consisting of but one species is said to be *monotypic*. *Phryma* (Phrymaceae), consisting of a single species, *P. leptostachya,* is monotypic (Cronquist, 1968). Some genera, such as *Senecio* of the Compositae family, contain 2000 to 3000 species.

Genera must be delimited following a worldwide study of their species. If the deciduous and the evergreen rhododendrons of North America are compared, they appear to be two distinct genera. Some taxonomists in the past have placed the deciduous species in the genus *Azalea*. With wider acquaintance of *Rhododendron* on a worldwide basis, the apparent discontinuities break down. As broadly and ordinarily interpreted, the genus *Rhododendron* includes both the deciduous azaleas and the evergreen rhododendrons of North America.

The collective group of species must be distinct from the other species groups to support two genera. If the species of two genera are not readily separable, most taxonomists would agree that it is more convenient to have larger units of classification. If necessary, the ranks of subgenus or section may be preferable to handle the classification and at the same time express relationships. When a genus is split into two or more genera, nomenclatural changes are required for certain of the species. Such changes may be avoided by the use of the infrageneric ranks because they do not cause changes in binomials. The present tendency in plant systematics is to have a broad, conservative view of generic concepts and limits. This is expressed in a relatively stable generic nomenclature. The category of genus is an old and useful concept which serves both the professional and the amateur botanist. Most colloquial or common names parallel generic concepts—e.g., oak, pine, rose, buttercup, sunflower.

### Families

Much of what has been said about the concept of the genus applies equally to the concept of the family. Ideally, families should be monophyletic and present both a biologically meaningful treatment and a practical taxonomy.

Both reproductive and vegetative features are used to characterize families

(Cronquist, 1968). The parts of the flower and fruits of flowering plants are numerous and varied, and they are subject to little environmental modification. Consequently, they provide more characters for the definition of families than do vegetative features. The category of family is more inclusive than either the genus or the species, and there may be exceptions to certain characters. Families are often delimited by ovary position, kinds of pistils and stamens, carpel number, fruit type, symmetry of the flower, leaf arrangement, leaf morphology, and habit. The use of certain characters to delimit families does not preclude the use of that same character to delimit taxa at higher or lower ranks in the hierarchy. For example, the type of fruit is often used to separate families, but in the Rosaceae it delimits only subfamilies. Some families are recognized easily even by amateurs because of certain distinguishing features including the head (inflorescence) of Compositae, the schizocarp (fruit) and umbel (inflorescence) of Umbelliferae, and the samara (fruit) of Aceraceae.

Some families are well defined and are relatively homogeneous, such as the Compositae, Cruciferae, Gramineae, Umbelliferae, and Sarraceniaceae. Leguminosae is a large family characterized by a unique type of fruit and seed, yet it is not homogeneous. The diversity of the family Leguminosae is reflected by the differing classifications that have appeared from time to time. Some divide the legumes into three families, while others maintain one family with three subfamilies. Taxonomists agree on where the lines should be drawn, but they disagree on the rank to be assigned.

A great diversity of both reproductive and vegetative features may be found in families such as Ranunculaceae, Berberidaceae, and Rosaceae. Families, like genera, do not have a fixed size. The Compositae, one of the largest families of flowering plants, has around 19,000 species; the Phrymaceae has but a single genus and one species. The size and diversity of plant families are the consequences of their evolutionary history, age, and radiation, coupled with our preception of the taxonomy of the group. Taxonomic schemes are by definition subjective; the whole point of taxonomy is to convey information through a derived scheme.

When distinctions are sharp among closely related families, there is no problem in defining the family. But as Cronquist (1968) wrote regarding whether *Phryma* should be placed in the Verbenaceae or in the Phrymaceae, "It is purely a matter of taste whether the evident relationship [with the Verbenaceae] or the obviously different gynoecium should be emphasized in determining the status of *Phryma*." Assignment of genera to poorly defined families is usually a matter requiring educated judgment.

Subjective classification is often criticized by biologists whose training does not include a knowledge of the principles of systematics or taxonomic methods. It may be defended, however, on the grounds of the *predictive value* of classifications developed using these principles. Predictive value suggests that because of their common ancestry the members of a monophyletic group will share some features not apparent at the time the classification was originally made. For example, if one wishes to look for new anti-tumor alkaloids similar to those found in certain species in the Rutaceae, the best place to look would be in other species of Rutaceae. Examples like this are common and have demonstrated the predictive value of classifications developed by systematists.

Most taxonomists favor broadly conceived family concepts which lend stability to classification. Although there is no clear discontinuity of features between the Labiatae and the Verbenaceae, the two are different, and custom maintains the two groups as separate families (Cronquist, 1968). In some instances, tradition is a taxonomic character. The Rosaceae may be broken into several families based upon good logical evidence. However, the overall similarities within the family, rather than its intrafamily differences, has caused most taxonomists to treat the Rosaceae as a single family. Again, there is no set definition for a family; yet, there is relatively little disagreement on family classification. This suggests satisfaction among systematists with the delimitation of most families.

## Orders

Orders include one or more families. The monophyletic requirement for establishing orders must be broadly interpreted and should not be too restrictive when developing a workable classification (Cronquist, 1968). Orders are much more difficult to define and delimit than families. The divergence in the lineages, as reflected in orders, occurred early in the evolutionary history of the flowering plants and left few fossil records. Orders and other higher categories are characterized by an aggregate of characters. Generally, there are greater numbers of exceptions at the level of order than at the lower ranks. Most taxonomists agree that orders should be monophyletic, but all do not agree as to their size or to the families included in each. Bessey (1915) used only a few orders with a large number of families. However, Hutchinson (1973) uses many orders that often contain only a few families. The classification used in this text (Cronquist, 1968) takes a somewhat broad approach by arranging the flowering plants into 74 orders. Kubitzki (1977) provides an excellent series of papers on the classification of higher categories.

## Typological Concepts of Taxa

Systematists often remember the essential characteristics of a species, genus, or family and represent that taxon in their mind by an image embodying its salient features. This symbolism exemplifies what is meant by typological concepts of taxa. Taxonomists familiar with the flora of an area or traveling in new areas unconsciously use typological methods in recognizing the hundreds of species, genera, and families of the flora. For example, acorns, five-angled pith, and buds clustered at tips of twigs provide a recognition pattern for oaks

(*Quercus*). Taxonomists also identify plant materials by comparison with previously identified specimens or by using keys, figures, and descriptions. Since these tools represent only a sample of the populations, a taxonomist builds up typological pictures of species, genera, and families from experience with the plants. As Davis and Heywood (1963) have observed, typology is a valid and necessary process in the identification and classification of plants, when intelligently used. Only when its use obscures the natural variation in populations is it a liability to systematics.

## CHARACTERS

The *characters* of an organism are all the features or attributes possessed by the organism which may be compared, measured, counted, described, or otherwise assessed. From this definition, it is clear that the differences, similarities, and discontinuities between plants or taxa are reflected in their characters. The characters of a taxon are determined by observing or analyzing samples of individuals and recording the observations, or by conducting controlled experiments. Herbaria contain millions of plant specimens providing a ready source of samples for descriptive plant systematics.

Plant taxonomy uses morphological and anatomical characters for the purposes of classification. Structures are observed with the eye, hand lens, or light microscope or by using scanning electron microscopy (SEM). Modern instrumentation allows comparative studies of physiological rates, such as photosynthesis, and analysis of chemical compounds produced by the plant. For classification purposes, there is increasing use of evidence from fields such as cytology, biogeography, paleobotany, palynology, phytochemistry, population biology, molecular biology, and ultrastructure. But as Cronquist (1975) noted, "For the present at least, comparative study of modern forms, from the eyeball to the SEM level, remains the mainstay of major taxonomic interpretation." He also concluded that because of the time and effort required to obtain molecular data ". . . morphology, in the broad sense, will continue to reign taxonomically supreme for some years to come." Turner (1977), however, disagrees and argues that Cronquist neglects the recent application of chemical data for purposes of classification.

Characters have many uses, including the following: providing information for the construction of taxonomic systems, supplying characters for construction of keys for identification, furnishing features useful in the description and delimitation of taxa, and enabling scientists to use the predictive value of classification.

Characters are such things as leaf width, stamen number, corolla length, locule number, and placentation. Some descriptive or numerical term must occur with the character to imply the *character state*. For example, a leaf width of 5 centimeters and basal placentation are character states. When used in description, delimitation, or identification, characters are said to be *diagnostic* or *key characters*. Characters of a constant nature, which are used to help define a group, are referred to as *synthetic characters*. *Quantitative characters* are those features that can be readily counted or measured, whereas *qualitative characters* are such things as flower color, leaf shape, or pubescence. Quantitative characters are normally easier to obtain and communicate than qualitative characters. Successful observations of qualitative features provide a challenge to plant taxonomists. Another term often used by systematists is *good character*. This implies that the character is genetically determined, largely unaffected by the environment, and relatively constant throughout the populations of the taxa. The more constant a character, the greater is its reliability and resulting importance.

In order to provide character states, there must be differences in the features of flowering plants. Vegetative parts of angiosperms, such as leaves, stems, and roots, are relatively large and easy to observe, but they generally provide fewer characters for classification than reproductive structures, such as sepals, petals, stamens, pistils, and so on.

Clones of genetically identical flowering plants, when grown in diverse environments tend to be relatively uniform in characters of the flower and fruit but highly variable in overall size of the plants and leaves (Clausen, Keck, and Hiesey, 1940). Vegetative features of flowering plants, especially size and shape, tend to be more influenced by environmental factors than are reproductive features. As a result, certain vegetative characters are less reliable and less useful than reproductive characters in taxonomy.

A character which is diagnostic for separating two genera in one family may be a key character for families in another order. Some species are delimited by leaf arrangement, while leaf arrangement is a synthetic character of certain families. Characters become important after they are proved valuable by experience. The relative value and utility of characters are empirically determined, and there is no such thing (in a practical sense) as a "fundamental" character.

A rule of long taxonomic precedence requires that taxa at all ranks be established on the basis of a correlation of several characters, thus avoiding the incongruities of classification constructed on the basis of a single character. Ideally, classifications should be based on an understanding of all characters from every population of each species under consideration. But on a practical basis, this is not possible. Some geographic areas may be better sampled than others, and some characters have simply never been examined. There is not enough time to examine every possible character.

One of the results of computer technology has been the development of *numerical taxonomy* (or *taximetrics*) (Sneath and Sokal, 1973; Gilmartin and Harvey, 1976). In numerical taxonomy, each character is generally given equal weight, and as many characters as possible are used for each taxon. Phylogenies are generated by multiple correlations of characters. The overall purpose of this approach is to lend objectivity to what has previously been the somewhat subjective methods employed by taxonomists. Most workers today agree that phylogenies developed by numerical methods are probably no better or no worse than phylogenies produced by the traditional taxonomist. One major advantage of numerical approaches is that it forces the taxonomist to provide definitions of the characters, as well as data that may be used by others in reassessing classifications. Computer techniques allow the taxonomist to incorporate many more character analyses than could be done by traditional "brain power." Taximetric techniques, therefore, are especially useful in creating classifications for exceptionally complex groups—e.g., certain cultivated plants.

### Parallelism and Convergence

If two organisms are alike in their features, the resemblance is usually attributed to either parallelism or convergence. *Convergence* is the resemblance between two or more distinct phyletic lines brought about through evolution by adaptation to similar environments or to similar reproductive biologies. Adaptation to similar environments is most often apparent in the vegetative characters of plants, but it may occur in reproductive parts. Convergence in vegetative features of unrelated taxa occurring in the Mediterranean vegetation of California and Chile show the similarities in morphology caused by similar climates (Mooney, 1977). Many desert species of Euphorbiaceae from Africa and Cactaceae from North and South America demonstrate convergence with their "cactus-like" appearance, although their flowers immediately suggest the proper family (Figures 4-3 and 4-4). This convergence in vegetative features presumably reflects adaptive responses to aridity. Adaptation of reproductive biology in pollination or seed-dispersal mechanisms causes convergence in flowers, fruit, or seed. The weed *Camelina sativa* (Cruciferae) has an ecotype that infests flax fields. The flax field ecotype mimics the flax plants in size and weight of the seeds. This adaptation results when *Camelina* seeds resembling flax seeds are selected in the harvest and seed separation process. Another example of convergence, seemingly related to their pollination biology, is the development of pollinia in both the orchids and milkweeds.

With *parallelism,* the resemblance is due to a common ancestry and genetic background. Two phyletic lines may diverge up to some stage and cease to diverge and then run parallel. Parallel evolution is apparently a common feature of flowering plants, but is difficult to document with certainty due to the poor fossil record. Parallel evolution may take place in related lineages at any stage in their separation, or at the level of infraspecific taxa, species, genera, families, or so on.

Parallelism and convergence are difficult concepts at best. If the angiosperms are monophyletic, the dividing line between parallelism and convergence is purely arbitrary. Given the sparse fossil record of the angiosperms, it may be impossible to distinguish between parallelism and convergence within closely related orders.

**Figure 4-3** A desert Euphorbiaceae from Africa with a cactus-like appearance.

**Figure 4-4** A New World Cactaceae. Many desert species of Euphorbiaceae from Africa and Cactaceae from North and South America show convergent evolution in their vegetative features although they represent distinct phylogenetic lineages.

## GUIDING PRINCIPLES OF ANGIOSPERM PHYLOGENY

The angiosperm phylogenist is faced with major problems of an inadequate fossil record, the prevalence of convergent evolution, and the extreme structural modifications of many flowering plants (Thorne, 1963, 1976). The relatively soft tissues of most angiosperm flowers and certain other plant parts provide nothing analogous to bones, exoskeletons, or shells for fossilization. Flowering plants do have resistant pollen, and some have hard woody tissue or waxy leaves suitable for fossilization. Even though the fossil record of flowering plants is meager, it is sufficient to indicate some general trends in the angiosperms. In addition, ancestral characteristics and the direction of trends of specialization can often be recognized in so called "living fossils." The classification of angiosperms is primarily based on the examination of contemporary flowering plants and establishing correlations of characters (Thorne, 1963).

The necessary use of contemporary plants to establish hypothetical lineages and inferred evolutionary relationships has inherent problems (Thorne, 1963). The prevalence of convergent evolution in flowering plants has made it difficult to develop monophyletic groups. The taxonomist is faced with the problem of deciding whether certain characters are due to common ancestry or to convergence. To overcome the problem of convergence, information from all possible sources regarding the plants must be assembled and evaluated. A voluminous literature of comprehensive comparative studies has been developed on groups thought to contain primitive characters (Bailey, 1957).

In 1915 Bessey presented a set of guiding principles of angiosperm phylogeny, or what he called "dicta," to facilitate piecing together data into a phylogenetic system of angiosperm classification. The information reflected in these dicta has largely been determined from studies of comparative morphology and paleobotany. These "Besseyan principles" have been modified through the years by numerous phylogenists as new information appeared in the literature (Cronquist, 1968; Davis and Heywood, 1963; Gundersen, 1950; Hutchinson, 1973; Takhtajan, 1969; Thorne, 1963, 1976; and others). Although clearly Besseyan in origin, the present principles have been refined and restated, and most modern classifications conform to them. Given the diversity of the angiosperms, there are exceptions to these principles. They are not dogma and should only be used as guides. The modified "Besseyan principles" are listed below.

*Morphological indicators of phylogeny*

**1** In most groups of flowering plants, trees and shrubs usually have preceded the herbs, vines, and climbers. Perennials gave rise to biennials, and annuals have been derived from both. Terrestrial seed plants usually preceded closely related aquatics or epiphytes, saprophytes, and parasites.

**2** Flowering plants with collateral vascular bundles arranged in a cylinder are more primitive than those with scattered vascular bundles. Dicots preceded monocots.

**3** Alternate leaves are more primitive than opposite or whorled leaves. In nearly all instances, simple, pinnately veined evergreen leaves preceded compound leaves.

**4** Bisexual flowers preceded unisexual flowers, and the dioecious situation is often derived from the monoecious condition. Flowers borne separately in the axils of subtending, leaf-like bracts are primitive.

**5** The many-parted, spirally imbricate flower is primitive, while those with few parts, which are whorled or valvate, generally are more advanced. Regular (actinomorphic) flowers preceded irregular (zygomorphic) flowers.

**6** Perianth parts which are separate and poorly differentiated into sepals and petals are more primitive than laterally fused perianth parts which are sharply differentiated. Usually, flowers with petals preceded apetalous ones.

**7** Hypogyny (superior ovary) is the primitive condition, and perigyny and epigyny (inferior) are derived.

**8** Generally, separate carpels represent a more primitive condition than united carpels.

**9** The primitive seed has a small embryo and contains endosperm. Nonendospermic seed is derived.

**10** Many separate stamens are more primitive than few or united stamens.

**11** Single fruits preceded aggregate fruits formed from several flowers. The capsule preceded the drupe or berry.

In summary, the trends in the evolution of the flower have come about by reduction in number, fusion, specialization of parts, and changes of symmetry. Simple structures are not necessarily primitive, but have become simple as a result of reduction from more complex parts. The rate of evolution is not always the same for all structures of the plant. Some parts of a plant may become more specialized than others, and some taxa may have both advanced and primitive features.

## CHAPTER SUMMARY

From the discussion of categories, characters, and phylogeny, a number of principles of taxonomy emerge.

**1** The goal of taxonomy is to develop a workable classification which reflects evolutionary relationships and provides for nomenclature and identification.

**2** Species represent lineages produced by evolution, and branching genetic relationships exist among the taxa of each group.

**3** Categories such as species, genera, families, and orders are not rigid, but are flexible and individually delimited for each group. Their sequence in the hierarchy is established by the International Code of Botanical Nomenclature. Categories are subjective and are definable only with respect to their position in the hierarchy.

**4** Taxa are based upon the correlation of characters and discontinuities in the variation pattern. Characters may be selected from any attribute of the plant and do not have a fixed value at all ranks.

**5** For delimiting taxa, characters should be constant and show little environmental variation.

**6** Taxa should be monophyletic. On a practical basis, the monophyletic requirements for higher ranking taxa must be broadly interpreted.

**7** Taxa may resemble each other because of either convergence or parallelism.

**8** The higher categories are based upon hypothesis and are delimited by synthetic characters, which may have exceptions.

**9** Taxonomic treatments should be practical and consistent in their use of the various categories.

**10** Whenever possible in the development of classifications, taxa should be sampled throughout their range, and all taxa at lower ranks should be examined.

**11** Ancestral features and trends of diversity may often be recognized in the structure of living angiosperms.

**12** Evolution may result in reduction or loss of parts. It can also proceed toward greater structural complexity.

**13** Morphological indicators of phylogeny may provide guidance to primitive versus advanced features and aid in developing phylogenies. Thinking should remain flexible so that modifications in classification may be made as new evidence becomes available.

## SELECTED REFERENCES

Bailey, I. W.: "The Potentialities and Limitations of Wood Anatomy in the Study of the Phylogeny and Classification of Angiosperms," *J. Arnold Arbor. Harv. Univ.,* **38:**243–254, 1957.

Bessey, C. E.: "Phylogenetic Taxonomy of Flowering Plants," *Ann. Mo. Bot. Gard.,* **2:**109–164, 1915.

Cain, A. J.: "The Post-Linnaean Development of Taxonomy," *Proc. Linn. Soc. Lond.,* **170:**234–244, 1959.

Camp, W. H., and C. L. Gilly: "The Structure and Origin of Species," *Brittonia,* **4:**323–385, 1943.

Clausen, J., D. D. Keck, and W. M. Hiesey: "Experimental Studies on the Nature of Species. I. Effect of Varied Environments on Western North America Plants," *Carnegie Inst. Washington Publ.,* No. 520, 1940.

Cronquist, A.: *The Evolution and Classification of Flowering Plants*, Houghton Mifflin, Boston, 1968.

———: "Some Thoughts on Angiosperm Phylogeny and Taxonomy," *Ann. Mo. Bot. Gard.,* **62:**517–520, 1975.

Davis, P. H., and V. H. Heywood: *Principles of Angiosperm Taxonomy,* Van Nostrand, Princeton, N.J., 1963.

Fosberg, F. R.: "Subspecies and Variety," *Rhodora,* **44:**153–157, 1942.

Gilmartin, A. J., and M. J. Harvey: "Numerical Phenetics in Routine Taxonomic Work," *Syst. Bot.,* **1:**35–45, 1976.

Graham, A. (ed.): *Floristics and Paleofloristics of Asia and Eastern North America,* Elsevier, Amsterdam, 1972.

Gray, A.: *The Elements of Botany for Beginners and for Schools,* American Book, New York, 1887. (Gray's Lessons in Botany reprinted by American Environmental Studies, 1970.)

Gundersen, A.: *Families of Dicotyledons,* Chronica Botanica, Waltham, Mass., 1950.

Hall, H. M.: "The Taxonomic Treatment of Units Smaller Than Species," *Proc. Int. Congr. Plant Sci.,* **2:**1461–1468, 1929.

Henning, W.: *Phylogenetic Systematics,* The University of Illinois Press, Urbana, 1966. (Translated by D. D. Davis and R. Zangerl.)

Hopwood, A. T.: "The Development of Pre-Linnaean Taxonomy," *Proc. Linn. Soc. Lond.,* **170:**230–234, 1959.

Hull, D. L.: "Certainty and Circularity in Evolutionary Taxonomy," *Evolution,* **21:**174–189, 1967.

Hutchinson, J.: *The Families of Flowering Plants,* 3d ed., 2 vols., Clarendon, Oxford, 1973.

Kapadia, Z. J.: "Varietas and Subspecies: A Suggestion Towards Greater Uniformity," *Taxon,* **12:**257–259, 1963.

Kubitzki, K. (ed.): "Flowering Plants—Evolution and Classification of Higher Categories," *Plant Syst. Evol.,* Suppl. 1:1–416, 1977. (A series of excellent papers on angiosperm classification.)

Lawrence, G. H. M.: *Taxonomy of Vascular Plants,* Macmillan, New York, 1951.

Mooney, H. A. (ed.): *Convergent Evolution in Chile and California: Mediterranean Climate Ecosystems,* Dowden, Hutchinson and Ross, Stroudsburg, Pa., 1977.

Persoon, C. H.: *Synopsis plantarum, seu Enchiridium botanicum,* 2 vols., Paris, 1805–1807.

Rollins, R. C.: "Cytogenetical Approaches to the Study of Genera," *Chron. Bot.,* **14:**133–139, 1953.

Ross, H. H.: *Biological Systematics,* Addison-Wesley, Reading, Mass., 1974.

Slobodchikoff, C. N. (ed.): *Concepts of Species,* Dowden, Hutchinson, and Ross, Inc., Stroudsburg, Pa., 1976.

Sneath, P.H. A., and R. R. Sokal: *Numerical Taxonomy,* Freeman, San Franciso, 1973.

Takhtajan, A.: *Flowering Plants—Origin and Dispersal,* The Smithsonian Institution, Washington, 1969.

Thorne, R. F.: "Some Problems and Guiding Principles of Angiosperm Phylogeny," *Amer. Nat.,* **97:**287–305, 1963.

———: "A Phylogenetic Classification of the Angiospermae," *Evol. Biol.,* **9:**35–106, 1976.

Turner, B.L.: "Chemosystematics and Its Effects upon the Traditionalist," *Ann. Mo. Bot. Gard.,* **64:** 235–242, 1977.

Valentine, D. H.: "The Taxonomic Treatment of Polymorphic Variation," *Watsonia,* **10:**385–390, 1975.

Chapter 5

# Sources of
# Taxonomic Evidence

Taxonomic evidence for the establishment of classifications and phylogenies is gathered from a variety of sources. Because all parts of a plant at all stages of its development can provide taxonomic characters, data must be assembled from many diverse disciplines. The use of information from studies on comparative anatomy, embryology, palynology, cytogenetics, chemistry, and so on has greatly improved the modern classifications of plants.

## MORPHOLOGY

The features of floral morphology are the most important characters in the classification of flowering plants. These features are easily observed, and they are practical for use in keys and descriptions. Morphology currently provides most of the characters used in constructing taxonomic systems. When considering the evolutionary lines of the flowering plants, each represented by perhaps thousands of species, it should not be expected that all morphological characters should occur uniformly among all species. The occasional absence of a synthetic or diagnostic character, or its rare occurrence in a group from which it is normally absent, may be expected.

    Terms used in identification of angiosperms are morphological, and these are illustrated or defined in Chapter 10. *Morphology of the Angiosperms* (Eames, 1961) provides a comprehensive review for the flowering plants. Additional information may be found in Radford et al. (1974).

Natural selection, associated with successful reproduction, maintains a basic similarity of the reproductive features of flowers, fruits, and seeds within the various species, genera, and families. This general constancy makes these structures ideal for characterizing taxonomic groups. In addition to being more constant than vegetative features, reproductive characters are generally more numerous and therefore provide more features to differentiate taxa. Floral features are often fundamental in defining natural groups.

Modifications in floral morphology usually can be related to the mode of pollination or specialized reproduction. Wind-pollinated taxa frequently have unisexual, reduced flowers which are individually inconspicuous. Insect-pollinated plants (which typically have large, colorful, bisexual flowers) are valuable to pollinators because of their nectar or pollen, and the showiness of the flowers helps attract pollinators. Corolla color, pollinator guides, and similar modifications of the corolla are related to insect pollination.

Other modifications of flowers (stamen number, anther position, ovary position, style length, stigma shape, number of carpels, number and fusion of perianth parts, inflorescence type, fruit and seed types) contribute to the reproductive success of the species. Although these characters play important functional roles in the plant's reproductive biology, the adaptive advantages are not always obvious. It is quite possible that some floral features have no adaptive advantage in themselves, but that either they had such advantages historically, or their elimination may require too much genetic manipulation, when the nonadaptive trait is linked to some essential feature.

The growth habit (herbaceous or woody) of plants may be of primary usefulness in classification. The condition may be constant within a genus or family, or it may be variable. All members of the family Cruciferae are herbaceous. Some families, such as the Compositae, have both woody and herbaceous members. Genera or families that contain both tropical and temperate taxa are more likely to have both woody and herbaceous growth habits. Some semitropical or tropical herbaceous perennial species vary in their tendency to woodiness.

Stem types, buds, thorns, spines, and so forth on woody plants are often useful in identification. Leaf arrangement, type, form, duration, and venation are frequently used in both classification and identification. In trees with small reduced flowers of short duration, vegetative features often assume great importance in plant taxonomy. Vegetative underground structures such as rhizomes, corms, and bulbs may sometimes characterize a group.

## COMPARATIVE PLANT ANATOMY

For over a century, taxonomy has used comparative plant anatomy to aid in classification, and several principles concerning the use of anatomical data have become established. These principles are the following: (1) anatomical features have the same inherent problems of other characters, i.e., sampling, reliability, parallelism, and convergence (see Chapter 4); (2) anatomical characters must be used in combination with other features; and (3) anatomical characters tend to be most useful in classification of the higher categories and less useful below the rank of genus. Further information on comparative plant anatomy may be found by consulting the following sources: Esau (1965) for a general review of plant anatomy; Metcalfe and Chalk (1950) and Metcalfe (1960, plus several additional books in this series) for anatomical information arranged by taxa; and Radford et al. (1974) for a list of pertinent citations and a summary of anatomical techniques for systematics.

Since the 1930s the value of the evolutionary trends of specialization of the secondary xylem has been clearly established in studies dealing with the evolution of angiosperms (Bailey, 1957). A progressive series from tracheids (commonly found in the gymnosperms) to specialized vessel elements occurs in the secondary xylem of angiosperms (Figure 5-1). All stages of specialization, from vesselless wood to highly specialized vessel elements, may be found in contemporary flowering plants. Those angiosperms with vesselless wood often have other features regarded as primitive. This evolutionary series of vessel elements has been used in combination with other morphological features to develop hypotheses regarding the phylogeny of the angiosperms.

Little information for classification has been obtained from anatomical studies of the phloem because of its lack of apparent variation (Dickison, 1975). However, types of sieve-element plastids have recently proved useful in classification (Behnke, 1977). Some comparative plant anatomists suggest that correlations of nodal anatomy with other features might provide significant information regarding angiosperm phylogeny (Dickison, 1975). Others find this of limited value because of convergence and reduction. Recently, nodal anatomy was used to help define the broad outline of dicot relationships among the primitive dicots.

Anatomical features of leaves frequently provide characters. For example, investigation of the anatomy of leaves, associated with $C_4$-photosynthetic pathways, has resulted in a revised classification for several genera in the grass family (Brown, 1975). $C_4$ species have prominent chlorenchymatous vascular bundle sheaths. Venation patterns of fossil leaves are beginning to yield significant information when examined with reference to the Cronquist-Takhtajan system of classification (Hickey, 1973; Doyle and Hickey, 1976). Petiolar vasculation has been helpful in classification of certain genera such as *Rhododendron,* but has not proved applicable to classification problems of the higher categories.

Variation patterns of epidermal hairs or trichomes may provide characters for classification at species-genus-family levels (Figure 5-2). For example, trichomes are diagnostic characters for certain species of *Vernonia* (Faust and Jones, 1973). Emphasis is placed upon the structure of the trichome, i.e., the size, shape, and arrangement of cells making up the hair. The presence and structure of trichomes, as well as their distribution patterns among taxa, are

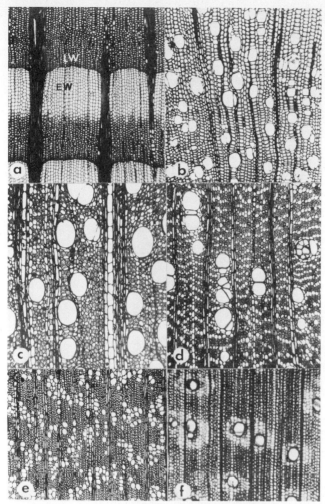

**Figure 5-1** Transverse sections of dicotyledonous woods. *(a) Trochodendron aralioides,* a primitive vesselless angiosperm with the secondary xylem composed of tracheids. *(b) Schumacheria castaneifolia,* with wood having solitary, annular vessels and mostly scanty axial parenchyma. *(c) Dillenia pentagyna,* with wood having a comparatively primitive structure but the vessels are more rounded in outline. *(d) Scytopetalum tieghemii,* with the vessels distributed as solitary and pores and pore multiples. *(e) Pittosporum tenuifolium,* having pores distributed in pore chains and radial pore multiples. *(f) Paulownia tomentosa,* a relatively advanced dicot; note the vasicentric parenchyma. *(Used with the permission of Dr. W. C. Dickison and the Ann. Mo. Bot. Gard.)*

taxonomically important. In the Compositae and several other families, trichomes are of value in analysis of suspected hybrids.

Stomatal types, produced by the characteristic arrangement of guard cells and the subsidiary cells, can be of taxonomic use. Perhaps one of the best examples is their use to characterize the subclasses of monocots (Cronquist, 1968).

**Figure 5-2**  SEMs of surface features of the shoot system of marijuana *(Cannabis sativa). 1,* Lobe of a young leaflet. *2,* Adaxial surface of a young leaf. *3,* Mature region of a leaf tip. *4,* Surface of a petiole showing nonglandular hairs, capitate glands, and bulbous glands. *5,* Abaxial leaf surface. *6,* Adaxial leaf surface. *7,* Adaxial leaf surface showing a capitate gland, long and short systolith trichomes, and a ring of cells surrounding the long trichome. The SEMs were made from fresh specimens without any treatment. *(From P. Dayanandan and P. B. Kaufman,* Amer. J. Bot., **63:** *578–591, 1971, with permission.)*

Some workers regard anatomical characters as conservative characters that are not easily modified by growing conditions. Closely related species or genera will often have the same anatomy, but there are exceptions. Leaves grown in sun may differ anatomically from those grown in shade. Anatomical features, like other characters, can be greatly influenced by selection. The stems of the hemp cultivar of marijuana *(Cannabis sativa)* have more fiber bundles than the stems from plants high in intoxicating resins. In most instances, variation such as this would not be expected in stem anatomy within a species. Anatomical features of vegetative structures have importance in separating gymnosperms from angiosperms and monocots from dicots. However, such features have proven of relatively minor value for the delimitation of subdivisions of these groups. Anatomical features tend to be most useful at the higher ranks or in determining relationships where there have been great structural modifications or reductions.

The concept of anatomical characters as conservative has been challenged by Carlquist (1969), who considers that anatomy is always related to adaptation and function. He stresses that data on the anatomy of floral structures should be viewed as part of a dynamic functioning system related to pollination biology and dispersal mechanisms.

## EMBRYOLOGY

Embryology includes micro- and megasporogenesis, development of gametophytes, fertilization, and development of endosperm, embryo, and seed coats (Palser, 1975). For general, systematic, and comparative information on plant embryology, the following reference books may be consulted: Bhojwani and Bhatnagar (1974), Davis (1966), and Maheshwari (1950).

Although there are a few minor differences in embryology among the flowering plants, there is strong embryological unity throughout the angiosperms, as exemplified by double fertilization. The major embryological character separating the dicots from the monocots is the number of cotyledons. The embryological features most common to all flowering plants are (1) four microsporangia per anther; (2) two-celled pollen grains; (3) eight-nucleate embryo sac; and (4) nuclear endosperm (Palser, 1975).

Embryological features are ordinarily constant at the family level in the angiosperms. In those families where variation has been found, the features are usually constant at the generic level. Embryology has had systematic significance in the grass family, where with several other characters, it has been used to revise the family's classification. Embryological features tend to be less useful as taxonomic characters at the rank of order, subclass, or class. When used judiciously and in combination with other characters, they may be useful in determining relationships within families, genera, and species. The technical work and time required to obtain sufficient embryological information for comparative purposes has limited its overall value as a taxonomic character.

## CYTOLOGY

Although cytology refers to the study of the cell, only information concerning the chromosomes—i.e., chromosome number, shape, or pairing at meiosis—is used for classification purposes (Stebbins, 1971). *Cytotaxonomy* refers to the use of chromosome number and morphology as a data source for classification.

*Cytogenetics* includes those studies dealing with observations of chromosome pairing or behavior at meiosis. For detailed discussions of cytotaxonomy and cytogenetics, consult Davis and Heywood (1965), Solbrig (1968), or Stebbins (1971).

The haploid number of chromosomes in angiosperms ranges from $n = 2$ in *Haplopappus gracilis* (Compositae) to around $n = 132$ in *Poa littoroa* (Gramineae). Most angiosperms have chromosome numbers ranging between $n = 7$ and $n = 12$. About 35 to 40 percent of the flowering plants are polyploids. *Polyploids* are organisms which have higher chromosome numbers due to multiplication of chromosome sets (i.e., complements). In an extensive review article, Raven (1975b) suggests that the original basic chromosome number for most angiosperms was probably 7. Raven speculates that an initial burst of polyploidy occurred during the early evolution of flowering plants, since many of the families with primitive features have high chromosome numbers.

There are several kinds of polyploid number relationships in flowering plants. Some genera such as *Pinus* with $n = 12$ have no deviation among the species. In other genera, a polyploid series is present, e.g., *Aster* (Compositae), with different species having $n = 9$, 18, or 27. There are genera with numbers which show no simple numerical relationship to each other—e.g., *Brassica* (Cruciferae), with $n = 6$, 7, 8, 9, or 10. *Vernonia* in the Old World has $n = 9$ or 10, with polyploids of $n = 18$, 20, or 30, whereas in the New World it has $n = 17$, 34, 51, or 68.

As with many other characters, the value of cytotaxonomic data depends upon the group or the category under consideration. A combination of information on chromosome number and chromosome morphology has proved useful for improving the classification of families, such as Agavaceae, or genera in the family Ranunculaceae. Base numbers and chromosome size have been very useful in understanding relationships in the grass family. In the family Onagraceae, information on chromosome number and morphology, in combination with experimental hybridizations and analysis of the chromosome set, provides clues for tracing evolution.

Because relatedness of taxa is often reflected in homology (similarity) of the chromosomes, determination of the amount of chromosome pairing at meiosis in hybrids of two species may aid in understanding relationships of closely related species. It is generally assumed that the more completely two chromosomes pair at meiosis, the more homologous they are. Or, if dealing with entire sets of chromosomes, the higher the percentage of pairing between the individual chromosomes of the two sets, the more closely the plants are presumed to be related. For additional information see Chapter 7.


## ELECTRON MICROSCOPY

Unlike its application to the study of lower plants, electron microscopy is a relatively new approach to the study of flowering plants. Features observable

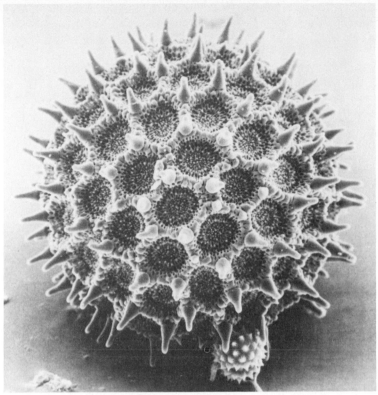

**Figure 5-3** SEM of a morning glory *(Ipomoea)* pollen grain and a smaller unidentified Compositae pollen grain. The sample was obtained from a fresh morning glory flower. Possibly the Compositae grain was brought into the morning glory flower by a pollinator. The Compositae grain is about 40 microns in diameter. Note the depth of field provided by the SEM.

only with the electron microscope are beginning to provide useful characters relevant to our knowledge of the phylogeny of the angiosperms (Cole and Behnke, 1975). *Scanning electron microscopy* (SEM) provides an image of unequaled depth of field, which is ideal for comparative studies of pollen grains and plant surfaces. An SEM micrograph of pollen grains is shown in Figure 5-3. SEM has a wide range of magnification and is relatively easy to use. Used to supplement light microscopy, SEM has been widely and readily adopted by plant systematists for pollen grains, small seeds, trichomes, and other surface features of plants.

Using *transmission electron microscopy* (TEM), Behnke (1972, 1977) demonstrated the taxonomic value of sieve-element plastids in flowering plant classification (Figures 5-4 and 5-5). Behnke found two distinct classes of plastids: the s-type that accumulates starch and the p-type that accumulates protein or protein and starch. This character was then successfully applied to taxonomic problems in the Caryophyllidae (Behnke, 1976) and to the angiosperms (Behnke, 1977). TEM studies of thin sections of pollen grain walls can yield re-

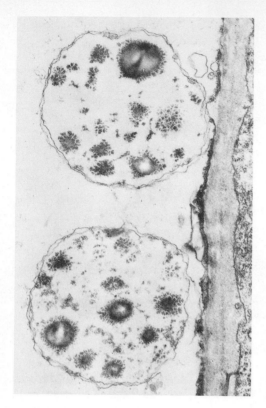

**Figure 5-4** TEM of sieve-element plastids, s-type, with many starch grains from *Cocculus trilobus* (the plastids are about 2 microns in diameter). *(Micrograph courtesy of Dr. H.-D. Behnke.)*

liable taxonomic information of certain groups (Skvarla and Turner, 1966). For example, some members of the family Compositae have differing types of wall structure and these have proved useful as characters. Dilated cisternae of endoplasmic reticulum are a characteristic feature of members of Cruciferae and Capparaceae of the order Capparales (Behnke, 1977).

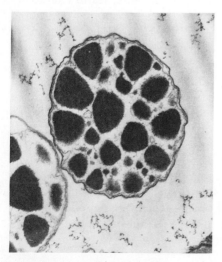

**Figure 5-5** TEM of sieve-element plastids, p-type, from *Dracaena hookeriana* (the plastids are about 1.5 microns in diameter). *(Micrograph courtesy of Dr. H.-D. Behnke.)*

In general, TEM has not been as successful as SEM in providing useful ultrastructural characters, for three reasons: (1) the general uniformity of cell organelles within each type of tissue; (2) the small number of taxa and tissues sampled; and (3) the greater difficulty and time required for specimen preparation. Ultrastructural characters may be found with TEM once identical tissues are compared and surveyed in groups that appear to have classification problems.

## PALYNOLOGY

Palynology is the study of pollen and spores. As a result of the stimulus and availability of SEM, taxonomists no longer overlook pollen as a source of characters (Figure 5-3). The availability of countless pollen samples from herbarium sheets and relatively rapid techniques for preparation allow a palynological survey of many taxa in a relatively short period of time. The taxonomic characters provided by pollen grains include pollen wall morphology, polarity, symmetry, shape, and grain size. Pollen grains do not differ within most genera; however, pollen has been very useful in determining patterns of species relationships in *Vernonia* (Keeley and Jones, 1977).

Two basic kinds of pollen grains are found in the angiosperms: monosulcate and tricolpate. *Monosulcate pollen grains* (Figure 5-6) are boat-shaped and have one long germinal furrow and one germinal aperture (Walker, 1976). Pollen of the monosulcate type is characteristic of the primitive dicots, the majority of the monocots, the cycads, and the pteridosperms. Palynologists agree that the first flowering plants probably had monosulcate pollen grains (Raven and Axelrod, 1974; Beck, 1976). *Tricolpate grains* (Figure 5-7) are globose-symmetrical, typically have three germinal apertures, and are characteristic of the advanced dicots. Walker and Doyle (1975) concluded that, with some exceptions, pollen morphology is consistent with the levels of advancement and the relationships shown in the Takhtajan and Cronquist systems of classification.

For further information on palynology, consult Wodehouse (1935) for terms; Erdtman (1966) for techniques and terminology; and Ferguson and Muller (1976) for a review of the evolutionary significance of the exine. (See Chapter 6 of this text regarding fossil pollen.)

**Figure 5-6** SEM of a monosulcate pollen grain of *Magnolia fraseri*, 1300X. *(J. W. Walker.)*

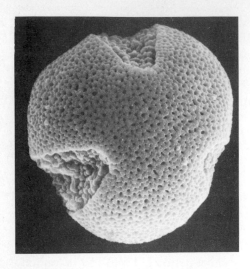

**Figure 5-7**  SEM of a tricolpate pollen grain of *Euptelea pleiosperma*, 3000X. *(J. W. Walker.)*

## PALEOBOTANY

Paleobotany uses microfossils, such as pollen, or macrofossils of leaves, stems, and other plant parts as sources of data. Paleobotanists attempt: (1) to elucidate the composition and the evolution of the floras of the past; (2) to trace these evolutionary developments through stratigraphic sequences; (3) to integrate paleobotanical data with comparative morphology; and (4) to determine past ecological conditions.

Students interested in additional readings should consult Andrews (1961) and Darrah (1960) for general surveys of the field, and Wolfe, Doyle, and Page (1975) and Beck (1976) for implications of paleobotany to angiosperm phylogeny. The U.S. Geological Survey has an extensive unpublished index to the fossil plant literature. It is housed in the Smithsonian Institution in Washington and must be examined there.

Until very recently, most systematists generally agreed that paleobotany could provide little evidence on the diversification and relationships of the major groupings of flowering plants. Evidence is accumulating that paleobotany can provide taxonomy with significant information regarding the angiosperms (Delevoryas, 1969). Investigations of fossil pollen have recently provided many new insights into the evolution of the angiosperms. The leading paleobotanists had stressed that the diversification of the angiosperms into orders, families, and genera had occurred during pre-Cretaceous time (see the geological time scale in Chapter 6), presumably in upland areas remote from the necessary basins required for deposition and fossilization, which therefore explained the lack of fossils. Now it is generally accepted that flowering plant fossils extend back only to the early Cretaceous, since no definite angiosperm fossils have been found in pre-Cretaceous sediments.

## CHEMOSYSTEMATICS

For centuries, humans have known that certain plants produce substances that can be used for specific purposes, e.g., poisons, drugs, stimulants, flavorings, starch, sugar, and so on. Little was known of the compounds except their economic value as plant products. By the first century after Christ, Dioscorides had recognized the aromatic mints, and this signified the beginning of chemosystematics. *Chemosystematics* is the application of chemical data to systematic problems. Chemosystematics developed as a hybrid between the chemistry of natural plant products and systematics. It offers a fertile field for investigation and is an increasingly important aspect of classification at various taxonomic levels.

With the development of modern methods and analytical equipment used in organic chemistry, information on natural plant products for phylogenetic comparisons is becoming increasingly available (Figure 5-8). Several recent books provide examples of the ways in which chemistry may be used in systematics (Gibbs, 1974; Harborne, 1970; Hawkes, 1968; Swain, 1973).

The chemical compounds of plants used in systematic studies are varied,

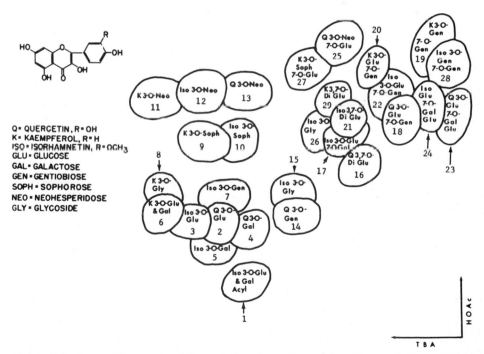

**Figure 5-8** Composite representation of the chromatographic patterns of flavonoids in *Nerisyrenia* showing the complementation of the flavonoids. *(Courtesy of T. J. Mabry and John Bacon.)*

but Turner (1969) classified them according to molecular size. Compounds of relatively low molecular weight—alkaloids, amino acids, cyanogenic glucosides, glucosinolates (mustard oil glucosides), pigments (anthocyanins, betalains, and so on), phenolics (flavonoids and so on), and terpenoids—are termed *micromolecules*. Compounds of high molecular weight, such as proteins, DNA, RNA, cytochrome c, and ferredoxin, are termed *macromolecules*.

Application of chemical data to systematics has shown that micromolecular data have practical utility in taxonomic problems at the generic level or below. They have been used to study variation caused by natural hybridization as well as other problems. In such studies, micromolecular data have been treated as characters used in classification. Generally, macromolecular data are most helpful when applied to problems of phylogeny above the generic level. Knowledge of biosynthetic pathways of natural plant products is useful in assessing relationships, because the pathways show the step-by-step sequence of the production of the compounds. The sequence of compounds can then be used in determining primitive or advanced compounds.

Chemosystematics provide a wealth of data on the chemical constituents of plants and have the potential to produce new insights into problems of angiosperm classifications. A number of illustrations are cited in the review article by Fairbrothers et al. (1975). In the subclass Caryophyllidae, the distribution of the betalain-producing families and those that produce only anthocyanins has led to reclassification of that subclass. The presence of betalains in the family Cactaceae is an example of chemical data aiding in determining relationships at higher taxonomic levels (Figure 5-9). Clearly, the Cactaceae should be placed in the Caryophyllidae, the only subclass with betalains. The presence of glucosinolates throughout the families of the order Capparales supports the monophyletic interpretation of the group. The distribution patterns of phenolics suggests a relationship of the monocots to primitive dicots such as the Magnoliales, Ranunculales, and so on.

Determination of the amino acid sequences of large protein molecules such as cytochrome c (Figure 5-10) and ferredoxin provides an approach to the study of phylogeny at the level of families and orders (Fitch and Margoliash, 1970). Amino acid sequences are precise, relatively stable, widespread, and independent of morphological characters. Technical problems of purification of these large protein molecules, however, are a major bottleneck in their use for establishing phylogenies (Fairbrothers et al., 1975). Cronquist (1976) considers the proper use of the sequences to be both complex and difficult because of problems of interpretation. Some workers suggest that the amino acid sequences for cytochrome c generally agree with our present morphologically based phylogenetic concepts (Boulter, 1972; Cronquist, 1976; Fairbrothers et al., 1975).

Theoretically, the technique of DNA and RNA hybridization has application to plant systematics because all organisms contain these compounds (Mabry, 1976). The method involves extracting denatured DNA strands and determining their affinity (or extent of pairing) with RNA or fragment DNA from another organism. The amount of pairing reflects similarity between the

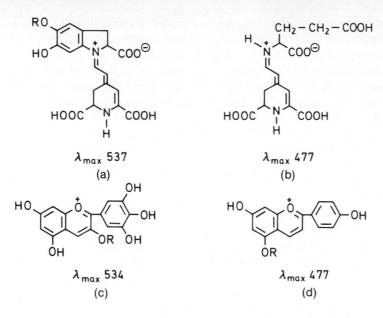

**Figure 5-9** The structure and visible absorption maxima of typical *(a)* red and *(b)* yellow betalains and *(c)* red and *(d)* yellow anthocyanins. The betalains are found in several closely related families in Caryophyllidae. Anthocyanins occur in most other nonbetalain families of flowering plants. *(Courtesy of T. J. Mabry.)*

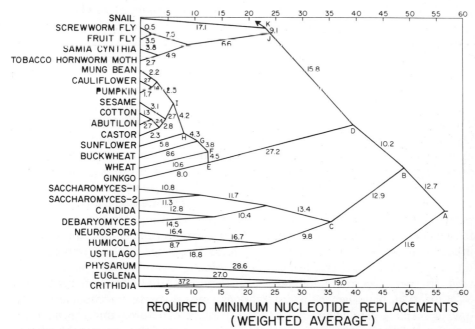

REQUIRED MINIMUM NUCLEOTIDE REPLACEMENTS
(WEIGHTED AVERAGE)

**Figure 5-10** Probable phylogeny for some plants and animals computed through the study of mutations in the cytochrome c gene. *(Used with the permission of Dr. W. M. Fitch.)*

two nucleic acids. Because of the complexity of the experiments and the difficulty in interpreting data, DNA and RNA hybridization has had little impact on the sytematics of angiosperms.

Electrophoretically determined variation in allozymes has had great impact in population biology and genetics (see the explanation in Chapter 7). Such data are helpful at the level of population, subspecies, species, or sometimes genera (Avise, 1974; Hamrick and Allard, 1975). Minor variation in soluble allozymes is compared between individuals within populations, or between populations, or between closely related taxa at low rank. Such data are not generally useful at higher ranks because of the wide distribution of similar allozymes in unrelated groups.

Immunological studies involving the injection of crude plant extracts into rabbits and analyses of the resultant antibodies have proved useful. Plant extracts are injected into the rabbit, which in turn produces the antibodies. The serum from the injected rabbit is mixed with protein extract from a presumably related taxon. The amount of immunological reactions obtained is used as an indication of phylogeny. Immunological techniques employed in studies on the comparative serology of flowering plants have contributed data useful in refining the taxonomic systems of Cronquist and Takhtajan (Fairbrothers et al., 1975). For example, serological data supported the separation of *Illicium* and *Schisandra* from the Magnoliaceae to the Illiciaceae and Schisandraceae as proposed by both Cronquist and Takhtajan. Presently the data are few, and additional taxonomic problems should be investigated by this method.

When all the technical problems are minimized, macromolecules will probably have a major impact on phylogenetic systematics and classification. Undoubtedly, this will be a rewarding field for workers trained in both chemistry and systematics.

## ECOLOGICAL EVIDENCE

Information on the ecology of flowering plants is basic to systematics in providing an understanding of (1) the distribution of taxa, (2) the variation within taxa, and (3) the adaptations of plants. In delimiting taxa, it is necessary for taxonomists to comprehend developmental responses of plants as distinguished from genetically fixed characters. Ecological studies have demonstrated that the character states of many morphological features are correlated with environmental factors such as light, moisture, and soil fertility. Ecology contributes to systematic interpretations of the evolutionary process by seeking environmental explanations for discontinuities in the structure, function, and distribution of plants (Kruckeberg, 1969). Plant ecologists examine ecotypic variation, edaphic specializations, pollination mechanisms, the effect of habitat on hybridization, plant-herbivore interactions, seed-dispersal mechanisms, ecology of seedling establishment, function of plant structures, and reproductive isolating mechanisms (Figure 5-11). These features are important in the adapta-

**Figure 5-11** Bumblebee *(Bombus)* queen pollinating flower of *Dodecatheon meadia* by vibrating pollen from the anther cone while stigma contacts residual pollen on the ventral side of the insect's thorax. *(Photograph furnished by Dr. L. W. Macior.)*

tion of populations to their environments and represent areas where ecologists make substantial contributions to systematics (Anderson, 1948; Clausen, 1951; Faegri and van der Pijl, 1966; Grant and Grant, 1965; Hadley and Levin, 1967; Harper et al., 1961; Janzen, 1966; Kruckeberg, 1969; Levin, 1971; Mooney, 1966; Muller, Muller, and Haines, 1964).

Much of the information derived from ecology has implications for classification below the level of genus. Yet ecological research has provided generalizations that may be applicable to the evolution of the flowering plants. For example, Stebbins (1974) presents the hypothesis that adaptive radiation of the angiosperms occurred in ecotones or marginal or transitional habitats under stresses of seasonal drought or cold, or in a mosaic of local edaphic conditions. Consequently, there may have been strong interactions between environmental factors and natural selection. He also favors a hypothesis of angiosperm origin in a climate having marked seasonal drought with a short season favorable for the formation of flowers. Under these conditions, natural selection might have favored the reduced angiosperm reproductive cycle.

Although these hypotheses can be neither proved nor completely rejected, they do reflect the role that environmental factors play in the development of concepts regarding the origin of flowering plants. Furthermore, they express the idea that ecological data must be considered in the classification of plants.


## PHYSIOLOGICAL EVIDENCE

Physiological and biochemical evidence is providing data of increasing importance to plant systematics. Of particular significance are data dealing with metabolic systems and biochemical pathways.

Recently, it has become apparent that a syndrome of anatomical and physiological features, related to a high-efficiency carbon fixation process, occurs in a large number of plants from tropical or warm temperate regions (Brown, 1975). This syndrome has been called the *Kranz syndrome,* $C_4$-photosynthesis, or the Hatch-Slack pathway. The most common pathway of carbon fixation is the *Calvin-Benson cycle,* or $C_3$-photosynthesis. In the algae, mosses, most ferns, gymnosperms, and many families of flowering plants, $C_3$-photosynthesis is the only known carbon fixation cycle. $C_4$-photosynthesis occurs in approximately 10 unrelated families of monocots and dicots (Downton, 1975). *Crassulacean acid metabolism* (CAM) has been found in four monocot families, in thirteen dicot families, and in one species of fern. CAM and $C_4$ both represent adaptations to arid climates.

The Kranz syndrome has proved useful in characterizing *Panicum* and other taxa in the grass family (Brown, 1975). In the dicot *Euphorbia* (Euphorbiaceae), $C_3$, $C_4$, and CAM species are known. Both $C_3$ and $C_4$ carboxylation occurs in *Zygophyllum* (Zygophyllaceae) and *Atriplex* (Chenopodiaceae). These findings suggest that the classification of the three genera should be reexamined, but these findings do not imply that the present classification is wrong.

Comparative studies of unique physiological processes of flowering plants might provide data of taxonomic significance. For example, the ability of halophytes to grow in highly saline soils seems to be restricted to relatively few families, but the phenomenon has not been well explained. There is much interest in the field of physiological plant ecology, but the major contributions to date have been in the physiology of ecotypes (Nobs, 1971). As more plants are examined with biochemical and physiological techniques, useful evidence will undoubtedly be found and applied to systematic problems.

## BIOGEOGRAPHY

The geophysical theory that land masses of the earth have been steadily moving during the course of geological time has been called *continental drift,* or more properly, *plate tectonics* (Dietz and Holden, 1968; McKenzie, 1972). The concept of plate tectonics provides a geophysical explanation for the isolation of land areas by sea floor spreading, the uplift of mountain ranges, the formation or disappearance of islands, and the shifting position of continents. Latitudinal changes, rafting of floras, and environmental stresses due to plate tectonics surely played a major role in the evolution of the angiosperms (Raven and Axelrod, 1974; Schuster, 1976). For these reasons, *biogeographers*, who study the distribution of plants and animals, have recently reexamined distribution patterns, fossil records, and evolutionary history of the flowering plants in respect to plate tectonic theory.

The earliest known angiosperm pollen has been found in lower Cretaceous strata (Raven and Axelrod, 1974). During this geologic period, the northern groups of continents (North America and Eurasia), which geologists call *Laurasia,* had largely separated from the southern group of continents (South America, Afric, Australia, Antarctica, and the Indian subcontinent), called

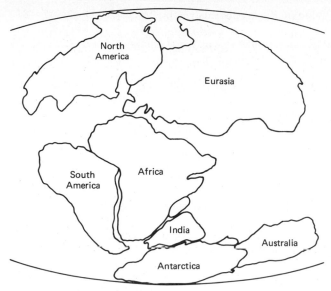

**Figure 5-12**   The relative positions of the continents at the beginning of the Cretaceous at the time the first angiosperms are believed to have evolved. *(Modified from R. S. Dietz and J. C. Holden,* The Breakup of Pangaea, *copyright © 1970; and D. P. McKenzie and J. G. Sclater,* The Evolution of the Indian Ocean, *copyright © 1973 by Scientific American Inc., all rights reserved, and used with permission.)*

*Gondwanaland.* Figure 5-12 shows that the components of Gondwanaland and Laurasia were adjacent to each other or in close proximity when the angiosperms first appeared on the earth near the beginning of the Cretaceous period. The pollen record strongly suggests that much of the radiation of the angiosperms occurred when the continents of Gondwanaland were essentially close together, allowing direct interchange of floristic elements. Taxa have subsequently radiated in various directions, occupying suitable habitats and forming their ranges.

Biogeography has had an enormous influence on the explanation of the evolutionary concepts of flowering plants. Explanation of distribution patterns and disjunct relationships or interruptions in ranges have fascinated and attracted the attention of plant systematists. For examples of recent publications, see Constance et al., 1963; Graham, 1972; Holt, 1970; Jones, 1976; Raven, 1972; and Raven, 1975a. Interpretations of distribution patterns of taxa have been one of the central problems of biogeography.

## SELECTED REFERENCES

Anderson, E.: "Hybridization of the Habitat," *Evolution,* 2:1–9, 1948.
Andrews, H. N.: *Studies in Paleobotany,* Wiley, New York, 1961.

Avise, J. C.: "Systematic Value of Electrophoretic Data," *Syst. Zool.*, **23**:465–481, 1974.

Bailey, I. W.: "The Potentialities and Limitations of Wood Anatomy in the Study of the Phylogeny and Classification of Angiosperms," *J. Arnold Arbor. Harv. Univ.* **38**:243–254, 1957.

Beck, C. B. (ed.): *Origin and Early Evolution of Angiosperms,* Columbia, New York, 1976.

Behnke, H. D.: "Sieve Tube Plastids in Relation to Angiosperm Systematics, an Attempt Towards a Classification by Ultra Structural Analysis," *Bot. Rev.*, **38**:155–197, 1972.

———: "Ultrastructure of Sieve-Element Plastids in Caryophyllales (Centrospermae), Evidence for the Delimitation and Classification of the Order," *Plant Syst. Evol.*, **126**:31–54, 1976.

———: "Transmission Electron Microscopy and Systematics of Flowering Plants," *Plant Syst. Evol. Suppl.* **1**:155–178, 1977.

Bendz, G., and J. Santesson (eds.): *Chemistry in Botanical Classification,* Proceedings of the 25th Nobel Symposium, Academic, New York, 1973.

Bhojwani, S. S., and S. P. Bhatnagar: *The Embryology of Angiosperms,* Vikas Publishing House, Delhi, 1974.

Boulter, D.: "The Use of Comparative Amino Acid Sequence Data in Evolutionary Studies of Higher Plants," *Prog. Phytochem.*, **3**:199–229, 1972.

Brown, W. V.: "Variations in Anatomy, Associations, and Origins of Kranz Tissue," *Amer. J. Bot.*, **62**:395–402, 1975.

Carlquist, S.: "Toward Acceptable Evolutionary Interpretations of Floral Anatomy," *Phytomorphology*, **19**:332–362, 1969.

Clausen, J.: *Stages in the Evolution of Plant Species,* Cornell, New York, 1951.

Cole, G. T., and H. D. Behnke: "Electron Microscopy and Plant Systematics," *Taxon*, **24**:3–15, 1975.

Constance, L., et al.: "Amphitropical Relationships in the Herbaceous Flora of the Pacific Coast of North and South America: A Symposium," *Q. Rev. Biol.*, **38**:109–177, 1963.

Cronquist, A.: *The Evolution and Classification of Flowering Plants,* Houghton Mifflin, Boston, 1968.

———: "The Taxonomic Significance of the Structure of Plant Proteins: A Classical Taxonomist's View," *Brittonia*, **28**:1–27, 1976.

Darrah, W. C.: *Principles of Paleobotany,* 2d ed., Ronald, New York, 1960.

Davis, G. L.: *Systematic Embryology of the Angiosperms,* Wiley, New York, 1966.

Davis, P. H., and V. H. Heywood: *Principles of Angiosperm Taxonomy,* Van Nostrand, Princeton, N. J., 1965.

Delevoryas, T.: "The Role of Paleobotany in Vascular Plant Classification," in V. H. Heywood and J. McNeil (eds.), *Phenetic and Phylogenetic Classification,* The Systematics Association, London, 1964, pp. 29–36.

———: "Paleobotany, Phylogeny, and a Natural System of Classification," *Taxon*, **18**:204–212, 1969.

Dickison, W. C.: "The Bases of Angiosperm Phylogeny: Vegetative Anatomy," *Ann. Mo. Bot. Gard.*, **62**:590–620, 1975.

Dietz, R. S., and J. C. Holden: "The Breakup of Pangaea," *Sci. Amer.*, **223**(4):30–41, 1968.

Doyle, J. A., and L. J. Hickey: "Pollen and Leaves from the Mid-Cretaceous Potomac Group and Their Bearing on Early Angiosperm Evolution," in C. B. Beck (ed.), *Origin and Early Evolution of the Angiosperms,* Columbia, New York, 1976, pp. 139–206.

Downton, W. J. S.: "The Occurrence of $C_4$ Photosynthesis Among Plants," *Photosynthetica (Prague),* **9:**96–105, 1975.

Eames, A.: *Morphology of the Angiosperms,* McGraw-Hill, New York, 1961.

Erdtman, G.: *Pollen Morphology and Plant Taxonomy,* vol. 1, *Angiosperms* (corrected reprint of 1952 edition with a new addendum), Hafner, New York, 1966.

Esau, K.: *Plant Anatomy,* 2d ed., Wiley, New York, 1965.

Faegri, K., and L. van der Pijl: *The Principles of Pollination Ecology,* Pergamon, New York, 1966.

Fairbrothers, D. E., T. J. Mabry, R. L. Scogin, and B. L. Turner: "The Bases of Angiosperm Phylogeny: Chemotaxonomy," *Ann. Mo. Bot. Gard.,* **62:**765–800, 1975.

Faust, W. Z., and S. B. Jones: "The Systematic Value of Trichome Complements in a North American Group of *Vernonia* (Compositae)," *Rhodora,* **75:**517–528, 1973.

Ferguson, I. K., and J. Muller (eds.): *The Evolutionary Significance of the Exine,* Linnean Society Symposium Series, no. 1, Academic Press, London, 1976.

Fitch, W.M., and E. Margoliash: "The Usefulness of Amino Acid and Nucleotide Sequences in Evolutionary Studies," *Evol. Biol.* **4:**67–109, 1970.

Gibbs, R. D.: *Chemotaxonomy of Flowering Plants,* 4 vols., McGill-Queens University Press, Montreal, 1974.

Good, R. O.: *The Geography of the Flowering Plants,* 4th ed., Longman Group Ltd., London, 1974.

Graham, A. (ed.): *Floristics and Paleofloristics of Asia and Eastern North America,* Elsevier, Amsterdam, 1972.

Grant, V., and K. A. Grant: *Flower Pollination in the Phlox Family,* Columbia, New York, 1965.

Grashoff, J., and B. L. Turner: "'The New Synantherology'—A Case in Point for Points of View," *Taxon,* **19:**914–917, 1970. ("Must" reading for those who think taxonomy is a dull subject.)

Hadley, E. B., and D. A. Levin: "Habitat Differences of Three *Liatris* Species and Their Hybrid Derivatives in an Interbreeding Population," *Amer. J. Bot.,* **54:**550–559, 1967.

Hamrick, J. L., and R. W. Allard: "Correlations Between Quantitative Characters and Enzyme Genotypes in *Avena barbata,*" *Evolution,* **29:**438–442, 1975.

Harborne, J. B. (ed.): *Phytochemical Phylogeny,* Academic, London, 1970.

Harper, J. L., J. N. Clatworthy, I. H. McNaughton, and G. R. Sagar: "The Evolution and Ecology of Closely Related Species Living in the Same Area," *Evolution,* **15:**209–227, 1961.

Hawkes, J. G. (ed.): *Chemotaxonomy and Serotaxonomy,* Academic, London, 1968.

Hickey, L. F.: "Classification of the Architecture of Dicotyledonous Leaves," *Amer. J. Bot.,* **60:**17–33, 1973.

Holt, P. C. (ed.): *The Distributional History of the Biota of the Southern Appalachians,* part II, *Flora,* Research Division Monograph 2. Virginia Polytechnic Institute and State University, Blacksburg, 1970.

Janzen, D. H.: "Coevolution of Mutualism Between Ants and Acacias in Central America," *Evolution,* **20:**249–275, 1966.

Jones, S. B.: "Cytogenetics and Affinities of *Vernonia* (Compositae) from the Mexican Highlands and Eastern North America," *Evolution, 30:*455–462, 1976.

Keeley, S. C., and S. B. Jones: "Taxonomic Implications of External Pollen Morphology to *Vernonia* (Compositae) in the West Indies," *Amer. J. Bot., 64:*576–584, 1977.

Kruckeberg, A. R.: "The Implications of Ecology for Plant Systematics," *Taxon, 18:*92–120, 1969.

Leppik, E. E.: "Origin and Evolution of Bilateral Symmetry in Flowers," *Evol. Biol., 5:*49–58, 1972.

Levin, D. A.: "Plant Phenolics: An Ecological Perspective," *Amer. Nat., 105:*157–181, 1971.

Mabry, T. J.: "Pigment Dichotomy and DNA-RNA Hybridization Data for Centrospermous Families," *Plant Syst. Evol., 126:*79–94, 1976.

Macior, L. W.: "Co-evolution of Plants and Animals—Systematic Insights from Plant-Insect Interactions," *Taxon, 20:*17–28, 1971.

McKenzie, D. P.: "Plate Tectonics and Sea-floor Spreading," *Amer. Sci., 60:*425–435, 1972.

—— and Sclater, J. G.: "The Evolution of the Indian Ocean," *Sci. Amer., 228*(5):63–72, 1973.

Maheshwari, P.: *An Introduction to the Embryology of Angiosperms,* McGraw-Hill, New York, 1950.

Metcalfe, C. R.: *Anatomy of the Monocotyledons,* vol. I, *Gramineae,* Clarendon, Oxford, 1960. (There are additional volumes in this series by E. S. Ayensu, D. F. Cutler, C. R. Metcalfe, and P. B. Tomlinson.)

—— and L. Chalk: *Anatomy of the Dicotyledons,* 2 vols., Clarendon, Oxford, 1950.

Mooney, H. A.: "Influence of Soil Type on the Distribution of Two Closely Related Species of *Erigeron,*" *Ecology, 47:*950–958, 1966.

Muller, C. H., W. H. Muller, and B. L. Haines: "Volatile Growth Inhibitors Produced by Aromatic Shrubs," *Science, (Wash. D.C.) 143:*471–473, 1964.

Nobs, M. A., et al.: "Physiological Ecology Investigations," *Carnegie Inst. Wash. Year Book, 69:*624–662, 1971.

Palser, B. F.: "The Bases of Angiosperm Phylogeny: Embryology," *Ann. Mo. Bot. Gard., 62:*621–646, 1975.

Radford, A. E., W. C. Dickison, J. R. Massey, and C. R. Bell: *Vascular Plant Systematics,* Harper and Row, New York, 1974.

Raven, P. H.: "Plant Species Disjunctions: A Summary," *Ann. Mo. Bot. Gard., 59:*234–246, 1972.

——: "Summary of the Biogeography Symposium," *Ann. Mo. Bot. Gard., 62:*380–385, 1975a.

——: "The Bases of Angiosperm Phylogeny: Cytology," *Ann. Mo. Bot. Gard., 62:*724–764, 1975b.

—— and D. I. Axelrod: "Angiosperm Biogeography and Past Continental Movements," *Ann. Mo. Bot. Gard., 61:*539–673, 1974.

Schuster, R. M.: "Plate Tectonics and Its Bearing on the Geographic Origin and Dispersal of Angiosperms," in C. B. Beck (ed.), *Origin and Early Evolution of the Angiosperms,* Columbia, New York, 1976, pp. 48–138.

Skvarla, J. J., and B. L. Turner: "Systematic Implications from Electron Microscope Studies of Compositae Pollen—A Review," *Ann. Mo. Bot. Gard., 53:*220–256, 1966.

Solbrig, O. T.: "Fertility, Sterility and the Species Problem," in V. H. Heywood (ed.), *Modern Methods in Plant Taxonomy,* Academic, London, 1968, pp. 77–96.

Stebbins, G. L.: *Chromosomal Evolution in Higher Plants,* Addison-Wesley, Reading, Mass., 1971.

————: *Flowering Plants, Evolution Above the Species Level,* Belknap Press of Harvard University Press, Cambridge, 1974.

Swain, T. (ed.): *Chemistry in Evolution and Systematics,* Butterworth, London, 1973.

Turner, B. L.: "Chemosystematics: Recent Developments," *Taxon,* **18:**134–151, 1969.

U.S. Geological Survey: *Compendium Index of Paleobotany,* Paleobotany Library, U.S. National Museum, Washington. (A literature index, unpublished, but may be consulted at the Smithsonian Institution in Washington.)

Walker, J. W.: "Comparative Pollen Morphology and Phylogeny of the Ranalean Complex," in C. B. Beck (ed.), *Origin and Early Evolution of Angiosperms,* Columbia, New York, 1976, pp. 241–299.

———— and J. A. Doyle: "The Bases of Angiosperm Phylogeny: Palynology," *Ann. Mo. Bot. Gard.,* **62:**664–723, 1975.

Wodehouse, R. P.: *Pollen Grains,* McGraw-Hill, New York, 1935.

Wolfe, J. A., J. A. Doyle, and V. M. Page: "The Bases of Angiosperm Phylogeny: Paleobotany," *Ann. Mo. Bot. Gard.,* **62:**801–824, 1975.

# Chapter 6

# The Origin and Classification of the Magnoliophyta (Angiosperms)

Among the plants inhabiting the earth, the angiosperms, or flowering plants, have the greatest number of species and occupy more types of habitats than any other group. Their habit includes trees, shrubs, perennials, annuals, and lianas. They range in size from tiny duckweeds to giant trees. Adaptive radiation of angiosperms has produced parasites, saprophytes without chlorophyll, and epiphytes. Insectivorous species such as pitcher plants, Venus flytrap, and sundew represent unusual leaf adaptations. Diversity in flower structure is another remarkable feature of flowering plants. The term *flower* usually refers to a structure containing sepals, petals, stamens, and carpels. However, flowers may consist only of a single ovary or a single stamen.

The origin of the angiosperms is unclear. Charles Darwin called the origin of this group an "abominable mystery" because little evidence was available to trace angiosperms back in time. The origin of the flowering plants is studied by comparisons of morphological characters of modern-day plants and fossil evidence.

## CHARACTERISTICS OF FLOWERING PLANTS

The angiosperms and gymnosperms constitute the seed plants. Generally, the difference between them is the extent to which the ovules are exposed at the time of pollination. The characteristics of angiosperms are the following: (1) vessels in the xylem, (2) sieve elements and companion cells in the phloem, (3) an embryo sac of eight nuclei (egg, two synergids, three antipodal, and two polar), (4) double fertilization, and (5) closed carpels. There are numerous exceptions to these characteristics.

Xylem vessels are not peculiar to angiosperms, because they occur among some gymnosperms and there are certain angiosperms which have vesselless wood. Many exceptions are known to the eight-nucleate embryo sac. The gametophyte generation of flowering plants is generally more simplified and reduced than that found in the gymnosperms. The male and female gametophytes of flowering plants are small and function efficiently with relatively few cell divisions. Angiosperms have *double fertilization,* with one sperm fusing with the egg nucleus and the other sperm fusing with the two polar nuclei in the embryo sac. The latter results in initiation of *endosperm,* the nutritive tissue of the seed. In angiosperms, the embryo is formed directly from the zygote. Double fertilization is not known to occur in other groups of plants. In the angiosperms, the specialized megasporophylls (or carpels) which enclose the ovules are unique and provide the name *angiosperm,* "covered seed." The carpels cover the ovules and facilitate adaptations for seed dispersal. Pollen collects and germinates on a stigmatic surface, forming a pollen tube that carries the sperm nucleus to the egg nucleus located within the embryo sac of the ovule. However, in gymnosperms, the pollen grains must have direct access to the ovule. The germination of pollen on the stigmatic portion of the carpel, and *not* directly on the ovule, is an important feature of angiosperms. Many of the unique reproductive characters of angiosperms, including an efficient and shortened life cycle, may have contributed to their success in the plant world.

The angiosperms share a number of common features. The relative positions of the stamens and carpels do not vary. The carpels are always in the uppermost position of the floral axis. The types of appendages on the floral axis—i.e., sepals, petals, stamens, and carpels—are uniform in the flowering plants. The structure of the angiospermous stamens is constant, with four sporangia in two pairs. The monosulcate pollen of certain dicots and all the monocots is analogous. Xylem vessels and phloem sieve tubes provide the angiosperms with a highly efficient conducting system resulting in their adaptability to arid regions. The leaves of many flowering plants have secondary and tertiary venation with isolated termination of veinlet endings. This is related to transpiration and permits the development of large, flat photosynthetic surfaces. Because of these and other adaptations, the angiosperms have been able to dominate successfully the terrestrial flora of the earth.

The angiosperms are a natural group, sharing characters which make them unique from all other vascular plants (Cronquist, 1968; Stebbins, 1974; Takhtajan, 1969). These common features suggest that the angiosperms be considered loosely monophyletic. This means that the angiosperms are derived not necessarily from a common ancestral species, but perhaps from a group of related organisms. Similar evolutionary changes may have taken place in several closely parallel lines to produce the features recognized today as angiospermous (Cronquist, 1965; Krassilov, 1977). For clues to a group that might have been ancestral to the angiosperms, it is necessary to examine the fossil record.

## FOSSIL RECORD: TIME OF ORIGIN AND DIVERSIFICATION

Paleobotanists have unsuccessfully attempted to determine the immediate ancestors of the angiosperms, and the following three hypotheses were formulated to explain this failure (Hughes, 1976): (1) incompleteness of the fossil record and the lack of fossil flowers; (2) a long and slow evolutionary history of flowering plants; and (3) evolution of early angiosperms in upland areas away from the necessary basins where fossils would have been preserved.

These so-called "escape" hypotheses led some botanists to suggest that the angiosperms originated in the early Mesozoic or even the late Paleozoic and that they underwent extensive diversification by the Aptian-Albian stages of the lower Cretaceous (see Table 6-1 for a geologic time scale). However, after reviewing the supporting evidence for this hypothesis, Wolfe, Doyle, and Page (1975) found ". . . no unequivocal evidence that indicates a pre-Cretaceous origin for the angiosperms." The best available evidence from the fossil record indicates that the angiosperms originated during the early Cretaceous about 130 to 135 million years ago.

In the older Cretaceous sediments, the angiosperm fossil records are overshadowed by those of ferns and gymnosperms, and it is not until the late part of the Cretaceous that the angiosperms become dominant. The first monosulcate angiosperm pollen grains, which are characteristic of the primitive dicots and the monocots, appear in the Barremian stage of the lower Cretaceous. Tricolpate pollen grains, which were found in the more advanced dicots, are first reported from slightly younger Aptian rocks. Pollen sequences have clearly shown a pattern of increasing diversity in the pollen record of the Cretaceous. The pollen of primitive angiosperms, however, is almost indistinguishable from gymnospermous pollen, creating uncertainty about the identification of the older fossil pollen. At an early date the angiosperms split into the monocots and dicots and very slowly gained prominence in the terrestrial flora which was largely dominated by ferns and gymnosperms. The progression of pollen from primitive types in older strata to more derived types in younger sediments indicates to paleobotanists that angiosperms underwent much diversification during the Cretaceous (Raven and Axelrod, 1974; Wolfe, Doyle, and Page, 1975). By the Turonian and Coniacian stages, angiosperm pollen had become more abundant than fern spores and gymnosperm pollen. By the Maestrichtian stage at the close of the Cretaceous, a number of modern families and genera are represented by fossil pollen and leaves. Members of such groups as the Magnoliales, Hamamelidales, Ranunculales, and Theales and some monocots had appeared by the end of the Cretaceous. Evolution and diversification of the angiosperms continued into the Cenozoic. Fossil flowers are known from

**Table 6-1   Geologic Time Scale**

| 10⁶ years | Era | Period | Epoch | Stage | Approximate Age 10⁶ years |
|---|---|---|---|---|---|
| | Cenozoic | Quaternary | Holocene Pleistocene | | 2.5 |
| | | Tertiary | Neogene — Pliocene | | 12 |
| | | | Neogene — Miocene | | 25 |
| | | | Paleogene — Oligocene | | 38 |
| | | | Paleogene — Eocene — Upper | | 46 |
| | | | Paleogene — Eocene — Middle | | 50 |
| | | | Paleogene — Eocene — Lower | | 54 |
| | | | Paleogene — Paleocene | | |
| 65 | Mesozoic | Cretaceous | Upper | Maestrichtian | 70 |
| | | | | Campanian | 80 |
| | | | | Santonian | 85 |
| | | | | Coniacian | 90 |
| | | | | Turonian | 100 |
| | | | | Cenomanian | 110 |
| | | | Lower | Albian | |
| | | | | Aptian | 122 |
| | | | | Barremian (Neocomian) | 125 |
| | | | | Hauterivian (Neocomian) | 127 |
| | | | | Valanginian (Neocomian) | 130 |
| | | | | Ryazanian (Neocomian) | 132 |
| 135 | | Jurassic | | | 135 |
| 180 | | Triassic | | | |
| 225 | Paleozoic | Permian | | | |
| 270 | | Carboniferous | | | |
| 350 | | Devonian | | | |
| 405 | | Silurian | | | |
| 440 | | Ordovician | | | |
| 500 | | Cambrian | | | |
| 600 | | | | | |

*Source:* Modified and adapted from Tschudy and Scott (1969) and Raven and Axelrod (1974).

Eocene strata (Crepet, Dilcher, and Potter, 1974). Although there is some evidence it may be older, pollen of the Compositae (considered one of the advanced families of dicots) appears at the end of the Oligocene (Muller, 1970). Pollen of a relatively advanced type similar to that found in present-day Vernonieae (a tribe in the Compositae) has been found in Miocene sediments.

Only a few records are available of putative dicotyledonous woods of early Cretaceous age. They represent primitive types of xylem and lack the features of advanced angiosperms. Investigations of Cretaceous leaves reveal a pattern of increasing diversity during the Cretaceous. These represent a transition from leaf types common to certain present-day Magnoliales to those morphologically similar to living Rosidae (Hickey and Doyle, 1977).

## ANCESTORS OF THE FLOWERING PLANTS

Morphological and paleobotanical evidence suggests that all the seed plants were derived from a broad but apparently cohesive group of ancient heterosporous ferns and fern-like plants. The monophyletic origin of seed plants is based upon evidence obtained from comparative studies. Both the gymnosperms and angiosperms may connect with the ferns through a gymnospermous seed-producing ancestor (Beck, 1966). The term *gymnosperm* describes a life habit and does not refer to a phyletically cohesive group; *gymnosperms* are conifers, cycads, seed ferns, and so on, which may be interrelated, but *gymnosperm* also refers to the great club mosses that produced naked seeds. Various groups of gymnosperms have been given serious consideration as possible progenitors of the angiosperms. The major groupings of gymnosperms are shown in Table 6-2.

There is practically no evidence to suggest that the subdivision Pinophytina was ancestral to the angiosperms. At one time, the gymnosperm subdivision Gnetophytina was linked to the flowering plants by the Amentiferae theory. This theory regarded the reduced unisexual flowers in the catkins of

**Table 6-2   Classification of the Gymnosperms**

I.  Cycadophytina
    1.  Lyginopteridatae
        a.  Lyginopteridales
        b.  Caytoniales
    2.  Bennettitatae
        a.  Bennettitales
    3.  Cycadatae
        a.  Cycadales
II.  Pinophytina*
III.  Gnetophytina*

*Classes and orders are not given.
*Source:* Adapted from Cronquist, Takhtajan, and Zimmermann, 1966.

such families as Salicaceae and Betulaceae as primitive. Superficially, the floral structures of the Amentiferae resemble the reproductive structures of subdivision Gnetophytina, but any similarities between their reproductive structures are merely examples of convergence.

The subdivision Cycadophytina is an ancient group extending back to the Carboniferous, and it has three classes: Lyginopteridatae, Bennettitatae, and Cycadatae. Class Cycadatae probably evolved from a line of pteridosperms (Lyginopteridatae) (Table 6-2). The Cycadatae, which are dioecious, are probably not the ancestors of the flowering plants. The Bennettitatae have stalked megasporophylls which are not homologous with angiosperm carpels. This excludes them from consideration as the ancestor. The Caytoniales of class Lyginopteridatae with ovules semienclosed in small pouches are regarded by some as too specialized to have been the ancestor (Thorne, 1976). However, Stebbins (1974) believes homologies exist between the angiosperm ovule and the structure of the pouches of the Caytoniales. The unusual two-integument structure of the angiosperm's ovules are explainable by the small pouches.

Since most other groups of gymnosperms have been eliminated, the Lyginopteridales (Pteridiosperms, or seed ferns), with their generally primitive characters and great diversity, remain as possible ancestors. It seems probable that some little-known, unspecialized group of Mesozoic seed ferns evolved very early in the Cretaceous into the first angiosperms (Thorne, 1976). Order Lyginopteridales was most abundant in the Carboniferous, but extended well into the Jurassic before becoming extinct.

The seed ferns had large pinnately compound, net-veined leaves. Their stems were simple or branched, with a vascular cambium. Seed ferns were monoecious, with distinct megasporophylls and microsporophylls. Their seeds were of a primitive type, apparently with a large female gametophyte. The ovules were probably borne on the margins of the leaves. Xylem vessels were not present, but this is not essential since vessels have arisen in many diverse groups of vascular plants due to convergent evolution. Extending well into the Jurassic, the seed ferns were available at the proper time to give rise to the first angiosperms, which might be called *proangiosperms*.

Although we have eliminated some groups as likely ancestors of angiosperms, Darwin's "abominable mystery" is not completely solved. It appears likely that plants having the characters of angiosperms could have evolved from a seed fern ancestral line (Cronquist, 1968). They were, perhaps, mesic pteridosperms which were under strong selective pressures for more efficient reproduction or shortened life cycle in a tropical climate with seasonal drought. The answer may be found in fossils of the late Jurassic or early Cretaceous.

## EARLY FLOWERING PLANTS

In the absence of fossil record, it is of interest to reconstruct hypothetically the early angiosperms. This speculation is subject to modification. Some botanists regard the herbaceous growth habit or form as derived and believe that the first flowering plants were woody shrubs or small trees. There are several lines of evidence used to support this: (1) the prevalence of the woody condition in the gymnosperms (regarded by some as ancestral); (2) the correlation of the woody habit with primitive reproductive characters; and (3) the presence of vesselless wood in some Magnoliales.

The leaves of the earliest flowering plants are thought to have been simple, evergreen, entire, alternate, pinnately veined, glabrous, and with stipules present. Recently, Hickey (1973) has begun to develop a classification of the architecture of dicot leaves which shows their evolutionary development.

The classical theory holds that the primitive flower was solitary, terminal, bisexual, actinomorphic, with numerous sepals and petals (or tepals), stamens, and carpels spirally arranged on an elongate axis. The perianth was made up of thick, leaf-like sepals. The stamens were broad and leaf-like. Carpels were large, lacking both well-developed styles and distinctive terminal stigmas. The ovules were anatropous, later developing into large seeds which were borne in carpels similar to those found in present-day Magnoliaceae (see Figure 6-1). When mature, they formed a cone-like structure of many folded conduplicate carpels, with each carpel maturing into a type of fruit called a *follicle* (Cronquist, 1968; Takhtajan, 1969).

According to Takhtajan (1969), the stamens of the earliest flowering plants were broad, leaf-like microsporophylls with deeply embedded microsporangia and three vascular traces (Figure 6-2a and b). He bases this hypothesis on the

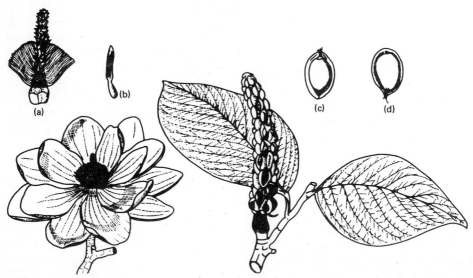

**Figure 6-1** *Magnolia campbellii* (Magnoliaceae). *(a)* Section showing arrangement of stamens and carpels. *(b)* Stamen. *(c)* Seed. *(d)* Section of seed (note tiny embryo). *(From J. Hutchinson,* The Families of Flowering Plants, *Oxford University Press, 1973; with permission.)*

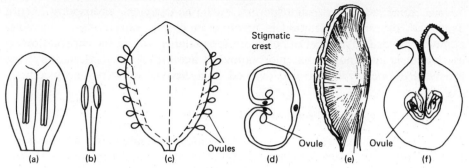

**Figure 6-2**   Stamens and carpels. *(a)* Leaf-like microsporophyll (stamen) believed by Takh-tajan to be primitive. *(b)* Stamen from Winteraceae. *(c)* Hypothetical megasporophyll with ovules and three vascular strands. *(d)* Hypothetical megasporophyll (cross section) closed to form a carpel. *(e)* Primitive conduplicate carpel of *Drimys* (Winteraceae); note stigmatic crest along suture. *(f)* Cross section of a conduplicate carpel of *Degeneria* (Winteraceae); cross section cut in about the position of the dashed line shown in *c.* (*e and f from Foster and Gifford, 1974; used with permission of W. H. Freeman Co.*)

series of intermediate and transitional stamens found in subclass Magnoliidae. Later, stamens evolved into the differentiated anther and filament known today as the typical stamen.

In plants ancestral to the flowering plants, ovules were borne on the margins or scattered over the surface of open megasporophylls. The first angiosperms may have had megasporophylls resembling young leaves, and these were folded along their midrib with the ovules inside (Figure 6-2*c–f*). The megasporophylls upon reduction gave rise to a carpel. A *carpel* is a mega-sporophyll, i.e., a modified leaf-like structure bearing ovules (megasporangia). In the first angiosperms, the margins of the carpels may not have been fused at the time of pollination. Thus, the stigmatic surface, or the area where the pollen must land for germination, was along the entire marginal suture. Later, a re-stricted stigmatic surface developed terminally to the style of a typical pistil, thus giving rise to a pistil with a distinct stigma, style, and ovary.

Takhtajan (1969) suggests that the anatropous ovule, in which the micropyle is bent back toward the funiculus, is primitive among flowering plants. He bases this conclusion on the distribution of anatropous ovules in families regarded as primitive and on the fact that it is widespread throughout the angiosperms. The first flowering plants are believed to have had two in-teguments or tissue layers surrounding the embryo, which later form the seed coat.

Ranalian families possessing primitive characters have an ovule with a large nucleus of many cells and an embryo sac of eight nuclei. A large endo-sperm is found in many of the families today regarded as having primitive char-acters. Some botanists have reasoned that the first flowering plants had a minute embryo embedded in copious oily endosperm. This is similar to the embryo and endosperm of *Magnolia*. Double fertilization is regarded as primi-tive because of its widespread distribution throughout the angiosperms.

In some advanced angiosperms, such as legumes, endosperm is not produced, or else its function is taken over by the cotyledons, which accumulate food. The number of cotyledons, or seed leaves, in the earliest flowering plants is not known, but morphologists agree that monocotyledony (the presence of one cotyledon) was derived from dicotyledony (two cotyledons).

### Alternative Views of the Primitive Angiosperm Flower

The prevailing ideas regarding the primitiveness of angiosperm flowers have changed in recent years (Carlquist, 1969; Gottsberger, 1974; Thorne, 1976). Morphologists and paleobotanists are questioning the concepts of the classical Cronquist-Takhtajan "*Magnolia* is primitive" school. They reject the idea that the long floral axis of *Magnolia,* with its many micro- and megasporophylls, is the prototype of flowers in the angiosperms. They object to the concept of leafy stamens and conduplicate (folded lengthwise) carpels as being primitive. Some paleobotanists are now even suggesting that the first angiosperm flowers were unisexual.

As an alternative hypothesis, some morphologists suggest that the most primitive flowers were probably middle-sized and grouped together into lateral clusters or inflorescences much like those in family Winteraceae. The Winteraceae have small flowers in lateral clusters, with several stamens and one to several carpels. The flowers of *Drimys* (Winteraceae) are shown in Figure 6-3. The Winteraceae are regarded as primitive because they have: (1) great similarity between their micro- and megasporophylls; (2) unifacial (one-front surface) carpels and stamens; (3) high chromosome numbers suggesting a long evolutionary history; (4) morphology similar to the pteridosperms; and (5) vesselless wood similar to that of the gymnosperms (Ehrendorfer et al., 1968). Additionally, the beetle pollination of *Drimys* is of a less specialized type than that of *Magnolia.* The latter has many functional modifications for attracting beetles (Gottsberger, 1974).

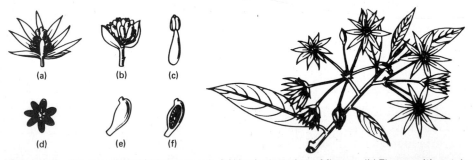

**Figure 6-3** *Drimys winteri* (Winteraceae). *(a)* Vertical section of flower. *(b)* Flower with petals removed. *(c)* Stamen. *(d)* Cross section of carpels. *(e)* One carpel. *(f)* Carpel in vertical section. *(From J. Hutchinson,* The Families of Flowering Plants, Oxford University Press, 1973; *with permission.)*

The botanists opposed to the "*Magnolia* is primitive" theory use the following characters to define a primitive flower: (1) flowers middle-sized, and bisexual; (2) flowers located laterally on the stem and perhaps clustered in an inflorescence; (3) flowers with relatively few stamens and carpels loosely arranged on a short floral axis; (4) similar unifacial micro- and megasporophylls; (5) flowers unspecialized for beetle pollination; and (6) seed dispersal by reptiles, a function later taken on by birds and mammals.

It should be noted that the "*Drimys* is primitive" view is not a drastic alteration from the classical position. Even the most dedicated partisan of either view will recognize that *Drimys* and *Magnolia* are relatively close to each other and that both represent ancient evolutionary assemblages. The question of what is primitive in the angiosperms continues to provide challenges for pollination biologists, plant morphologists, paleobotanists, and taxonomists. Until additional fossils are located and new information evaluated, no one can be sure about the nature of primitive angiosperms.

## INTERRELATIONSHIPS OF ANGIOSPERMS AND ANIMALS

Biologists now realize that the flowering plants are closely interrelated with animals in the evolutionary processes. Noting the amazing correlation between host plants and certain animals (particularly insects), biologists have proposed that patterns of adaptations are generated by a process of reciprocal interactions. Krassilov (1977) believes angiosperm differentiation was directly related to mammalization and other processes in the development of Cenozoic flowering plants. One of the important areas of interaction and adaptation has been that of pollination and flower development.

### Pollination

The ancestors of the flowering plants may have been wind-pollinated. Wind pollination requires plants to release enormous quantities of pollen into the air and is relatively inefficient in terms of energy expended on pollen or effective seed set.

Beetles were plentiful when the proangiosperms appeared and may have been the first animal pollinators. Today they assist with the pollination of *Magnolia, Drimys,* and numerous other flowering plants with primitive characters (Figures 6-4 and 6-5). The beetles may have been attracted by the nutritious qualities of pollen and the sticky sap droplets exuding from the micropyle of the ovules. Beetles feeding on the protein-rich pollen grains carried the pollen to the sticky micropyles. The accidental beetle pollination provided a more efficient method of pollination than chance pollination by wind. The insects—e.g., bees, butterflies, and so on—which we usually think of as pollinators, had not yet appeared when the first angiosperms evolved.

Pollen distribution was more effective, and therefore seed production

**Figure 6-4**   First-day flower of *Magnolia grandiflora,* with beetles *(Trichiotinus piger)* chewing on stigmatic papillae. Note that the inner petals partially cover the gynoecium. Beetles enter the flower just as petals open. Later, as petals open completely, the stigmas turn brown and are not receptive to pollen (about 1X). In the process of eating pollen the insects become coated with pollen, and subsequent visits to receptive stigmas result in pollination. *(Photograph by L. B. Thein.)*

increased when plants efficiently attracted insects. Through natural selection, the ancestral angiosperms developed adaptations to encourage beetle pollination. The development of a bisexual flower with adjacent carpels and stamens made beetle visits more efficient. Plants that provided food in the form of nectar, pollen, sap, and succulent tidbits of tissue attracted more beetles, giving these plants an advantage over plants that attracted fewer insects. This successful attraction of insects created a problem. The ovules of many early flowers were probably exposed on megasporophylls and required protection to prevent them from being eaten. The hollow closed carpel functioned to protect the ovules from insect predation. The protection afforded by the closed carpels allowed the ovules to become smaller and capable of faster development. The development of a shorter life cycle and reduced energy requirement was advan-

**Figure 6-5**  Late second-day flower of *Magnolia grandiflora*. The stigmas have turned brown and the dehisced stamens have fallen from the stalk of the gynoecium. When the stigmas are receptive or the pollen dehisces from the stamens, the petals are partially or completely closed to preserve food for beetles ($^3/_4$ X). *(Photograph by L. B. Thein.)*

tageous under conditions of stress and seasonal drought—conditions under which the angiosperms had an advantage over the gymnosperms and true ferns.

Before the evolution of the closed carpel, each ovule collected its own pollen by way of the micropyle droplet. After the carpel closed, the pollen-collecting function had to shift from each individual ovule to a central stigmatic area. Eventually, during the evolution of the flowering plants, a style and stigma developed which required the germination of pollen some distance from the ovule.

By the beginning of the Cenozoic era, the fossil record shows that the higher insects (including moths, butterflies, bees, wasps, and flies of the orders

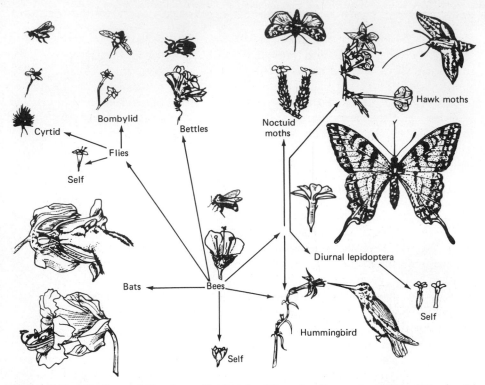

**Figure 6-6** Adaptive radiation for pollination by different pollen vectors in the *Phlox* family Polemoniaceae. *(From Stebbins, 1974; after Grant and Grant, 1965. Used with permission of Belknap Press of Harvard University Press and Dr. Verne Grant.)*

Lepidoptera, Hymenoptera, and Diptera) were undergoing diversification. The rise of the higher insect orders was coincidental with the radiation of flowering plants, which provided energy sources for the insects. The higher insects profoundly influenced the evolutionary diversification of the angiosperms.

The angiosperms underwent a transition from promiscuous pollination by unspecialized beetles to restricted pollination by insects and other animals specialized for flower pollination. By means of natural selection, many plant species developed floral features to accommodate their insect visitors (see Figure 6-6). The evolution of structural change in flowers permitted more effective pollination, resulting in the production of more of its kind. Diversification of the flowering plants also caused natural selection in insects for features adapted for flower visitation. Thus, there was a mutual influence of plants on insects and insects on plants. Transfer of pollen by bees, wasps, moths, butterflies, flies, birds, and bats is generally considered advanced as compared with the primitive but effective pollination by beetles. Of course, there have been reversals in this trend. Secondary wind pollination has reappeared a number of different times in unrelated families, e.g., ragweed (*Ambrosia*) Compositae, and walnuts (*Juglans*) Juglandaceae.

## Seed Dispersal

In the lower Cretaceous, angiosperm seed dispersal probably occurred as the abundant reptile population moved about their habitat. Then, in the late Mesozoic, birds and mammals replaced reptiles as the primary dispersers of seeds. Today flowering plants have many adaptations that promote seed dispersal. Some birds are attracted to the fleshy fruit of wild cherries, and the seeds are scattered about large areas after passing through the digestive tracts of these birds. This is merely a single example of angiosperms spreading over animal ranges.

## Biochemical Coevolution

Some groups of flowering plants have evolved various natural products such as alkaloids, tannins, flavonoids, terpenes, and even insect hormones. Many of these chemicals are regarded as defensive compounds which tend to protect the plants from selected herbivores. Still other herbivores may be found associated with these "protected" plants. Ehrlich and Raven (1964) coined the term *coevolution* to refer to any situation in which a pair of organisms act as selective agents for one another. According to their theory, plants and phytophagous insects have undergone adaptive radiation in a stepwise manner, with plant groups evolving new and highly effective chemical defenses against herbivores and the herbivores continually evolving means of overcoming these defenses. Probably part of the evolutionary success of angiosperms is linked to the chemical defenses they have erected. Ehrlich and Raven (1964) suggest that "... the plant-herbivore 'interface' may be the major zone of interaction responsible for generating organic diversity."

Mustard oils, characteristic of the family Cruciferae, are toxic to many animals. Yet the volatile mustard oils attract other herbivores to plants of that family. The white adult of the imported cabbage worm (Lepidoptera), often seen flying around cabbage and broccoli, is guided to the cabbage by mustard oils. The larvae cause major damage to cultivated cabbages. The imported cabbage worm has exploited the chemical defenses of cabbage and uses the mustard oils to locate a host plant for laying its eggs. Present in the genus *Hypericum,* the chemical hypericin repels almost all herbivores. An exception is the beetle genus *Chrysolina,* which can detoxify hypericin and use it to locate the host plant (Whittaker and Feeny, 1971).

According to coevolutionary theory, not all herbivores will respond to a chemical defense in the same way. For example, laboratory feeding tests with five species of insects and two species of mammals demonstrated that a bitter sesquiterpene lactone from *Vernonia* is an effective deterrent to some *but not all* herbivores (Burnett et al., 1974). Defensive compounds act as filters, screening only a portion of potential herbivores (Burnett, Jones, and Mabry, 1974).

Some natural plant products ingested by insects aid in protecting the insects against predators or microbial pathogens. For example, the monarch butterfly ingests cardiac glucosides from milkweed plants (*Asclepias*). These poisons accumulate in the larval stages of the monarchs and are retained by the adults. The ingestion of monarch butterflies by blue jays causes the jays to become violently sick (Brower and Brower, 1964). Blue jays learn to recognize the toxic, brightly colored monarchs. Thus the monarchs gain protection from predators using the cardiac glucosides of milkweed. (Consult Gilbert and Raven, 1975, for additional references concerning coevolution.)

## CLASSIFICATION OF THE ANGIOSPERMS

Truly phylogenetic classifications are the ultimate aim of taxonomy. Such classifications are highly useful to systematists, other biologists, and lay people. Numerous systems have been developed, such as those of de Jussieu, de Candolle, Bentham and Hooker, Engler and Prantl, Bessey, Hutchinson, and more recently Takhtajan, Cronquist, and Thorne.

Once widely accepted, the Englerian system (Engler and Prantl, 1887-1915) has received considerable criticism from a phylogenetic viewpoint. This criticism is especially directed to the question of relative primitiveness of various groups and the linear sequence of families. One fault is the positioning of the monocots before the dicots, a placement reversed in the most recent edition of the *Syllabus der Pflanzenfamilien* (Engler, 1964). The major problem concerns the position of the Amentiferae, or catkin bearers, and other apetalous groups. These families making up the so-called "Amentiferae" (Betulaceae, Fagaceae, Juglandaceae, and so on) were regarded by Engler and his followers as the most primitive. The Amentiferae (which are now considered to be a phylogenetically advanced, heterogenous group and not a phylogenetic unit) were therefore placed before the petaliferous families such as Ranunculaceae and Magnoliaceae and were regarded as evolutionarily primitive. The flowers and inflorescences of the Amentiferae are now regarded as the product of phenomena similar to those associated with the miniaturization of flowers and inflorescences (Abbé, 1974). The critical weakness of the Englerian system is the failure to recognize the significance of reduction, and for this reason "simple" was equated with "primitive" (Cronquist, 1965).

Since all families were treated in the Englerian system, it became the standard way to arrange herbaria and manuals in the United States. The practice of using this outmoded linear sequence continues mainly because it is a convenient filing system.

Although largely original, the Besseyan system of classification (Bessey, 1915) is rooted in the arrangement of de Jussieu's work (1789), further expanded by de Candolle in 1813, and again later expanded by Bentham and Hooker (1862–1883). Bessey also drew information from the Engler and Prantl classification scheme. The Besseyan school regards the "ranalian complex" (those plants having flowers with many, free, equal, and spirally arranged parts) as primitive. Hence, the phylogenetic system developed by Bessey, often

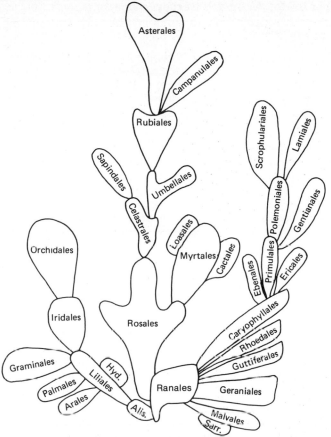

**Figure 6-7**  Bessey's cactus. The chart shows the relationship of orders recognized by Bessey. (Ann. Mo. Bot. Gard., **2**:118, 1915. *Used by permission.*)

called the *ranalian concept of evolution,* is essentially the same as the "*Magnolia* is primitive" school. Diagramed, it took on the form of a cactus plant which came to be called "Bessey's cactus" (see Figure 6-7).

Bessey considered the angiosperms monophyletic and derived from a cycadeoid ancestor with bisexual strobili. Bessey began his scheme with the group considered the least modified from the ancestral prototype. Believing the primitive angiosperms were insect-pollinated, Bessey concluded that the wind-pollinated Amentiferae had resulted from reduction and much evolutionary change. He disagreed with Engler and Prantl, who regarded the Amentiferae as primitive. A comparison of Englerian and Besseyan concepts is presented in Table 6-3.

Bessey's major contribution was the first intentionally phylogenetic classification scheme; he attempted to create a taxonomy that would reflect evolution. In the process, Bessey created his "dicta" (see Chapter 4) to provide the scientific guideline for understanding the construction of his system. In Bessey's

**Table 6-3    Comparison of the Englerian and Besseyan Concepts**

| Features | Englerian school | Besseyan school |
|---|---|---|
| Primitive flower | Apetalous; unisexual | Polypetally; perianth of many, free, equal parts; bisexual |
| Primitive pollination mechanism | Wind pollination | Insect pollination |
| Dicots began with | Amentiferae | Ranales |
| Monocots derived from | Gymnosperous stock | Primitive dicots |
| Ancestors | Coniferoid or gnetoid gymnosperms | Cycadicae gymnosperms, most likely, Pteridosperms |
| Philosophy | Simple flowers are primitive | Many parted flowers are primitive; Aggregation, fusion, reduction, and loss are advanced trends |

cactus the implication of direct ancestry of the orders from one another is open to criticism. Likewise, certain evolutionary lines proposed by Bessey are not accepted today. Regardless of these criticisms, the most recent taxonomic schemes of Cronquist, Takhtajan, and Thorne remain close to Bessey's system.

Hutchinson's classification was a modification of the Bentham and Hooker–Bessey tradition (Figures 6-8 and 6-9). Like Bessey, Hutchinson (1973) considered the flowering plants monophyletic, derived from a cycadeoid ancestor. Hutchinson, however, made the unfortunate error of separating the dicots into two lines: one line primarily herbaceous and the second line primarily woody (see Figure 6-8). This unnatural division caused families with close affinities to be separated. For example, the family Labiatae was separated from the closely related Verbenaceae and the Araliaceae from the related family Umbelliferae. Evidence indicates these pairs of families share many common features. Hutchinson also failed to provide evidence for his taxonomic decisions. However, he was extremely knowledgeable concerning flowering plant families on a worldwide basis, and his family treatments provide a major contribution to the understanding of angiosperms.

Takhtajan (Figure 6-10) is an adherent of the Besseyan philosophy and has been strongly influenced by some of the more progressive German workers. Many consider his division of flowering plants into 18 subclasses a major conceptual advance over previous systems (Takhtajan 1959, 1966, 1969). Published in Russian, the earlier versions of Takhtajan's work were not readily appreciated in the West. This disadvantage was largely overcome with the publication in 1969 of an English translation and summary prepared by Charles Jeffrey. A chart of the Takhtajan system is presented in Figure 6-11. Takhtajan has used much recent information and many diverse characters in the development of his classification scheme. One criticism of his work is that his taxa are so narrowly defined that unnecessary splitting of related groups has occurred. He is a proponent of the "*Magnolia* is primitive" school.

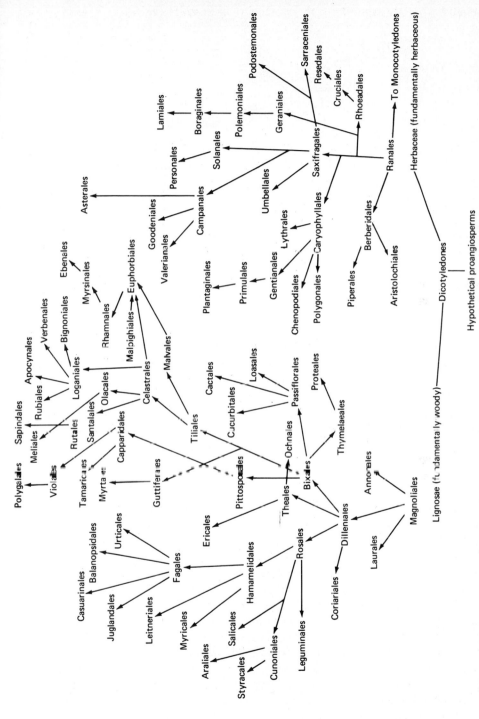

**Figure 6-8** Hutchinson's diagram showing the probable phylogeny and relationships of the orders of dicots. Note the division into the largely woody and largely herbaceous lines. *(From J. Hutchinson, The Families of Flowering Plants, Oxford University Press, 1973; with permission.)*

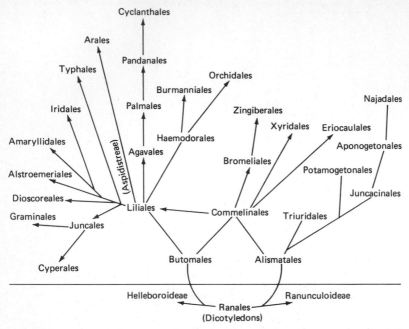

Figure 6-9   Hutchinson's diagram showing the probable phylogeny and relationships of the orders of monocots. Note origin of the monocots from dicot lines. *(From J. Hutchinson,* The Families of Flowering Plants, *Oxford University Press, 1973; with permission.)*

Figure 6-10   Dr. Armen Takhtajan of the Botanical Institute, Academy of Sciences of the Soviet Union, Leningrad.

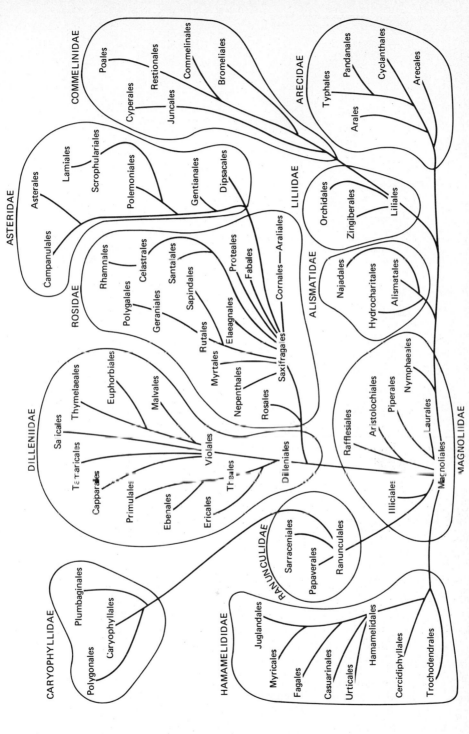

**Figure 6-11** Takhtajan's classification—a chart showing the probable relationships of the flowering plants. *(Published in the Great Soviet Encyclopedia and used with the permission of the Copyright Agency of the Soviet Union and Dr. Takhtajan.)*

**Figure 6-12**  Dr. Arthur Cronquist of the New York Botanical Garden. *(Photograph, New York Botanical Garden.)*

Thorne (1968, 1976) proposed a putatively phylogenetic classification of the angiosperms. His system, also in the Besseyan tradition, divides flowering plants into superorders, orders, and suborders (see Appendix 4). Other systems have been proposed by Gundersen (1950), Pulle (1952), Soo (1953, 1961), Novak (1954), Kimura (1956), and Benson (1957). These are conveniently summarized and compared in Engler (1964).

Cronquist (Figure 6-12) published a classification conceptually similar to Takhtajan's system, but differing in detail (Cronquist, 1968). A modified diagram of Cronquists's classification is shown in Figure 6-13. Since it has been widely distributed in book format and offers detailed explanations, the system of Cronquist is widely used and referred to in America. A comparison of the placement of families in Takhtajan, Thorne, Engler, Hutchinson, and Bentham and Hooker in relation to the scheme of Cronquist is provided by Becker (1973). This has been reprinted in Radford et al. (1974). A brief survey of the arrangement of Cronquist is presented below. Cronquist considers the angiosperms to merit the rank of division (Magnoliophyta). (A list of angiosperm families according to Cronquist's system is provided in Appendix 3.)

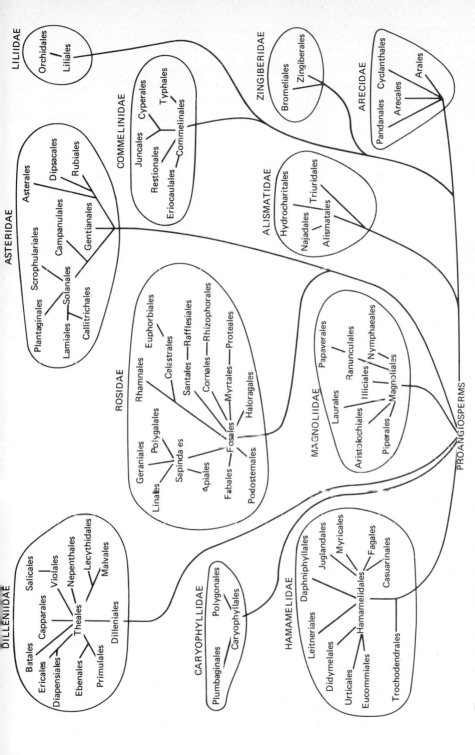

**Figure 6-13** Cronquist's classification—a diagram showing relationships of the flowering plants. The subclasses and orders are those of Cronquist *(1968 and personal communication)*. The chart distance from the center bottom is in rough proportion to prevalence of advanced characters. No living subclass has been derived from any other living subclass. The diagram is designed to be compared with Takhtajan's classification shown in Figure 6-11.

**The Two Classes of Magnoliophyta: Magnoliopsida (Dicots)
and Liliopsida (Monocots)**

Division Magnoliophyta includes all the angiosperms. This natural group consists of two major subgroups: the monocotyledons and dicotyledons. To conform to the International Code of Botanical Nomenclature, formal names with
Latinized endings have been applied to each of the categories, Magnoliopsida
for the dicots and Liliopsida for the monocots. The English names are useful
and continue to be employed. The features by which the dicots and monocots
differ are compared in Table 6-4.

In general, the dicots are much more diverse in habit than the monocots
are. About 50 percent of the present-day species of dicots are woody. Few of
the monocots are woody, and most of these are found in family Palmae. The
Palmae are regarded as relatively advanced in the monocots. Most woody
dicots are highly branched, but the woody monocots are generally unbranched.

Most systematists agree that monocots were phyletically derived from the
dicots early in the evolutionary history of the angiosperms. Cronquist (1968)
believes the monocots arose from a primitive vesselless ancestor resembling
present-day Nymphaeales. If this viewpoint is correct, the monocots had an
aquatic origin profoundly channelizing the evolutionary pathways available to
them. If aquatic in origin, the first monocots probably had apocarpous flowers
with undifferentiated tepals and monocolpate pollen, and they lacked a functional cambium and xylem vessels. Others, such as Cheadle (1953), using evidence from xylem vessels, have rejected the idea of an aquatic origin. Instead,
they suggested that monocots arose from terrestrial woody dicots that lacked
xylem vessels. Regardless of these arguments, the characters (Table 6-4)
separating the monocots from the dicots are usually regarded as secondarily
derived with but one exception. The monocolpate pollen of monocots is shared

**Table 6-4   A Comparison of the Features of Dicots and Monocots**

| Characters | Dicots (class Magnoliopsida) | Monocots (class Liliopsida) |
|---|---|---|
| Cotyledons | 2 (rarely 1, 3, or 4) | 1 (embryo sometimes undifferentiated) |
| Leaves | Usually net-veined | Usually parallel-veined |
| Intrafascicular cambium | Usually present | Absent, mostly without cambium |
| Primary vascular bundles | In a ring | Scattered or in two or more rings |
| Pollen | Monocolpate in primitive families, mostly tricolpate | Monocolpate |
| Floral parts (except carpels) | Sets of 5s or 4s | Sets of 3s, seldom 4s |
| Root system | Primary and adventitious | Adventitious |
| Habit | About 50% are woody | About 10% are woody, mostly in Palmae |

with certain living dicots, as well as with their gymnospermous ancestors. This one exception really poses no problem, since a lineage of pollen grains can be extended from the monocots back through the primitive dicots into gymnospermous stock. In addition to the morphological evidence, data from comparative biochemistry point to an origin of monocots from dicots (Harborne, 1977).

## CHARACTERISTICS OF THE SUBCLASSES OF THE CRONQUIST SYSTEM OF CLASSIFICATION

### Division: Magnoliophyta

Angiosperms or flowering plants (two classes, 11 subclasses, 81 orders, 380 families, and about 220,000 species).

**I   Class: Magnoliopsida**   Dicots (six subclasses, 63 orders, 315 families, and about 165,400 species). (See Table 6-5.)

*1   Subclass: Magnoliidae*   (Eight orders, 39 families, and about 11,000 species.) This subclass includes what are probably the most primitive living angiosperms. Magnoliidae generally have a many-parted, well-developed perianth of tepals, often differentiated into sepals or petals, but sometimes apetalous. The stamens are numerous and mature in a centripetal manner. The pollen typically is binuclcate and monosulcate. The gynoecium is typically apocarpous with bitegmic ovules and crassinucellulate endosperm. All familes in the subclass have endosperm with the exception of the Lauraceae. Magnoliales is the largest order, with about 5600 species.

*2   Subclass: Hamamelidae*   (Eleven orders, 23 families, and about 3400 species.) Hamamelidae is the smallest subclass of dicots. With the exception of some taxa in the order Urticales, the families are typically woody and often contain relatively few species. The subclass is characterized by reduced, often unisexual flowers, which in the advanced groups are arranged in catkins. The perianth is either absent or poorly developed, perhaps in response to wind pollination. The mature fruits contain a single ovule. Evidence is accumulating that this subclass, held together mainly by the bonds of reduced floral structure, probably should be regarded as only loosely monophyletic. Many familiar deciduous trees of eastern North America are members of subclass Hamamelidae.

*3   Subclass: Caryophyllidae*   (Three orders, 14 families, and about 11,000 species.) Several of the families in this mostly herbaceous subclass contain succulents or halophytes. The perianth of the group is morphologically complex and diverse. The primitive members have only a single perianth whorl, and from this, various modified perianths developed with apparent sepals and petals. The stamens are centrifugal in maturation sequence and produce trinucleate pollen. Placentation is free central to basal. The ovules are bitegmic with a crassinucellate endosperm. The ovules are either campylotropous or amphitropous, and when mature the embryos are often surrounded by peri-

**Table 6-5  Some of the General Characteristics of the Dicot Subclasses**

| Subclass | Advancement | Carpels | Flowers | Stamens | Pollen |
|---|---|---|---|---|---|
| Magnoliidae | Relatively primitive | Typically apocarpous | Well developed, with an evident perianth; polypetalous | Numerous; centripetal | Monosulcate, with one germ pore |
| Hamamelidae | Relatively advanced in one or more features | Mostly syncarpous, a few apocarpous | Reduced and often unisexual; perianth poorly developed; wind-pollinated | Often in staminate catkins | Mostly tricolpate, never with only one germ pore |
| Caryophyllidae | Relatively advanced in one or more features | Mostly syncarpous, a few apocarpous | Well developed; usually polypetalous | When numerous, centrifugal | Mostly tricolpate, never with only one germ pore |
| Dilleniidae | Relatively advanced in one or more features | Mostly syncarpous, a few apocarpous | Well developed; usually polypetalous | When numerous, centrifugal | Mostly tricolpate, never with only one germ pore |
| Rosidae | Relatively advanced in one or more features | Mostly syncarpous, a few apocarpous | Well developed; polypetalous (rarely apetalous or sympetalous) | When numerous, centripetal | Mostly tricolpate, never with only one germ pore |
| Asteridae | Relatively advanced in one or more features | Syncarpous | Well developed; sympetalous | The same number or fewer than the number of corolla lobes and alternate with them | Mostly tricolpate, never with only one germ pore |

sperm. Betalains, a distinctive class of pigments, are found in many of the families in order Caryophyllales. Often called the Centrospermae, order Caryophyllales contains around 10,000 species representing the bulk of the subclass.

**4 Subclass: Dilleniidae** (Thirteen orders, 78 families, and about 24,000 species.) With the notable exception of the apocarpous order Dilleniales, subclass Dilleniidae is distinguished from typical members of subclass Magnoliidae by syncarpy. Members of subclass Dilleniidae have a centrifugal maturation of stamens and binucleate pollen, except family Cruciferae, which is trinucleate. Ovules are either unitegmic or bitegmic and have crassinucellate to tenuinucellate endosperm. This subclass contains many woody species.

**5 Subclass: Rosidae** (Eighteen orders, 113 families, and about 60,000 species.) This is the largest subclass, containing about one-third of the dicotyledonous species. The flowers have numerous stamens that mature in centripetal sequence. The ovules are bitegmic or unitegmic and have crassinucellate or tenuinucellate endosperm. The flowers have a polypetalous corolla, although a few apetalous and sympetalous flowers occur.

**6 Subclass: Asteridae** (Ten orders, 48 families, and about 56,000 species.) Subclass Asteridae is the second largest subclass of dicotyledons. Flowers of this subclass are sympetalous and only rarely apetalous or polypetalous. The stamens are few and alternate with the petals. Members of subclass Asteridae usually have two carpels with unitegmic ovules and tenuinucellate endosperm. Of the dicots, subclass Asteridae is probably the best-defined. Present evidence strongly suggests that the Asteridae were likely derived from the evolutionary Rosidae line.

**II Class: Liliopsida** Monocots (five subclasses, 18 orders, 65 families, and about 54,000 species.) (See Table 6-6.)

**1 Subclass: Alismatidae** (Four orders, 15 families, and about 500 species.) Alismatidae characteristically inhabit aquatic or wetland habitats and are herbaceous. Most are apocarpous and have trinucleate pollen. When mature, the seeds lack endosperm. Two subsidiary cells surround the stomates. This subclass is regarded as a relic side branch of the monocots. Subclass Alismatidae has retained a number of primitive characters.

**2 Subclass: Arecidae** (Four orders, six families, and about 6400 species.) In habit, the members of subclass Arecidae vary from tiny macroscopic duckweeds to huge woody palms. About 50 percent of the species are arborescent. The flowers tend to be numerous, small, and often aggregated into a spadix subtended by a spathe (see Chapter 10). The stomatal subsidiary cells are mostly four, but may be two or three. Many species have atypical monocot features of broad, petiolate, net-veined leaves. All but the order Arales contain vessels.

**3 Subclass: Commelinidae** (Six orders, 16 families, and about 16,200 species.) The vast majority of the species of subclass Commelinidae are herbaceous. They inhabit sites ranging from aquatic to terrestrial or even epiphytic. The flowers may have either well-defined sepals and petals or a perianth

**Table 6-6  Some of the General Characteristics of the Monocot Subclasses**

| Subclass | General information | Carpels | Flowers | Stomatal subsidiary cells |
|---|---|---|---|---|
| Alismatidae | Herbaceous, often aquatic; a relic side branch of the monocots | Mostly apocarpous | With many primitive features; stamens numerous | Two |
| Arecidae | Highly variable, ranging in habit from tiny duckweeds to woody palms | Mostly syncarpous | Small, aggregated into a spadix, subtended by a spathe | Mostly 4, but may have only 2 |
| Commeliniidae | Mostly herbaceous; may be terrestrial, aquatic, or epiphytic, but generally terrestrial | Mostly syncarpous | Perianth highly variable, well developed to chaffy, bristly, or lacking; those lacking a perianth are usually wind-pollinated | Several |
| Zingiberidae | A largely tropical group; terrestrial or epiphytic, many xerophytic species | Mostly syncarpous | Regular to irregular, with septal nectaries | Several |
| Liliidae | Largely herbaceous; terrestrial or epiphytic | Mostly syncarpous | With petaloid sepals and petals; highly evolved for insect pollination | Usually absent, but sometimes two or more |

which is chaffy, bristly, or lacking. Primitive Commelinidae have showy insect-pollinated flowers, whereas advanced members with reduced perianths are wind-pollinated. Commelinidae pollen is either trinucleate or, less commonly, binucleate. Family Gramineae, the grasses, supplies much of the food required for human consumption.

 **4  Subclass: Zingiberidae**   (Two orders, nine families, and about 2800 species.) The vast majority of the members of Zingiberidae are tropical and are either terrestrial or epiphytes with regular to irregular flowers, septal nectaries, and an inferior ovary. The two orders are distinctive, and although associated here, many of their features appear to have developed independently. They differ from the other monocots in having retained the floral nectaries and in having achieved epigyny.

 **5  Subclass: Liliidae**   (Two orders, 19 families, and about 28,000 species.) The members of the monocotyledonous subclass Liliidae are syncarpous, with petaloid sepals and petals. They are highly developed for insect pollination. The majority are terrestrial or epiphytic herbs. Leaves are linear and parallel-veined to broader with net venation. The ovary in subclass Liliidae is frequently inferior. Stomatal subsidiary cells are usually absent, but sometimes two or more are present.

## SELECTED REFERENCES

Abbé, E. C.: "Flowers and Inflorescences of the 'Amentiferae'," *Bot. Rev.*, **40**:159–261, 1974.

Beck, C. B.: "On the Origin of Gymnosperms," *Taxon*, **15**:337–339, 1966.

Becker, K. M.: "A Comparison of Angiosperm Classification Systems," *Taxon*, **22**:19–50, 1973.

Benson, L.: *Plant Classification*, Heath, Boston, 1957.

Bentham, G., and J. D. Hooker: *Genera plantarum*, 3 vols., London, 1862–1883.

Bessey, C. E.: "Phylogenetic Taxonomy of Flowering Plants," *Ann. Mo. Bot. Gard.*, **2**:109–164, 1915.

Brower, L. P., and J. V. Z. Brower: "Birds, Butterflies and Plant Poisons: A Study in Ecological Chemistry," *Zoologica*, **49**:137–159, 1964.

Burnett, W. C., S. B. Jones, and T. J. Mabry: "Evolutionary Implications of Sesquiter-pene Lactones in *Vernonia* (Compositae) and Mammalian Herbivores." *Taxon*, **26**:203–207, 1977.

——, ——, —— and W. G. Padolina: "Sesquiterpene Lactones—Insect Feeding Deterrents in *Vernonia*," *Biochem. Syst. Ecol.*, **2**:25–29, 1974.

Candolle, A. P. de: *Théorie élémentaire de la botanique*, Paris, 1813.

Carlquist, S.: "Toward Acceptable Evolutionary Interpretations of Floral Anatomy," *Phytomorphology*, **19**:332–362, 1969.

Cheadle, V. I.: "Independent Origin of Vessels in the Monocotyledons and Dicotyle-dons," *Phytomorphology*, **3**:23–44, 1953.

Constance, L.: "The Systematics of the Angiosperms," in *A Century of Progress in the*

*Natural Sciences, 1853–1953,* California Academy of Science, San Francisco, 1955, pp. 405–483.

Crepet, W. L., D. L. Dilcher, and F. W. Potter, "Eocene Angiosperm Flowers," *Science (Wash. D.C.)* **185:**781–782, 1974.

Cronquist, A.: "The Status of the General System of Classification of Flowering Plants," *Ann. Mo. Bot. Gard.,* **52:**281–303, 1965.

———: *The Evolution and Classification of Flowering Plants,* Houghton Mifflin, Boston, 1968.

———: "Broad Features of the System of Angiosperms," *Taxon,* **18:**188–193, 1969.

———, A. Takhtajan, and W. Zimmermann: "On the Higher Taxa of Embryobionta," *Taxon,* **15:**129–134, 1966.

Eames, A.: *Morphology of the Angiosperms,* McGraw-Hill, New York, 1961.

Ehrendorfer, F., F. Krendl, E. Habeler, and W. Sauer: "Chromosome Numbers and Evolution in Primitive Angiosperms," *Taxon,* **17:**337–353, 1968.

Ehrlich, P. R., and P. H. Raven: "Butterflies and Plants: A Study in Coevolution," *Evolution,* **18:**586–608, 1964.

Engler, A.: *Syllabus der Pflanzenfamilien,* 12th ed., vol. 2, H. Melchior and E. Werdermann (eds.), Gebrüder Borntraeger, Berlin, 1964.

———, and K. Prantl: *Die natürlichen Pflanzenfamilien,* 23 vols., Leipzig, 1887–1915.

Foster, A. S., and E. M. Gifford: *Comparative Morphology of Vascular Plants,* 2d ed., Freeman, San Francisco, 1974.

Gilbert, L. E., and P. H. Raven (eds): *Coevolution of Animals and Plants,* The University of Texas Press, Austin, 1975.

Gottsberger, G.: "The Structure and Function of the Primitive Angiosperm Flower—A Discussion," *Acta Bot. Neerl.,* **23:**461–471, 1974.

Grant, V., and K. A. Grant: *Flower Pollination in the Phlox Family,* Columbia, New York, 1965.

Gundersen, A.: *Families of Dicotyledons,* Chronica Botanica Company, Waltham, Mass., 1950.

Harborne, J. B.: "Flavonoids and the Evolution of the Angiosperms," *Biochem. Syst. Ecol.,* **5:**7–22, 1977.

Hickey, L. J.: "Classification of the Architecture of Dicotyledonous Leaves," *Amer. J. Bot.,* **60:**17–33, 1973.

——— and J. A. Doyle: "Early Cretaceous Fossil Evidence for Angiosperm Evolution," *Bot. Rev.,* **43:**3–104, 1977.

Hughes, N. F.: *Paleobiology of Angiosperm Origins,* Cambridge, London, 1976.

Hutchinson, J.: *The Families of Flowering Plants,* 3d ed., 2 vols., Clarendon, Oxford, 1973.

Jussieu, A. L. de: *Genera plantarum secundum ordines naturales disposita,* Paris, 1789.

Kimura, Y.: "Systéme et Phylogénie des Monocotyledones," *Mus. Natl. Hist. Nat: Not. Syst.,* **15:**137–159, 1956.

Krassilov, V. A.: "The Origin of Angiosperms," *Bot. Rev.,* **43:**143–176, 1977.

Kubitzki, K. (ed): "Flowering Plants—Evolution and Classification of Higher Categories," *Plant Syst. Evol.,* Suppl. **1:**1–416, 1977. (Contains papers by Cronquist, Gottsberger, Dahlgren, Kubitzki, Sporne, Thorne, Wagenitz and others; excellent.)

Lawrence, G. H. M.: *Taxonomy of Vascular Plants,* Macmillan, New York, 1951.

Melchior, H. See Engler, A.

Muller, J.: "Palynological Evidence on Early Differentiation of Angiosperms," *Biol. Rev. Cambridge Phil. Soc.,* **45:**417–450, 1970.

Novák, F. A.: "System Angiosperm," *Preslia, (Prague),* **26:**337–364, 1954.

Pulle, A. A.: *Compendium van de Terminologia, Nomenclatuur en Systematiek der Zaadplanten* 3d ed., Utrecht, 1952.

Radford, A. E., W. C. Dickison, J. R. Massey, and C. R. Bell: *Vascular Plant Systematics,* Harper and Row, New York, 1974.

Raven, P. H., and D. I. Axelrod: "Angiosperm Biogeography and Past Continental Movements," *Ann. Mo. Bot. Gard.,* **61:**539–673, 1974.

Soó, R.: "Die Modernen Grundsätze der Phylogenie im Neuen System der Blütenpflanzen," *Acta Biol.,* **4:**257–306, 1953.

———: "Present Aspect of Evolutionary History of Telomophyta," *Ann. Univ. Sci. Budap. Rolando Eotvos Nominatae Sect. Biol.,* **4:**167–178, 1961.

Sporne, K. R.: *The Morphology of Angiosperms,* St. Martin's, New York, 1975.

Stebbins, G. L.: *Flowering Plants; Evolution Above the Species Level,* The Belknap Press, Harvard University Press, Cambridge, 1974.

Takhtajan, A.: *Die Evolution der Angiospermen,* Gustav Fischer Verlag, Jena, 1959.

———: *Systema et Phylogenia Magnoliophytorum,* Soviet Publishing Institution, Nauka, 1966. (In Russian.)

——— *Flowering Plants—Origin and Dispersal,* Smithsonian Institution Press, Washington, 1969.

Thein, L. B.: "Floral Biology of *Magnolia,*" *Amer. J. Bot.,* **61:**1037–1045, 1974.

Thorne, R. F.: "Synopsis of a Putatively Phylogenetic Classification of the Flowering Plants," *Aliso,* **6**(4):57–66, 1968.

———: "A Phylogenetic Classification of the Angiospermae," *Evol. Biol.,* **9:**35–106, 1976.

Whittaker, R. H., and P. P. Feeny: "Allelochemics: Chemical Interactions Between Species," *Science (Wash. D.C.)* **171:**757 770, 1971.

Wolfe, J. A., J. A. Doyle, and V. M. Page: "The Bases of Angiosperm Phylogeny: Paleobotany," *Ann. Mo. Bot. Gard.,* **62:**801–824, 1975.

# Chapter 7

# Evolution and Biosystematics

By observing the diversity, structural complexity, and adaptative nature of plants, systematists have described taxa, developed classifications, and unraveled the process of evolution. Biosystematics attempts to discover the mechanisms and processes in flowering plants that: (1) direct their evolution, (2) influence their variation patterns, and (3) cause speciation. Classical taxonomists emphasize the evolutionary end products and differences between taxa.

*Organic evolution* is the series of transformations of the genetic materials of populations through time. This is primarily caused by natural selection in response to changing environmental interactions (Dobzhansky et al., 1977). Evolution is not necessarily synonymous with speciation. Evolution is the gradual accumulation of small genetic changes through time which sometimes leads to speciation (Stebbins, 1950).

The term *biosystematics* is derived from biosystematy and was introduced by Camp and Gilly (1943). As originally conceived, biosystematics was an attempt to produce a system of nomenclature to express explicit relationships. Since the 1940s the term has come to mean the application of experimental, genetic, cytologic, and population approaches to systematic problems, especially at the level of species and infraspecific taxa.

112

## SOURCES OF VARIATION

The complexity and diversity of plant life is apparent among species and between individuals of the same species. This variation forms the basis of both
evolution and classification. An understanding of the evolution of plants may
be discovered by careful analysis of this variation. Systematists are confronted
with variability in populations possessing three components: (1) developmental
variation, (2) environmentally induced variation, and (3) genetic variation.

### Developmental Variation

Adult plants are often strikingly different from the immature seedlings. This
developmental variation is genetically determined. The adult leaves of gum tree
(*Eucalyptus*) are alternate and usually pendulous, whereas those on young
seedling shoots of many species are opposite and horizontal to the erect axis.
The seedling leaves of ash (*Fraxinus*) are simple, but the mature leaves are pinnately compound. Because the seedling stage is the most critical in a plant's
life, the characters present during this period surely have survival value. The
needle-like seedling leaves of red cedar (*Juniperus*) and arborvitae (*Thuja*) may
carry on more photosynthesis than the small scale-like leaves on the mature
branches. The biosystematist studies these developmental differences in the
garden and greenhouse, observing them at all stages of their life cycle. Developmental variation has been primarily of interest in understanding morphology
and evolution.

### Environmental Variation

Plants are quite variable, and some species may vary by altering their pattern of
growth in response to environmental changes. Phenoplasticity (changes in appearance) is caused by such variables as light, water, nutrients, temperature,
and soil. A classic example may be found among some species of the aquatic
plant arrowhead (*Sagittaria*). These plants have arrow-shaped leaves when
emergent and long strap-shaped leaves when submerged. The aquatic water-
buttercup (*Ranunculus aquatilis*) has dissected submerged leaves, but on the
same stem will have lobed emergent leaves (Figure 7-1). Mature plants of many
annuals vary in size. Chickweed (*Stellaria media*) will reach only 3 centimeters
in height in poor, dry soil, but will exceed 20 centimeters in rich, moist soil.
Systematists must detect environmentally induced variation. This is
usually done by growing cloned (genetically identical) plants under several different environmental conditions. When experimenting with wild plant material,
the phenotype (the appearance of the plant) is usually rather plastic when grown
in different environments (Clausen, Keck, and Hiesey, 1940). Characters
which are easily modified are size of plant and individual organs, and number of
branches, leaves, or flowers. Those characters which are less easily modified
are lobing, serration, and pubescence of leaves, and the size, shape, and

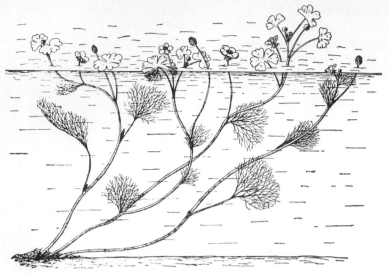

**Figure 7-1** Habit sketch of a water crowfoot *(Ranunculus aquatilis)* with submerged and emergent leaves. *(From Weaver and Clements, 1938; used with permission.)*

number of floral parts. Plasticity may have survival value and contribute to the success of an organism because it allows the plant to adjust to its environment.

### Genetic Variation

The third type of variation in plants is genetic variation (genotypic variation). Genotypic variation, which is heritable, has two sources: (1) mutation and (2) gene flow and recombination. *Mutation* is a transmissible change in the hereditary material. It is the ultimate source of variation in species and replenishes the supply of genetic variability. When broadly used, the term *mutation* includes both genic or point mutations (nucleotide substitutions) and chromosomal mutations (Stebbins, 1971a).

**Gene Mutations**   A gene mutation is an alteration in the sequence of nucleotides which brings about a change in the action of the gene (Grant, 1975). In higher plants, genic mutations are known to affect many features of plants, e.g., the hairless peach or nectarine. The role of genic mutations in evolutionary change depends upon several factors: (1) the amount of their affect upon the organism, (2) the adaptive advantage or disadvantage of the mutant individuals, and (3) the role of the mutant gene in population-environment interactions. The environment in which a mutation occurs will have a great effect on its incorporation into the local populations. Unless the environment is changing, mutations will most likely lower the level of adaptation of the species.

If a species is adjusted well to its environment, small mutations might possibly allow it to inhabit new or changing environments, but drastic changes are almost certain to make it function poorly. Hybrids between two distinct ecotypes of *Potentilla glandulosa* demonstrated that their features were controlled by a series of genes called *multiple genes* (Clausen, 1951). With multiple genes, the hereditary differences in quantitative characters are usually determined by two or more independent genes having similar and cumulative effects (Grant, 1975). Each of these local populations of *Potentilla* had become specialized or adapted for living under one particular set of conditions. Adaptedness in *Potentilla,* therefore, was achieved by the accumulation of many small changes through time rather than by large mutations (Grant, 1963).

**Chromosomal Mutations**    These include: (1) *polyploidy,* or multiplication of chromosome sets; (2) *aneuploidy,* or loss or gain of chromosomes from the set; and (3) gross structural changes in the chromosomes themselves. Structural changes may be any of the following: deficiencies or loss of segments from a chromosome; duplications of chromosome segments; translocations, or interchange of chromosome segments among nonhomologous chromosomes; or inversions of segments of chromosomes. Structural changes in chromosomes do not produce new genes but function in providing unique gene combinations, new gene arrangements, and different linkage groups (Stebbins, 1971b).

**Gene Flow and Recombination**    The movement and exchange of genes between breeding populations is described as *gene flow. Recombination* is the result of new combinations of previously existing genes. Gene flow and recombination involve genes already existing in nature. These processes are the main, immediate sources of variability in nature and are brought about by cross-fertilization between individuals, populations, and species. Recombination produces new gene arrangements by cross-fertilization and by the interchange (crossing over) of segments of homologous chromosomes followed by independent assortment of the chromosomes at meiosis. Many genotypic variations in a population may be due to recombination of genic differences which have existed in the population system for many generations. Of major importance is that the potential for this variability is already present in the populations (Grant, 1963). Biochemical polymorphisms have clearly demonstrated the existence of great stores of variability in natural populations of various organisms (Avise, 1974; Hamrich and Allard, 1975).

**Mode of Reproduction**    The modes of reproduction in plants must be understood in order to evaluate recombination. These include: (1) *cross-fertilization,* (2) *self-fertilization,* or *autogamy,* (3) *vegetative propagation* from buds, and (4) *agamospermy* from seeds developed without fertilization (Grant, 1971). If vegetative propagation or agamospermy tend to replace sexual reproduction, the process is referred to as *apomixis* (Stebbins, 1950). Apomixis is commonly found in plants associated with interspecific hybridization and high polyploidy. It is a reproductive strategy that overcomes problems of chromosome distribution at meiosis (Grant, 1971). Autogamy, or self-fertilization, tends to reduce recombination, but cross-fertilization enhances recombination.

## NATURAL SELECTION

In order for evolutionary change to occur, there must be a source of genetic variation and a driving force. The prime force is called *natural selection,* and it is the most important of the evolutionary forces or processes. The consequence of natural selection is *adaptation,* or adjustment to the environment. Natural selection is the mechanism by which populations become modified in response to the environment (Grant, 1963).

Natural selection operates by differential reproduction of certain genotypes. The favored genotypes are those which have higher reproductive success and consequently contribute a disproportionate share of individuals to the succeeding generation. The adaptations of plants which permit them to live successfully in given environments generally involve a combination and coordination of characters. Natural selection establishes and preserves favorable gene combinations ranging from simple evolutionary changes to changes involving the habit of the plant.

Evolutionists have described three kinds of selection (see Figure 7-2): (1)

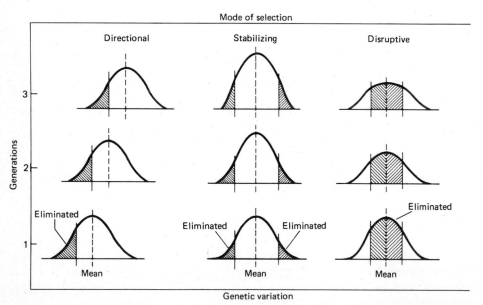

**Figure 7-2** The three kinds of selection: directional, stabilizing, and disruptive. Directional selection operates in a changing environment, stabilizing selection in a stable environment, and disruptive selection in a mosaic environment. *(From Grant 1977; used with permission.)*

stabilizing, (2) directional, and (3) disruptive (Grant, 1977). These are determined by the organism's environment. Stabilizing selection favors "normal" individuals—i.e., it eliminates variants, or deviations from the norm—and operates in a more or less constant environment. Directional selection operates in a unidirectional changing environment, i.e., changes from a moist to dry climate following a modification of the weather pattern. Divergent selection pressures in a heterogeneous environment produce disruptive selection. This breaks up a uniform population. An example is the formation of ecotypes adapted to their own local environmental conditions.

## RANDOM EVENTS

Theoretically, in the absence of natural selection, gene frequencies will remain relatively constant in natural populations. However, gene frequencies may fluctuate owing to the random sampling of genes transmitted in each generation. This effect is greatest in small, isolated populations. It may be verified by tossing coins; i.e., a few tosses may give a disproportionate number of heads or tails, but in a large sample the number of heads and tails will be almost equal. This random sampling of genes is known as *genetic drift*. The overall importance of genetic drift in the evolution of flowering plants has long been debated, but it undoubtedly occurs in nature. It is difficult, if not impossible, to find good examples of drift in natural populations because of the interaction of drift with natural selection (Stebbins, 1971a). Some of the best examples are with human blood groups, but even here recent evidence indicates selection may be involved.

The *founder principle* is a random event related to the chance distribution of individual seeds or propagules that colonize new areas. For example, a colonizing seed which arrives on an island or some other isolated area obviously cannot bring a complete sample of the genetic diversity of that species to the new area. Simply by chance, that seed could carry some genetic variant and establish a new population distinctly different from the original populations.

The gene frequencies of a population system may also be altered by a *genetic revolution* associated with a sudden reduction in the size of a population. By chance, the few survivors may possess systems of variability different from those found in the original population.

There has been an unresolved controversy regarding the adaptive nature of much biochemical or molecular variation (Ayala, 1974). One school of population genetics holds that gene frequency changes in biochemical traits have occurred solely through random processes. Another school holds that these changes have been governed by natural selection. The implications of this disagreement are yet to be fully understood, but quite possibly chance may be shown to play a greater role in evolution than is currently believed to be the case.

## VARIABILITY IN POPULATIONS AND RACIAL DIFFERENTIATION

Many years of experimental work have clearly demonstrated that local populations of species of flowering plants are highly variable. This variation is in morphological features and in cryptic differences such as allozymes, chromosomes, natural plant products, and so on (Heslop-Harrison, 1964). Both *clinal variation* (gradual changes) and mosaics of distinctive ecotypes may be recognized. The type of pattern depends upon the geographical and ecological distribution of the species concerned. Forest trees, prairie grasses, and other plants found over large areas in which environmental conditions change gradually tend to have clinal variation. However, plants occupying territories that are broken into mosaics of sharply distinct ecological conditions tend to have *mosaic variation*. Mosaic patterns are often related to soil and moisture conditions.

*Experimental cultivation* is a major technique for exploring phenotypic plasticity of species and for detecting differences between ecotypes (Langlet, 1971). In the early 1800s A. P. de Candolle advocated cultivation of critical forms side by side. In the middle 1800s the French botanist A. Jordan used comparative culture to study many species native to France. Finding much ecotypic variation, Jordan split many good Linnaean species. For example, he divided *Erophila verna* into 200 entities which are now referred to as "Jordanons." This was the ultimate in splitting.

In the 1920s G. Turesson pioneered the use of experimental cultivation to show the genetic differentiation of populations. Turesson coined the word *ecotype* for genetically adapted ecological races and proposed the term *genecology* for studies dealing with the ecology of ecotypes (Langlet, 1971). Turesson investigated several common species of plants from southern Sweden which grow in diverse habitats. When samples of these species were grown in a uniform garden, Turesson found they differed in a number of features, indicating genetic differentiation. He was able to demonstrate climatic and edaphic (soil) ecotypes in a number of herbaceous species. Shortly afterwards at Edinburgh, J. W. Gregor, working with *Plantago,* was able to demonstrate an *ecocline*. This shows genetically based gradual changes in variation of a species, correlated with an observable gradient in environmental conditions.

The classic work in North America was by a team of scientists: Jens Clausen, a cytologist; David Keck, a taxonomist; and William Hiesey, an ecologist. In the 1930s they established experimental plots at three locations in California which differed greatly in climate and altitude (Clausen, 1951). One experimental plot was at Stanford University at an elevation of 10 meters; another at Mather on the western side of the Sierra Nevadas, at 1400 meters; and the third at Timberline at an elevation of 3000 meters, near the summit of the Sierra Nevada. For their studies, they selected perennial plants that grew along the transect from sea level to alpine conditions. They selected plants which grow in clumps. These could be divided to produce genetically identical clones which were transplanted into each of the three gardens (see Figures 7-3 and 7-4). Thus, genetically identical plants were compared in the differing envi-

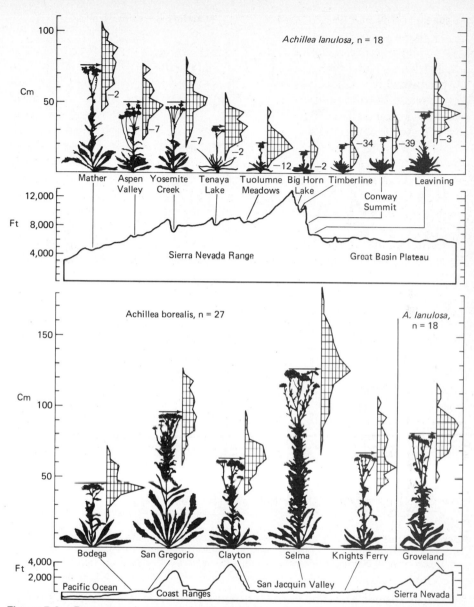

**Figure 7-3.** Response of races of *Achillea* when grown in the garden at Stanford. Plants were taken from an east-west transect across central California. The arrows indicate the origin of the populations, frequency diagrams indicate variation in height, and the plant represents the mean. *(From Clausen, 1951; with permission.)*

ronments of each location. Their carefully documented work clearly demonstrated that local populations are distinctly adapted to their own environments. The experimental work of the California transplant experiments paved the way for much recent genecological work and had great influence on plant taxonomists.

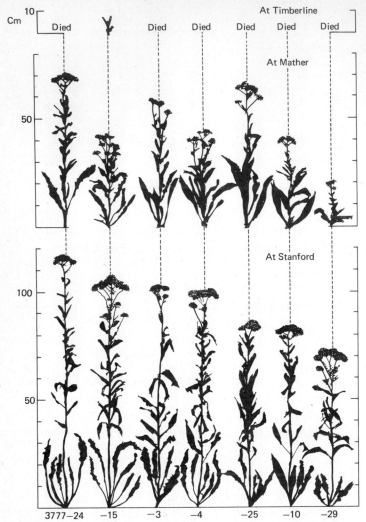

**Figure 7-4** The response of cloned individuals of *Achillea borealis* from the coastal region of California when transplanted to the gardens at Stanford (near sea level), Mather (4000 feet), and Timberline (10,000 feet.). *(From Clausen, 1951; with permission.)*

Most species are subdivided by their distribution and breeding patterns into small local populations (Grant, 1963). This provides an ideal situation in which natural selection may operate on the genetic variation without interference from gene flow from nearby populations. These local populations may differentiate ecologically and geographically and eventually form genetically different ecotypes. Such ecotypes or races may be the basis of a taxonomic subspecies or variety. Pure races in the sense of homogeneous population systems do not exist in nature (Grant, 1963). Instead, geographical and ecological gradients are found in the variation pattern (Figure 7-5). Races normally

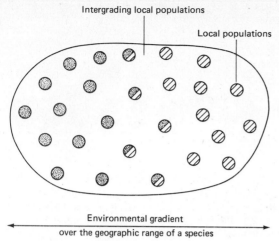

Figure 7-5 A diagrammatic drawing of the range (large circle) of a species and its local populations showing geographical differentiation in morphological and ecological features and intergrading local populations. A species is generally not continuous throughout its range but occurs in local populations in suitable habits.

intergrade if they continuously inhabit territory which provides intermediate ecological zones. If races are to be described and given names, they are usually delimited by sharp discontinuities in their variation pattern.

New developments in population genetics have made it possible to survey a random sample of the genotype of an organism and to detect variation in individual gene loci (Dobzhansky et al., 1977). Genetic information is encoded in the nucleotide sequence of the DNA molecule and translated into the amino acids making up a protein or enzyme. Therefore, variation in proteins suggests variation in the gene.

*Gel electrophoresis* is used by population geneticists to detect the amino acid substitutions in individual proteins or enzymes. Proteins and enzymes are extracted from plant parts. This extract is placed in a gel. The gel, along with the extract, is subjected to an electric current. Since each protein in the sample migrates in the electrical field of the gel at rates depending upon their molecular size and charge, it is possible to visualize their position by using stains that are specific for certain proteins or enzymes (Figure 7-6). For details of these procedures see the appendix at the end of this chapter. Such techniques permit detection of otherwise hidden genetic variation. Around 15 or 20 individual enzymes or proteins studied in a suitably sized sample are considered sufficient to estimate the amount of genetic variability in a population. Since not all variants are detectable by the use of gel electrophoresis, the variation is considered to be underestimated.

Although the technique of gel electrophoresis has some limitations, such studies have demonstrated that many organisms have a large store of genetic variation. Wild oats (*Avena barbata*) collected on a hillside transect from moist (bottom of hill) to dry (top of hill) differed appreciably in the frequency of genes

**Figure 7-6** Photograph of illustrative gel of 16 individual oysters with genetic variation at the PGM (phosphoglucomutase) locus. Oysters 1 and 2 (at the left) have alleles 2 and 3; oyster 3 has alleles 1 and 2; and oyster 4 has alleles 2 and 4. The gel indicates the genotype of the oyster for the PGM locus. *(Photograph furnished by Wyatt Anderson.)*

at six loci (Hamrich and Allard, 1975). Although this technique has been applied to animals more than to plants, it is finding increasing favor with plant taxonomists.

## ISOLATION AND THE ORIGIN OF SPECIES

In his now classic *Variation and Evolution in Plants,* Stebbins (1950) noted that the most important principle derived from evolutionary studies was that evolution and the origin of species are not synonymous as implied by the title of Darwin's classic book. The present definitions of species stress the importance of genetic and morphological continuity within species and the discontinuity among closely related species. Yet systematists realize that an array of genetic variation may exist within the species. Infraspecific taxa are based upon incomplete morphological separation as reinforced by ecological and geographical isolation. These two factors are the keys to recognizing genetically based distinctions. Therefore, taxa are maintained by limited gene exchange and adaptations to local environmental conditions.

### Reproductive Isolation

Isolating mechanisms are phenomena which prevent interbreeding of closely related taxa. They are based upon the following factors: (1) spatial distance, (2) environmental factors, and (3) reproductive biology. Isolating mechanisms are usually placed in one of two categories: *prezygotic* mechanisms, which prevent fertilization, or *postzygotic* mechanisms, which function after fertilization. A classification of reproductive isolating mechanisms is given in Table 7-1. In general, most instances of isolation in nature involve combinations of one or more mechanisms rather than a single mechanism.

**Table 7-1   A Classification of Reproductive Isolating Mechanisms**

**Prezygotic mechanisms**

> *Geographical.* Spatial distance prevents interbreeding because the taxa do not occur in the same area.
> *Ecological.* The plants live in the same region but grow in distinctly different habitats.
> *Seasonal.* The plants occur in the same region but flower at different times.
> *Ethological.* The plants are isolated from each other by pollinator behavior patterns.
> *Mechanical.* Fertilization is prevented by structural differences in flower parts.

**Postzygotic mechanisms**

> *Hybrid Inviability.* The hybrid zygote dies after fertilization, or inviability occurs at some developmental stage before flowering.
> *Hybrid Sterility.* First-generation hybrids are produced which flower, but the stamens abort or the gametes are infertile.
> *Hybrid Breakdown.* First-generation hybrids are normal and produce fertile pollen, but second- or later-generation hybrids are weak and have reduced fertility.

*Source:* Grant, 1963; Stebbins, 1971a.

*Geographical isolation* exists between two species whose respective geographical ranges are separated by gaps greater than the normal radius for dispersal of their pollen and seed. *Platanus occidentalis* and *P. orientalis* occur naturally in eastern North America and in the Mediterranean region. They are morphologically distinct and effectively isolated by spatial distance. Yet, when brought together in the garden, they form fully fertile hybrids.

When the ranges of two taxa do not overlap, as in the previous example, the species are said to be *allopatric*. When the ranges of two species overlap, the species are described as being *sympatric*.

*Ecological isolation* results from differentiation in habitat requirements. Although plants might live in the same region, interbreeding is prevented because they grow in different habitats. Two species of *Vernonia* occur in middle Georgia: *Vernonia noveboracensis* in sunny, moist roadsides and meadows; and *V. glauca* in shady, well-drained, mature hardwood forests. Where the two species are sympatric, hybrids are rarely found, although the cross may be easily made in the greenhouse.

Genetically controlled differences in reproductive features and habits may also block gene exchange. Sympatric species may be *seasonally isolated* and flower at different times. The ecological isolation between *V. noveboracensis* and *V. glauca* is reinforced by differences in flowering time. One species blooms in midsummer and the other in late summer.

*Ethological isolation,* or incompatible behavior patterns, is a very important isolating mechanism in animals. When interpreted to include plant reproduction, it involves the flower-visiting behavior of some insects and birds. Certain hummingbirds, bees, and hawkmoths as well as other insects and birds are species-specific. They have sensory perceptions enabling them to distinguish different flowers on the basis of odor, color, form, and so on (Grant, 1963). For example, hummingbirds are attracted to red flowers and hawkmoths to white flowers. Certain pollinators are flower-constant and have instincts which cause them to feed preferentially on one species during a succession of feeding visits. The advantage of flower-constant behavior to the insect is a matter of efficiency and energy (Heinrich and Raven, 1972). Once a colony of plants in flower is located and the floral mechanism learned, an insect can obtain more nectar and pollen in a shorter time by continuing to visit the same species.

The structural differences in flower parts may prevent cross-pollination. This *mechanical isolation* occurs in families with complex floral mechanisms, such as the Asclepiadaceae and Orchidaceae. In both of these families, pollen is released in pollen sacs or pollinia of various sizes and shapes. The proper pollinia must be inserted by the pollinator into slits in the gynoecium. The location of the pollinium on the insect, whether on the head or thorax, may be very important for successful insertion of the pollinium (Pijl and Dodson, 1966).

Interspecific pollination does not necessarily lead to the formation of successful and fertile hybrids. Many developmental steps occur between pollination and production of mature fertile hybrids. Developmental blocks may occur in various stages in the production of sexually mature plants: e.g., the pollen tubes fail to reach the ovules, the sperm fails to unite with egg, the hybrid zygote dies, the endosperm disintegrates causing death of the embryo, the fruit fails to set, the number of seeds is reduced, seeds fail to germinate, seedlings die, anthers abort, or gametes are sterile (Grant, 1963).

In *hybrid sterility,* first-generation hybrids are produced which flower, but have aborted stamens, or more typically, infertile pollen. Hybrid sterility may be in part developmental, involving the development of reproductive structures. It may also result from abnormal segregation of chromosomes at meiosis. Doubling the chromosome number may partly overcome problems with meiosis but has no effect on developmental sterility. Segregational sterility is caused by translocations, inversions, and other chromosomal rearrangements, as well as by cryptic structural differences in the chromosomes. Hybrids between two species may exhibit any degree of reduction in fertility. Some hybrids will be fully fertile, others semifertile, and still others completely sterile.

The final barrier to gene exchange between species, *hybrid breakdown,* occurs in second- or later-generation progeny. In some interspecific crosses in *Vernonia,* the first-generation hybrids are completely fertile, but the second-generation progeny exhibit continuous variation from fully fertile and vigorous to completely inviable. The causes of hybrid breakdown are not fully understood but are probably related to chromosome structure, gene content, and also to cryptic structural differences in the chromosomes. These cause developmental and segregational difficulties which interfere with the physiology and the formation of gametes.

Isolation mechanisms develop as a by-product of evolutionary divergence. As parts of population systems differentiate ecologically, morphologically, and biochemically, barriers to gene exchange begin to develop. For example, the presence of a particular pollinator in large numbers in one part of the range of a species might, through natural selection, channelize the adaptation of floral structures toward those best-suited for that insect. The gradual accumulation of structural differences in the chromosomes in subpopulations will eventually lead to the origin of chromosomal barriers. The origin of isolating mechanisms is no different from any other type of revolutionary change.

### Speciation

*Speciation* is the evolutionary divergence and differentiation of a formerly homogeneous population system into two or more separate species. Speciation requires genetic variations, which are universal in population systems, plus isolation. In the late 1860s the German zoologist Wagner (1889) and the Austrian botanist Kerner (1896–1897) formulated the idea that geographical and ecological radiation of populations is paralleled by morphological and physiological differentiation. Kerner also suggested that new races originate in areas marginal to the main population. The American zoologist Mayr and other evolutionists have stressed the need for spatial distance as a prerequisite for speciation. This concept is designated the *geographical theory of speciation*.

Basic to this idea is the principle that before populations can differentiate, there must be divergent pressures of natural selection. These conditions generally must be met for any divergence, whether it is the formation of isolating mechanisms or that of races or species (Grant, 1963). Furthermore, the divergent element must be isolated enough so that the initial genetic divergence will not be overwhelmed by gene exchange with neighboring populations; e.g., on lead mine tailings there is selection for lead tolerance in grasses within 1 square meter, but speciation does not occur because of the proximity of nontolerant plants of the same species. In related animal groups, protein and enzyme differences closely correspond to levels of morphological divergence into races, subspecies, species, or genera (Avise, 1974).

There are basically two modes of evolutionary change: phyletic evolution and divergent evolution (Figure 7-7). A change from one state to another state

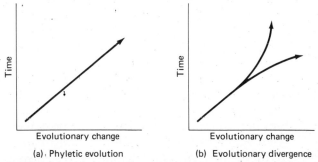

(a) Phyletic evolution          (b) Evolutionary divergence

**Figure 7-7** The two modes of evolutionary change: (a) phyletic evolution and (b) evolutionary divergence. In phyletic evolution, the species moves in time from one state to another state, but in evolutionary divergence one species diverges through time into two species. *(From Grant, 1963; with permission.)*

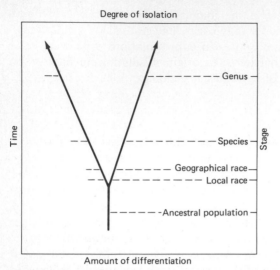

**Figure 7-8** States of divergence. As the divergence increases, the amount of isolation increases. *(Modified from Grant, 1977; used with permission.)*

through time in which a species might evolve into something different from its ancestor is called *phyletic evolution.* In contrast, a single population system which differentiates into two evolutionary lines is called *divergent evolution.* As populations become genetically differentiated, isolating mechanisms also begin to develop (Figure 7-8). Partial reproductive isolation is commonly present between local populations of a species owing to spatial distance, season of flowering, pollinating agents, ecological requirements, and so on. Differentiation of ecotypes in response to their environment reinforces the development of isolating mechanisms. As an ecotype becomes specialized for its environment, isolation mechanisms may build up, reducing gene exchange. As divergence increases, the elements of the populations become less and less able to exchange genes. Divergence leads to isolation, and isolation promotes speciation; i.e., one event reinforces the other.

There has been unresolved debate as to whether divergence without geographic isolation, or *sympatric speciation,* can occur. Sympatric speciation involves the differentiation of closely related populations that coinhabit a geographical area—i.e., are sympatric. The answer to this is not yet clear. Theoretically, the possibility of sympatric speciation seems more likely if the plants are self-fertilizing, because this would tend to restrict gene flow. Although many evolutionists have minimized sympatric differentiation, it might happen under strongly divergent ecological differentiation, reduction of outbreeding, or structural changes in the chromosomes. In one example, there was rapid (40 to 60 years) differentiation of populations of *Anthoxanthum odoratum,* an outbreeding grass, on small 20 by 35 meter plots in response to pH (Snaydon, 1970). Variation in chromosomes not correlated with ecogeography has been reported in *Lasthenia* in California.

## HYBRIDIZATION

The role of hybridization in evolution has been a controversial topic in systematics (Grant, 1963). Linnaeus suggested a hybrid origin for several plant species. In the late 1800s and early 1900s, it was believed hybridization could be the starting point for new species. Until the early 1950s many floristic botanists were reluctant to recognize hybrids in nature. Today plant systematists regard interspecific hybrids to be common in certain genera. The sunflower (*Helianthus*) and oak (*Quercus*) have species that are difficult to identify, because there has been interspecific hybridization blurring species lines.

Natural hybridization is more frequent in perennial plants than in annuals. The perennial nature ensures that occasional hybridization between species will not interfere with their existence, since the genotypes are preserved by the perennial habit. Therefore, there has not been strong selection against hybridization in perennials. Interspecific hybrids are more common in plants than in animals. There are several reasons for this: (1) ethological isolation is more effective with animals; (2) zoologists may have overlooked some hybridization because animals are not as easily collected as plants; and (3) because of the complex structure and sequence of development of animals in relation to plants, hybridization is more likely to cause problems in gene interaction and development in animals than in plants.

First-generation hybrids generally have characteristics which are intermediate between the parental species. Such hybrids represent crosses between normally well-adapted parents. First-generation hybrids are also frequently intermediate in their habitat requirements. They are usually at a competitive disadvantage unless the environment is changing or unless intermediate habitats are available. Two sympatric species may be ecologically isolated, with one species growing in a wet place and the other in a dry place. They may hybridize when intermediate habitats are located between adjacent populations of the two, such as along a wet to dry transect. An example of this occurs between *Vernonia fasciculata* and *V. baldwinii*.

Following natural hybridization, first-generation hybrids typically backcross to one of the parental species. There are several reasons for this. More plants with the parental genotype are present, increasing the chances of crosses with a member of the parental group. Fertility of the hybrids may be lower, so that they produce less-viable pollen than do the parents. Therefore, the most common effect of interspecific hybridization is successive backcrossing of hybrids with the parents. This results in the reversion of hybrid offspring toward the parental types. Backcross hybrids are usually more successful because they tend to approach the adaptive peaks of their parents. Backcrossing may result in the movement of genes from one species to another via the hybrids and backcrosses. This is referred to as *introgressive hybridization* (Anderson, 1949). Introgressive hybridization may lead to changes in the variation pattern of the parents, because of gene flow.

The overall effect of hybridization depends largely on the environment in which it takes place. In stable environments, it is likely to have no important effects. In unstable, new, or changing environments, hybrid segregates may be better-adapted than are the parental taxa. Because hybrids are often associated with disturbed habitats, Edgar Anderson (1949) coined the phrase *hybridized habitat*.

## POLYPLOIDY

The multiplication of the chromosome set, *polyploidy,* is one of the most widespread and distinctive cytogenetic processes affecting the evolution of flowering plants (Stebbins, 1971a). Around 40 percent of the species of flowering plants and a higher percentage of the ferns are polyploids. The morphological and physiological effects of polyploidy may include an increase in cell size, a reduction in number of cell divisions, changes in shape and texture of plant parts, lowered fertility, and physiological imbalance.

Traditionally, polyploids derived from one species are known as *autopolyploids* and those from two parental species as *allopolyploids* or *amphiploids*. However, the situation in nature is much more complex than this, and natural polyploids often have various combinations of chromosomes. Unreduced gametes (2n), resulting from meiosis failure, may combine to yield a 4n polyploid (Harlan and de Wet, 1975). Polyploidy may also arise from mitosis failure during vegetative growth which later could develop a polyploid flower (4n).

Like hybridization, polyploidy is often related to the growth habit. Annuals tend to have the lowest frequency of polyploidy, and herbaceous perennials have higher frequency than woody plants. Interspecific hybridization is often involved in polyploidy, with crossing taking place between parental species and hybrid progeny at various ploidy levels, blurring distinctions between taxa.

## APOMIXIS

The replacement of sexual by asexual reproduction is referred to as *apomixis* (Grant, 1971). This process, which is widespread among angiosperms and ferns, sometimes involves vegetative buds or propagules, but more often seeds. When embryos are produced without fertilization, the apomictic process is referred to as *agamospermy*. Agamospermy has several advantages: (1) it allows asexual reproduction of hybrid genotypes which may possess adaptively valuable gene combinations; (2) it facilitates the perpetuation of a successful maternal genotype by allowing the hybrid to reproduce its own kind by seeds; and (3) seeds have dispersal and adaptive advantages over vegetative reproduction.

Apomixis is often associated with polyploidy, but it does appear in a few diploids such as *Citrus*. Many apomictic plants are adapted to microhabitats, as

8X

7X

6X

5X

4X

3X

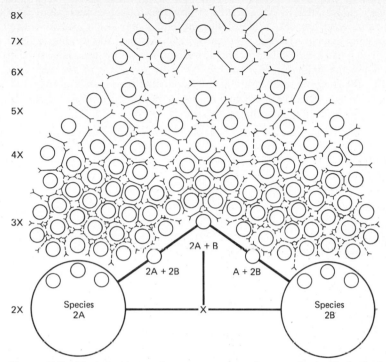

2A + B

2A + 2B          A + 2B

2X          Species          X          Species
             2A                           2B

**Figure 7-9** The structure of an agamospermic complex with both hybridization and polyploidy (2X, 3X, and so on). The small circles represent morphologically distinct, isolated, apomitic populations that provide a continuum of variation between the two parental species. This causes much difficulty in establishing a classification for the populations. *(Modified from Grant, 1977, with permission.)*

was clearly demonstrated by Solbrig and Simpson (1974) with dandelion. Some apomictic groups are capable of both sexual and asexual reproduction, e.g., *Poa* and *Potentilla*. Some have geographical variation in sexual and asexual populations, e.g., *Taraxacum* and *Bouteloua* (Grant, 1971). With apomixis, irregular meiosis is no barrier to seed production, and triploids, pentaploids, and other odd-chromosome-numbered apomictic clones may be found (Figure 7-9).

## APPENDIX TO CHAPTER 7: Techniques

### COUNTING CHROMOSOMES BY THE SQUASH TECHNIQUE*

Freshly collected flower buds are used for studies of chromosomes at meiosis. Many beginners tend to collect bud material which is too old. For most plants, an easy way to tell when bud material is past the appropriate stage is to examine the color of the

---

*Adapted from a handout of B. L. Turner; used by permission.

anthers. If the anther sacs have already turned yellow, the pollen grains have been formed and the material is too old. When collecting bud material, collect an entire inflorescence, including buds just forming to those with anthers already turning yellow. The buds are dropped into a killing solution. Small, wide-mouth, screw-top vials are best for collecting buds in the field. *Voucher specimens must be made of the material collected.* Root tips for studying mitosis are usually collected from newly germinated seed or from rapidly growing roots of pot-bound plants. A modified *Carnoy's solution* is used by many botanists as a killing and fixing solution:

*For meiosis (4:3:1)—*

4 parts chloroform
3 parts absolute alcohol
1 part glacial acetic acid

*For mitosis (1:3:1)—*

1 part chloroform
3 parts absolute alcohol
1 part glacial acetic acid

## Bud Squashing

The bud material in the 4:3:1 solution should be poured into a petri dish. The material need not be washed in water or other solutions. With tweezers or needles, select a promising bud and place it on a glass slide. If the buds are small, select several of approximately the same size. Add 1 or 2 drops of *acetocarmine* to stain the chromosomes. To make acetocarmine, boil 2 to 4 grams of carmine dye in 100 cubic centimeters (cc) of 45 percent acetic acid for several minutes, and then cool and filter. The resulting saturated solution is diluted with 45 percent acetic acid to obtain the required strength. For best results, use a 1 percent acetocarmine solution for meiotic material.

Squash the buds with the smooth end of a glass rod. Observe the material under the low-power objective of a compound microscope, and make a rapid search. It is quite easy to see both tetrads and diads and even chromosomes in pollen mother cells (PMCs) under low power without a coverslip. If diads are observed, then some PMCs will be found to have metaphase divisions and perhaps a few cells at late prophase.

If, in superficial observation, the material is found to be either too young or too old, then the macerated material is wiped off immediately and a new bud is selected and placed in the acetocarmine, and the whole process is repeated. It is quite possible to look at 20 to 30 unpromising buds in a period of 15 or 20 minutes. If the bud material is adequate, the proper meiotic stages should be found soon.

If metaphase figures or diads are observed, remove the slide and macerate the tissue with a little more care. Using a pair of flattened needles, remove any remaining large floral tissues. Place a No. 1 coverslip over the squashed material. A blotter is placed over the top of the coverslip (to absorb the extruded stain following pressure), and with the thumb, pressure is applied so that the cells are flattened. The degree of pressure that must be applied to the coverslip varies and must be learned by experience.

## Interpretation

Interpretation is the greatest difficulty in counting chromosomes from meiotic material. Beginners may feel frustrated when trying to make an accurate count from meiotic cells.

The bivalents (paired chromosomes) are sometimes torn apart, and it is often difficult to determine the exact number present in any given cell. Another complication is that some species may be apomictic and show asynapsis with the chromosomes existing as univalents. If one chromosome is superimposed upon another chromosome, it may appear as a single unit. Chromosomes that are superimposed will be about twice as dark as those in which the arms are separated and flattened in a single plane. Fine-adjustment manipulation can reveal the true nature of such chromosomes.

Some species will form chains or rings of four and six chromosomes at meiosis. These formations are best detected at diakinesis when their ring structure becomes apparent. At metaphase it is much more difficult to detect rings of four, since the extreme shortening will often make them appear as a single large bivalent. Chromosome fragments can usually be recognized by their very small size and their irregular pattern in the cell.

The best stage at which to count chromosomes in PMCs of nearly all species is that of diakinesis. While the chromosomes at diakinesis do not take such a dark stain as those at metaphase, they are widely separated and do not tend to stick, and their bivalent nature can be easily ascertained. However, in some species, particularly those with small chromosomes, metaphase is the best stage for chromosome counts.

## EXPERIMENTAL HYBRIDIZATIONS

Experimental crosses are used to help define and position taxa. Such experiments require time and greenhouse and garden space. Plants must be maintained daily. Six months to several years may be required to obtain flowering progeny. Some plants require environmental conditions that are difficult, if not impossible, to maintain in a greenhouse or garden. Other species flower for a brief period each year. The various species in a genus may be seasonally isolated, requiring timing on crosses. Synchronization of crossing may become critical in hybridization experiments. These problems are mentioned to provide information for planning, not to discourage one from performing crossing experiments.

If collections of living plant material are not available, travel to the area where the plants grow may be required. Before the collecting trip, it must be determined where populations may be located and when they flower and produce seed. Growth habit—i.e., annual, biennial, perennial, or woody—will influence the methods used to gather living material. This information may often be obtained from herbarium specimens. Annuals are best started with seed. Rootstocks of perennials and cuttings from woody plants are useful in propagation. Horticulturists who have worked with native plant material may give pointers for handling the plants.

Care must be taken with living material to ensure survival during movement from the field into the greenhouse and garden. Seed should never be dried with heat. It is usually best to cut a large plant back to ground level. This keeps the root and shoot system in balance, since most roots are lost in digging. Rootstocks can be washed, tagged, and placed in heavy plastic bags. Never allow bags of living plant material to be directly exposed to the sun or to stay in a hot automobile.

Voucher specimens should be collected from all populations sampled. The plant material can be identified with a collection number. Seed or plants obtained from other collectors may be given an accession number and recorded in a bound data book. One such system used is "79–33," for the thirty-third accession of 1979.

In some instances, collecting permits are needed and laws regarding endangered species must be considered. An import permit number must be obtained to import living material into the United States. The permit can be obtained from the Permit Section, Plant Importation Office, 209 River Street, Hoboken, New Jersey 07030.

The germination of seeds or establishment of rootstocks is most critical, and competent advice about these processes should be sought from horticulturists. Once the plants are established and flowering, the first step is to determine whether the taxa are self-incompatible, self-fertile, or apomictic. Individual plants should be isolated in a screened greenhouse or enclosed in insect-proof cages. Bagging of individual flowers or flower clusters to prevent pollination often causes problems with plant diseases due to high temperature and relative humidity. If seeds are not set in selfing tests, it is likely that the plants are self-incompatible, which can make crossing rather easy. If isolated plants set seed, they are either self-fertile or apomictic. This result should be followed with a progeny test. If the progeny are homogeneous, apomixis might be expected—especially if it has been reported in other studies of the genus. Apomixis can be verified by detailed embryo sac studies. Another approach might be to emasculate the flowers and to check for seed set. Self-fertile flowers pose difficulties in crossing experiments because they must be emasculated. The degree of difficulty depends on the size and morphology of the flower.

Experimental crosses should be carefully planned, and $F_1$s, $F_2$s, and backcrosses should be made, if time permits. The experiments should be designed with controls available for comparison with the experimental individuals. When fruit from the crosses is mature, estimates can be made of seed set percentages in the hybrids versus the controls. Seed set, however, does not in itself indicate if the seed will produce a viable hybrid.

Seeds from the hybrids should be planted and controls established to determine their viability. The squash technique may be used to estimate sterility by checking chromosome pairing at meiosis in pollen mother cells. Another estimation of fertility can be obtained by staining pollen grains. For this, the pollen grains are stained for 24 hours in a drop of 1 percent aniline blue in lactophenol on a microscope slide with a coverslip. Two hundred grains are counted and scored as either stained or not stained. The stained grains are used to estimate the frequency of fertile grains.

## LOCAL POPULATION SAMPLES AND CHARACTER ANALYSIS

To study variation in plants, populations must be sampled and their characters analyzed. Considerations must be given to the objectives of the study, the proposed techniques, methods for obtaining the sample, the part of the plant to be sampled, and the characters to be analyzed.

Collecting a local population sample involves pressing and drying a number of whole specimens or the critical parts of individual plants. Although not usually feasible, the best sample of a local population would be all the plants in the colony (Jones and Wilkins, 1971). Since this is not practical, a sample is taken that gives an estimation of the variation in the colony. *Local population samples* consisting of 25 to 75 plants will yield an estimate of the variation within and between populations and taxa. The number of plants sampled depends upon the variability and size of the population. With little other than normal variation, 25 individuals will be adequate; but if two species are hybridizing, 50 to 75 individuals are required. Care should be taken not to exterminate the colony. A sampling procedure should be used to ensure that the population sample is a random one representative of the local plants. This precaution is often difficult to implement in the field, because the terrain or vegetation may restrict the mechanics of sampling.

Collect only the parts of the plant essential to the study, along with a voucher specimen. This saves press and storage space and eases the drying problem. A voucher specimen is collected from each population. It is given a collection number, and data are recorded in the field book (see Chapter 8). Each local population sample receives a collection number.

There are several basic approaches used to analyze local population samples. First, an intensive study may be made of one or two characters, especially those that show geographical or ecological variation, using a large number of population samples. Second, an analysis of the interrelationships of 8 to 20 characters on fewer samples may be performed (Stebbins, 1950). The samples may be compared by means of diagrams, statistics, or numerical methods (Benson, 1962; Jones and Wilkins, 1971; Sneath and Sokal, 1973). Each approach has its own advantages and disadvantages. With few characters and large numbers of samples, they may be analyzed with ease, but great care is required in selecting characters. The use of many characters in the second approach helps with the critical problem of character selection. Interrelationships of characters can be shown, and it is sometimes possible to distinguish populations and taxa using a combination of characters.

Visual presentations of data are helpful when communicating with others. These include polygons, bar diagrams, scatter diagrams, ideographs, line drawings, graphs, dice grams, and so on. Benson (1962) and Davis and Heywood (1963) provide excellent examples. Frequently, population data are visually presented in relation to a background map or physiographic profile in order to convey not only the features of the population but also geographical relationships.

## GEL ELECTROPHORESIS

The technique of *gel electrophoresis* is used to estimate the amount of genetic variation in populations. When combined with the appropriate staining techniques, it may be used to visualize directly proteins or enzymes. The procedure for gel electrophoresis is schematically shown in Figure 7-10. Fresh tissues of individual plants from a local population sample are individually homogenized in a buffer solution. This releases the proteins and enzymes. Pieces of filter paper are used to absorb the homogenate supernatants. The saturated pieces of filter paper are placed in a slit cut in a gel (usually made of starch) 1 centimeter thick. The gel and samples are then subjected to a direct electric current from a power supply to electrodes in a buffer solution. Moist contact bridges carry the current from the buffer solution to the gel. The gels then are run (subjected to the electric current) for a given length of time. During the run, they must be cooled, since they tend to warm up because of the electric current. The voltage used is high, and care must be taken to prevent shocks.

The different allelic forms of the proteins and enzymes each migrate different distances in the gel, depending upon their net electric charges. After completion of electrophoresis, the gels are sliced into 3-millimeter thicknesses so that three gels can be obtained from each original gel. This permits the application of three different stains to this gel. Specific stains are presently available for over 40 enzymes. A slice of gel is placed in

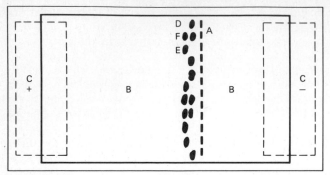

**Figure 7-10**   A diagrammatic representation of the technique of starch gel electrophoresis. *A*, samples and origin, each of 11 samples represents one individual; *B*, gel; *C*, bridge (carries current to gel); *D*, stained enzyme homozygous for allele 1 (Enzymes and proteins are actually later stained in a staining dish. The bands are not visible while the gels are in the apparatus.); *E*, stained enzyme homozygous for allele 2; *F*, stained enzyme heterozygous for alleles 1 and 2. Migration is from negative to positive.

a staining tray and covered with a stain. The gels are usually stained for one to three hours in a warm incubator. The stains cause bands to appear which visualize proteins or enzymes. The frequency of the allelic forms of the enzyme is determined by examining the gel. This must be done relatively soon because the stains are often not permanent. The genes studied are chosen by the availability of appropriate staining techniques and not with regard to the functions they specify, so presumably they are a nearly random sample from the entire set of genes.

### Preparation of Starch Gels

    **1**   Weigh 100 grams of electrostarch and place it in a 2-liter side-arm vacuum flask.

    **2**   Add 832 milliliters of dehydrogenase buffer solution to the flask.

    **3**   Insert a large magnetic stirring bar, place the flask on the automatic stirrer, adjust stirring speed to maximum, and turn heat control to high.

    **4**   Heat the starch until it thickens and becomes transparent.

    **5**   Remove the flask, cork it, and aspirate with a vacuum line until all small bubbles disappear.

    **6**   Pour the starch into two gel molds using a spatula to prevent the stirring bar from falling into the mold.

    **7**   Allow the gels to cool to room temperature. Cover with plastic wrap. They may be refrigerated or left out overnight. The gels will then have the consistency of very thick gelatin.

### Preparation of Tissues

    **1**   Cut plant tissues into small sections with a razor blade.

    **2**   Place the tissue in a small mortar in a tray of ice.

    **3**   Grind the tissue as finely as possible with 2.0 milliliters of dehydrogenase buffer.

    **4**   Place this crude extract in a small labeled vial.

    **5**   The extract may be used immediately or frozen in an ultrafreezer until needed.

### Loading and Running Gels

**1**  Cut a slit at about one-fourth the gel length for the placement of samples.
**2**  Absorb the plant extract into filter paper strips (3 by 9 millimeters) and place up to 20 of these samples in the slit.
**3**  Put buffer into each side of the buffer tray.
**4**  Place the gel across the buffer tray with the sample end near the negative platinum wire.
**5**  Put a sponge wick into each buffer compartment and over each end of the gel.
**6**  Spray the exposed plug with silicone spray to prevent fouling. Place the tray into the low-temperature incubator and connect the appropriate plugs.
**7**  Turn on the power, allowing a 1-minute warm-up time. Adjust power so that 30 milliamperes of DC current move through the gels.
**8**  The gels are run for five hours at 4°C.

### Slicing and Staining Gels

**1**  To indicate the orientation of the gel, punch a hole in the upper left corner with the end of a disposable pipette.
**2**  Slice the gel into three layers with a gel slicer.
**3**  Place the slices into the staining trays, add the appropriate staining solutions, cover with plastic wrap, and place in the 37°C incubator. The bands representing the allozymes should begin to appear within a half hour.

### Interpreting the Gels

The bands for each individual plant are scored according to the distance migrated.

Further instructions on electrophoretic techniques and stains are provided in Brewer and Sing (1970).

### SELECTED REFERENCES

Anderson, E.: *Introgressive Hybridization,* Wiley, New York, 1949.

Avise, J. C.: "Systematic Value of Electrophoretic Data," *Syst. Zool.,* **23:**465–481, 1974.

Ayala, F. J.: "Biological Evolution: Natural Selection or Random Walk," *Amer. Sci.,* **62:**692–701, 1974.

Babbel, G. R., and R. K. Selander: "Genetic Variability in Edaphically Restricted and Widespread Plant Species," *Evolution,* **28:**619–630, 1974.

Baker, H. G., and G. L. Stebbins (eds.): *The Genetics of Colonizing Species,* Academic, New York, 1965.

Bartholomew, B., L. C. Eaton, and P. H. Raven: "*Clarkia rubicunda:* A Model of Plant Evolution in Semiarid Regions," *Evolution,* **27:**505–517, 1973.

Benson, L.: *Plant Taxonomy: Methods and Principles,* Ronald, New York, 1962.

Bradshaw, A. D.: "Evolutionary Significance of Phenotypic Plasticity in Plants," *Adv. Genet.,* **13:**115–155, 1965.

Brewer, G. J., and C. F. Sing: *An Introduction to Isozyme Techniques,* Academic, London, 1970. (Basic information on gel electrophoresis.)

Camp, W. H., and C. L. Gilly: "The Structure and Origin of Species," *Brittonia,* **4:**323–385, 1943.

Carson, H. L.: "The Genetics of Speciation at the Diploid Level," *Amer. Nat.,* **109:**83–92, 1975.

Chase, V. C., and P. H. Raven: "Evolutionary and Ecological Relationships Between *Aquilegia formosa* and *A. pubescens* (Ranunculaceae), Two Perennial Plants," *Evolution,* **29:**474–486, 1975.

Clausen, J.: *Stages in the Evolution of Plant Species,* Cornell, Ithaca, N.Y., 1951.

———, D. D. Keck, and W. M. Hiesey: *Experimental Studies on the Nature of Species. I. Effect of Varied Environments on Western North American Plants,* Carnegie Institution of Washington Publication, no. 520, 1940.

Davis, P. H., and V. H. Heywood: *Principles of Angiosperm Taxonomy,* Van Nostrand, Princeton, N.J., 1963.

de Wet, J. M. J., and H. T. Stalker: "Gametophytic Apomixis and Evolution in Plants," *Taxon,* **23:**689–697, 1974.

Dobzhansky, T., F. J. Ayala, G. L. Stebbins, and J. W. Valentine: *Evolution,* Freeman, San Francisco, 1977.

Ehrendorfer, F., et al.: "Biosystematics at the Cross-roads; Proceedings of a Symposium Held at the XI International Botanical Congress, Seattle 1969," *Taxon* **19:**137–214, 1970. (Papers by various authors.)

Fryxell, P. A.: "Mode of Reproduction in Higher Plants," *Bot. Rev.,* **23:**135–233, 1957.

Gadgil, M., and O. T. Solbrig: "The Concept of r- and K-selection: Evidence From Wild Flowers and Some Theoretical Considerations," *Amer. Nat.,* **106:**14–31, 1972.

Gottlieb, L. D.: "Biochemical Consequences of Speciation in Plants," F. Ayala (ed.), *Molecular Evolution,* Sinauer Associates Inc. Publishers, Sunderland, Mass., 1976, pp. 123–159.

Grant, K. A., and V. Grant: "Mechanical Isolation of *Salvia apiana* and *Salvia mellifera* (Labiatae)," *Evolution,* **18:**196–212, 1964.

Grant, V.: *The Origin of Adaptations,* Columbia, New York, 1963.

———: *Plant Speciation,* Columbia, New York, 1971.

———: *Genetics of Flowering Plants,* Columbia, New York, 1975.

———: *Organismic Evolution,* Freeman, San Francisco, 1977.

Gregor, J. W., and P. J. Watson: "Ecotypic Differentiation: Observations and Reflections," *Evolution,* **15:**166–173, 1961.

Hadley, E. B., and D. A. Levin: "Habitat Differences of Three *Liatris* Species and Their Hybrid Derivatives in an Interbreeding Population," *Amer. J. Bot.,* **54:**550–559, 1967.

Hamrick, J. L., and R. W. Allard: "Correlations Between Quantitative Characters and Enzyme Genotypes in *Avena barbata,*" *Evolution,* **29:**438–442, 1975.

Harlan, J. R., and J. M. J. de Wet: "On Ö. Winge and a Prayer: The Origins of Polyploidy," *Bot. Rev.,* **41:**361–390, 1975.

Heinrich, B., and P. H. Raven: "Energetics and Pollination Ecology," *Science (Wash. D.C.),* **176:**597–602, 1972.

Heiser, C. B.: "Introgression Re-examined," *Bot. Rev.,* **39:**347–366, 1973.

Heslop-Harrison, J.: "Forty Years of Genecology," *Adv. Ecol. Res.,* **2:**159–247, 1964.

Hiesey, W. M., and M. A. Nobs: "Genetic and Transplant Studies on Contrasting Species and Ecological Races of the *Achillea millefoliun* Complex," *Bot. Gaz.,* **131:**245–259, 1970.

Jones, D. A., and D. A. Wilkins: *Variation and Adaptation in Plant Species,* Heinemann, London, 1971.

Jones, S. B.: "Cytogenetics and Affinities of *Vernonia* (Compositae) from the Mexican Highlands and Eastern North America," *Evolution,* **30:**455–462, 1976.

Kerner, A.: *The Natural History of Plants, 1831–1898,* 2 vols., Blackie, London, Glasgow, Edinburgh, and Dublin, 1896–1897. (Translation from the German by F. W. Oliver.)

Langlet, O.: "Two Hundred Years of Genecology," *Taxon,* **20:**653–721, 1971.

Lerner, I. M.: "The Concept of Natural Selection," *Proc. Amer. Philos. Soc.,* **103:**173–182, 1959.

Levin, D. A.: "Reinforcement of Reproductive Isolation: Plants Versus Animals," *Amer. Nat.,* **104:**571–581, 1970.

———: "The Origin of Reproductive Isolating Mechanisms in Flowering Plants," *Taxon,* **20:**91–113, 1971.

——— and H. W. Kerster: "Natural Selection for Reproductive Isolation in *Phlox,*" *Evolution,* **21:**679–687, 1967.

Lewis, H.: "Speciation in Flowering Plants," *Science (Wash. D.C.),* **152:**167–172, 1966.

Long, R. W.: "Artificial Interspecific Hybridization in Temperate and Tropical Species of *Ruellia* (Acanthaceae)," *Brittonia,* **27:**289–296, 1975.

Mayr, E.: "Isolation As an Evolutionary Factor," *Proc. Amer. Philos. Soc.,* **103:**221–230, 1959.

Ornduff, R.: "Reproductive Biology in Relation to Systematics," *Taxon,* **18:**121–133, 1969.

Pijl, L. van der, and C. H. Dodson: *Orchid Flowers: Their Pollination and Evolution,* University of Miami Press, Coral Gables, Fla., 1966.

Snaydon, R. W.: "Rapid Population Differentiation in a Mosaic Environment. I. The Response of *Anthoxanthum odoratum* Populations to Soils," *Evolution,* **24:**257–269, 1970.

Sneath, P. H. A., and R. R. Sokal: *Numerical Taxonomy,* Freeman, San Francisco, 1973.

Solbrig, O. T.: *Principles and Methods of Plant Biosystematics,* Macmillan, New York, 1970.

——— and B. B. Simpson: "Components of Regulation of a Population of Dandelions in Michigan," *J. Ecol.,* **62:**473–486, 1974.

——— et al.: "IOPB Symposium on Biosystematics," *Taxon,* **16:**253–333, 1967. (Papers by various authors.)

Stebbins, G. L.: *Variation and Evolution in Plants,* Columbia, New York, 1950.

———: "The Inviability and Weakness of Interspecific Hybrids," *Adv. Genet.,* **9:**147–215, 1958.

———: "The Significance of Hybridization for Plant Taxonomy and Evolution," *Taxon,* **18:**26–35, 1969.

———: *Processes of Organic Evolution,* 2d ed., Prentice-Hall, Englewood Cliffs, N.J., 1971a.

———: *Chromosomal Evolution in Higher Plants,* Addison-Wesley, Reading, Mass., 1971b.

Straw, R. M.: "Isolation in *Penstemon,*" *Amer. Nat.,* **90:**47–53, 1956.

Thompson, H. J.: The Biosystematics of *Dodecatheon, Contrib. Dudley Herb.,* **4:**73–154, 1953.

Wagner, M.: *Die Entstehung der Arten durch räumliche sonderung,* B. Schwabe, Basel, 1889.

Weaver, J. E., and F. E. Clements: *Plant Ecology,* 2d ed., McGraw-Hill, New York, 1938.

Wooten, J. W.: "Experimental Investigations of the *Sagittaria graminea* Complex: Transplant Studies and Genecology," *J. Ecol.,* **58:**223–242, 1970.

Chapter 8

# Specimen Preparation and Herbarium Management

Collections of plant specimens are essential for taxonomic research. They circumscribe species and document their variability, they are the prime sources for floristic studies, and they are vouchers for experimental investigations. Plant materials must be carefully selected, prepared, and preserved, since herbarium specimens become a permanent record for later investigators to examine.

The taxonomist in the field observes plant locations and records habitat information. The variation within a single population or among populations of the same species in different environments may be measured. Observations may be made of patterns produced by hybridization or by changes in soil, moisture, slope, light, and so on. Features may be evident in living plants which are not easily observed in dried materials, e.g., flower color or fragrance. Species response to environmental disturbances, such as fire, grazing, and clearing, may be noted. There is no substitute for firsthand knowledge of plants as living entities through field observations.

Student collections are fundamental to study and training in plant systematics. As a beginner's collection grows, methods change from those of a collector interested in natural history to those of a researcher interested in the taxonomic problems of a particular group.

Ideally, the best specimen for identification and research is an intact and complete plant. Attempts to identify a specimen from a single flower or leaf usually fail. Such specimens have little or no scientific value. It is possible to collect entire plants of small annual species and some herbaceous perennials. However, one would not attempt to press an entire tree. Underground parts of herbaceous perennials such as rhizomes, roots, and bulbs should be collected. Representative leaves and reproductive structures are essential. The flowers, fruits, and seed of flowering plants are especially important, since most keys for identification use reproductive characters. With large herbs, shrubs, and trees, the different kinds of foliage will be helpful. Individuals should be selected which are representative of all phases of the natural population. Insect-damaged plant material or monstrosities should be avoided. When pressed and dried, the specimen should yield the maximum amount of information concerning the living plant species and be representative of the population.

## PRESSING AND DRYING PLANT SPECIMENS

The materials needed to press plants consist of five items: the press, straps or ropes, blotters, corrugate ventilators, and torn newspapers (Figure 8-1). The *press* may be purchased from a biological supply house or may be inexpensively constructed from $^3/_8$-inch plywood cut into two 12 by 18 inch pieces to be used for either end of the press. *Straps or ropes* may be constructed from webbing with claw buckles or from window sash cord. The straps or ropes should be at least 5 feet long and are used to tighten the press. *Blotters* are heavy blotting papers that will absorb moisture from the plant specimen. They are 12 by 18 inches and may be purchased from paper companies or biological supply houses. The *corrugate ventilators* are sheets of corrugated 12 by 18 inch cardboard. They provide space for air passage through the press to remove water vapor. The corrugations must run the short distance rather than the long distance of the cardboard. This is because the press is usually dried with the 18-inch side down over the heat source, and the 12-inch corrugations act as vents. The *torn newspaper* receives and contains the specimen while in the press and until the time it is mounted. A double-page sheet of newspaper is torn or cut in half and then folded over. The newspaper should be just slightly smaller than the press.

Useful tools include a heavy-duty trowel for digging roots, pruning shears for clipping tough plants, and pole pruners for collecting specimens from tall trees (see Table 8-1).

Plant specimens should be pressed as soon as possible after they are collected. The best preserved specimens are obtained by using a field press (see Table 8-2). This is arranged to hold 100 sheets of torn newspaper, allowing the pressing of 100 specimens. At the end of the day or by the next morning the specimens must be transferred (still in their own sheets of newspaper) to a drying press (see Table 8-2). By this time, the specimens will have relaxed and may be slightly rearranged to improve the quality of the specimens.

**Figure 8-1** Field equipment: a plant press, pruning shears, digging tool, field book, plastic bag, and other items.

It is also possible to obtain suitable specimens by wrapping and rolling the freshly collected plants in newspaper while in the field. Be sure that the specimens are totally covered by the wrapping, or damage will result. The bundle should be tagged or labeled and tied with string. Water is poured through the open ends of the bundle and the excess drained away. Plastic milk jugs are excellent containers for taking water to the field. The ends of the bundle should be closed and the bundle placed upright in a large, heavy plastic bag. A second bag is placed over the open end of the first bag to keep the plants moist. Specimens collected by this method will keep up to 20 hours. Specimens not pressed in the field are usually pressed directly in the drying press. Taxonomists formerly used a metal collecting can called a *vasculum* to collect plants, but these have largely been replaced by heavy-duty plastic bags such as a 50-pound fertilizer bag.

Plants pressed in the field will yield specimens of higher quality than those wrapped, bagged, and pressed later. The disadvantage of the field press is that

## Table 8-1   Checklist of Field Equipment and Supplies

I  **For both beginners and advanced students**
1  Bound field notebook (for collection data)
2  Field press (complete)
3  Drying press (complete)
4  Digging tool (such as trowel, geologist pick, bricklayer's hammer, dandelion digger)
5  Pruning shears (avoid cheap shears; Wiss or Seymour Smith give years of service)
6  Heavy-duty plastic bags (50-lb. fertilizer bags are excellent)
7  Newspaper (for wrapping bundles of specimens and for pressing)
8  Plastic milk jug with water (for moistening specimens)
9  10X hand lens
10  Pocket knife
11  Soft lead pencils

II  **Additional items that may be needed for advanced students and professionals**
1  String tags
2  Plastic pot labels (for making waterproof labels for living material)
3  Collecting vials and jars
4  FAA and Carnoy's killing and fixing solutions
5  Seed envelopes
6  Plastic bags for living material
7  Highway, topographic, and geologic maps; county road maps; aerial photographs
8  Camera and film
9  Portable plant drier
10  Insect repellant
11  Pole pruner (for trees and large shrubs)
12  Semipermanent aluminum tag labels (for tagging plants in the field)
13  Plastic flagging (to relocate areas)
14  Altimeter
15  Compass
16  Entrenching tool (war surplus)
17  Folding pruning saw
18  Paper sacks

## Table 8-2   Organization of the Field Press and the Drying Press

| Field press | Drying press |
|---|---|
| *Plywood press* | *Plywood press* |
| Corrugate | Corrugate |
| Blotter | Blotter |
| 10 sheets of torn newspaper (each sheet will be used for one specimen) | Newspaper (with specimen) |
|  | Blotter |
|  | Corrugate |
| Blotter | Blotter |
| Corrugate | Newspaper (with specimen) |
| Blotter | Blotter |
| 10 sheets of torn newspaper | Corrugate |
| Blotter | (Continue the sequence |
| Corrugate | until the press |
| (Continue the sequence | reaches a maximum of |
| until 10 groups of 10 | 3 feet in height.) |
| newspapers are present.) | *Plywood press* |
| *Plywood press* |  |

fewer specimens may be collected in a day's time. Around 100 specimens may be collected in a day using the field press, but over 300 could be collected using the bagging technique.

The *placing and arrangement* of the specimens in the torn newspaper is a matter of great importance and requires careful attention to details. The final appearance of the specimen depends upon how it is pressed and dried. Each specimen should be arranged to look more or less natural and show the essential botanical details. Unnecessary overlapping of leaves and other plant parts must be avoided, since this slows drying and lowers specimen quality. Whenever possible, at least one leaf (or parts of a compound leaf) should be arranged with the lower side uppermost. This will allow observation of the lower leaf surface even when the specimen is mounted. All soil and trash should be removed from underground parts of plants before pressing.

Maximum efficiency of the plant press is obtained when the surface of the newspaper is loosely covered with plant materials, but the plant specimen should never be overcrowded. Only one collection should be pressed in each newspaper to avoid the confusion of data. It is possible and desirable with small plants to have more than one plant of the same species per sheet if they are not overcrowded.

Slender plants may be folded in the shape of a V or W. Large plants require cutting into segments. Occasionally, two sheets for a single specimen may be required, but this should be avoided because of the problems of keeping the two parts together in the herbarium.

Bulky organs such as fruit may be reduced in thickness by slicing. Both cross and longitudinal sections of fruits are useful. Pads of folded newspaper or foam rubber may be placed on the leaves to keep them flat and to avoid crushing the bulky stems, fruits, or other materials. Cones may be tagged with the collection number on a stringed label and set aside. Cacti and succulents should be split and the inner parts removed before pressing. Some collectors salt the cut surface to aid drying. In humid tropical regions specimens are slow-drying, and possible decomposition of specimens may be a problem. This may be avoided by brushing or soaking the specimens with formaldehyde or alcohol to control decomposition.

Tall grasses, sedges, and rushes are difficult to press and will not stay bent during the pressing operation. A fold may be maintained in such material by slipping the bent end of the plant specimen through a 1-inch slit in a card. The card may later be removed and reused. Because of pressing and size considerations, specimens should never extend beyond the press.

Filamentous or filmy aquatic plants may be arranged by floating them in a pan of water over an $8^1/_2$ by 11 inch sheet of bond paper. Slowly lift the paper out of the water with the specimen on it. The specimen will assume a fairly natural shape and will make an attractive specimen. The sheet of paper with the plant resting on it is then placed in the newspaper and the specimen is pressed and dried. The sheet of paper with the specimen adhering to it is often mounted on the herbarium paper.

Plants with mucilage or with delicate flowers which stick to newspapers may be pressed between waxed or tissue paper. Some plants with large flowers should have a few flowers split open and pressed. Actually, it is helpful to collect a few extra flowers and fruits and fill in vacant spaces to maximize the efficiency of the press and to increase the value of the specimen.

Once the plant specimens are arranged, the press must be tightly closed with ropes or straps to prevent wrinkling of the specimens. Place the press on the floor or ground and kneel on one end while tightening the ropes on that end. Repeat the process several times until the press is tight. The press is now ready for the drier.

For best results, the specimens should be dried as rapidly as possible. When traveling in arid regions, the press may be placed on the luggage rack of a vehicle with the corrugations oriented to funnel the dry air through the press. In humid regions, a drier must be used. A wooden box 18 inches wide and 3 feet long made of 1 by 10 inch boards with five 60-watt light bulbs as a heat source will usually dry one press in 12 hours. Some small openings should be present in the bottom of the box to allow air to enter. The air will be heated and rise by convection through the corrugates. The press becomes loose when the plants are dry.

## THE FIELD NOTEBOOK

A permanently bound field notebook with horizontal rulings is an indispensable item for the collector. Full data on each collected plant should be recorded in the field at the time the collection is made. The data should be so complete that the label may be prepared directly from the field notebook. An example of an entry from a field notebook is shown in Table 8-3.

Data recorded at the time of collection should include locality, elevation, habitat, information about the plants, date, and *collection number*. If the specimens were collected in the same location on a given day, the locality data may be recorded only once. Additional data for each species can be added down the page.

The collection number is a numerical series starting with number 1 and

**Table 8-3   An Example of a Typical Format\* for Data Entry in a Field Notebook**

Union Co. Georgia        8 July 1977
Near summit of Brasstown Bald Mountain, ca. 8 miles ENE of Blairsville and ca. 5.5 miles SW of Hiawassee. Herb 1.5 m tall, corollas scarlet, locally abundant in openings in scarlet oak-black oak-*Rhododendron-Kalmia* forest. Elevation ca. 4000 ft.

23155†*Monarda fistulosa* L.

\*This format can be modified if more than one species is collected so as to have only one location heading and individual notes on each plant continued down the page.
†23155 is the collection number.

continuing throughout the collector's lifetime. Duplicate collections of one species which are taken on the same date at the same place are given the same collection number. Otherwise, the numbers are not duplicated. Taxonomists routinely collect duplicate specimens in sets of 5 or 10 or 20 for exchange to other herbaria. At the time the collection is pressed, the collection number should be recorded on the long margin of the newspaper opposite the fold. Later, when the specimen is identified, its complete scientific name (genus, specific epithet, author) is also written in the long margin. Additional notes regarding the specimen are sometimes written on the margin. The specimen will stay in this same newspaper until it is mounted, so the collection number serves as an identification number associating the specimen with data in the field book.

If not pressed in the field, the specimens must be tagged or labeled so that errors of associating specimens with incorrect collecting localities will not be made. Pocket-size tape recorders are sometimes useful in the field. The tapes are then used to edit the field notes later that day. Never delay recording data in the field notebook or editing the field notes.

## COLLECTING, CONSERVATION, AND THE LAW

Increasing human population and changing land-use patterns in North America during the past century have reduced or destroyed some habitats. This has had a serious impact on the size of the population systems of species which have specific habitat requirements. Those critical habitats which remain are often under constant threat from development, forestry practices, agriculture, actions of governmental agencies, and pressures from the increasing population. Many plant species which were once common are now rare. It is essential that collectors be thoughtful conservationists and use good judgment when collecting, and not collect rare or uncommon plants. The wise collector will not dig the only plant of a species at a locality. Some species have become extinct at particular sites because of thoughtless collecting. Some conservationists advocate a ban on all collecting, which would severely limit specimen collection.

There are many laws which protect native plant species. The Federal Endangered Species Act of 1973 protects plants determined to be threatened or endangered, and federal and state agencies have obligations under this act. Collectors should be aware of laws that apply to the areas where they are gathering specimens. Collecting is generally prohibited in city, county, state, and national parks. States may have laws giving further protection to specific endangered plants. Obtain the necessary permissions, including collecting permits in advance, if at all possible. For some foreign countries, this may require six months to a year and probably the assistance of a local contact. Consult the local district ranger before collecting in national forests or on lands managed by the Bureau of Land Management. Whenever possible, it is advisable to obtain permission in writing before collecting on private lands.

## IDENTIFICATION

Professionals usually identify their specimens after they have been pressed and dried. Beginners should attempt to identify their collections by using fresh material. If it is necessary to make dissections, herbarium material may be softened by boiling or by using a wetting agent. One such wetting agent is prepared by the following formula: aerosol O T (dicotyl sodium sulfosuccinate), 1 percent; distilled water, 74 percent; methanol, 25 percent (vol/vol). Specimens are identified by the techniques discussed in Chapter 9. Once identified, the names are typed on the labels and the specimens are mounted.

## HERBARIUM MANAGEMENT

A collection of pressed and dried plants arranged in some order and available for reference or study is known as a *herbarium* (plural, *herbaria*). Many large research and educational institutions serving as basic resources for systematic botany had their beginning as gardens which included herbaria. Thus, the definition has been extended to include such institutions. A herbarium may contain a few hundred specimens collected locally or millions of gradually accumulated specimens which document the flora of one or more continents. The overall goal of herbarium management is to collect and preserve plant specimens with adequate label notes and to collect the literature of taxonomy in the herbarium library.

Herbaria began early in the sixteenth century in Italy as collections of dried plants sewn on paper. Luca Ghini probably initiated the first collection and may have passed the concept along to his students and associates. Its value was appreciated by those interested in plants, and the technique quickly spread over Europe. By the mid 1500s the Englishmen John Falconer and William Turner are reported to have had collections. Up to the early part of the nineteenth century, plants were sewn or pasted on plain pages and bound into volumes. Linnaeus apparently popularized the current practice of mounting specimens on single sheets of paper and storing them horizontally. From such simple beginnings, herbaria have developed into facilities housing millions of specimens, usually in steel cases. A list of the herbaria of the world and their standard abbreviations can be found in Holmgren and Keuhen (1974). The size of the larger herbaria in North America and elsewhere is shown in Table 8-4. The larger herbaria tend to be national institutions supported by governmental funds. University herbaria generally are smaller because of limited resources.

## FUNCTIONS OF HERBARIA

Herbaria are permanent repositories of plant specimens and are sources of information about plants and vegetation. Practically all taxonomic research in-

**Table 8-4   Some of the Major Herbaria of the World and the Number of Their Specimen Holdings**

**Outside of North America**

| | |
|---|---:|
| Museum of Natural History, Paris | 6 million |
| Royal Botanic Gardens, Kew | 4 to 5 million |
| Komarov Botanical Institute, Leningrad | over 5 million |
| Conservatory and Botanical Garden, Geneva | 5 million |
| British Museum of Natural History, London | about 4 million |
| University of Lyon, Lyon | 3.8 million |
| Natural History Museum, Vienna | 3 million |
| University of Uppsala, Uppsala | 2 million |
| Botanical Survey of India, Calcutta | 2 million |
| Botanical Garden and Botanical Museum, Berlin (Largely destroyed in World War II.) | 1.8 million |
| National Botanical Garden of Belgium, Brussels | 1.7 million |
| National Herbarium of Melbourne, Victoria | 1.5 million |
| Royal Botanic Garden, Edinburgh | 1.2 million |

**North America**

| | |
|---|---:|
| Combined Herbaria, Harvard University, Cambridge | 4.2 million |
| New York Botanical Garden, Bronx | 4 million |
| U.S. National Herbarium (Smithsonian), Washington, D.C. | 3.9 million |
| Field Museum of Natural History, Chicago | 2.5 million |
| Missouri Botanical Garden, Saint Louis | 2.4 million |
| University of California, Berkeley | 1.5 million |
| University of Michigan, Ann Arbor | 1.3 million |
| Academy of Natural Sciences, Philadelphia | 1.2 million |
| University of Texas, Austin | 750 thousand |
| Vascular Plant Herbarium, Ottawa | 625 thousand |
| University of Montreal, Montreal | 600 thousand |
| University of North Carolina, Chapel Hill | 500 thousand |

*Source:* From Holmgren and Keuken, 1974.

volves the use of preserved specimens which yield vast amounts of data when properly prepared. For practical purposes, only in a herbarium can all the related species of a genus or family be gathered together for study. The classification of the world's flora is primarily based on herbarium material and the literature associated with it (which in turn, was derived from herbarium materials).

Herbaria supplement the limited individual field studies by utilizing the results of travel and collecting of many botanists. Monographers of genera and families study not only the specimens of their own herbaria but those in other herbaria. This is accomplished through loans of specimens or by visits to other institutions.

No longer are taxonomists content to study a single or a limited number of specimens of each species. The importance of studying a multitude of specimens has become evident as the complexity of species has become better-understood. Ideally, specimens studied from several herbaria will include most of the range of the geographic and ecological variations of the species and will reveal the trends (constancy or instability) of characters. Many distinct populations are represented since herbaria usually contain a series of specimens collected in different places by many individuals.

Herbaria are *the* repository of "original documents"—i.e., specimens upon which all our knowledge of the taxonomy, evolution, distribution, and so on of the flora rests. All manuals, monographs, and wild flower books eventually stem from herbarium resources. Questions about the nature of species, taxonomy, identification, and nomenclature ultimately lead to the herbarium. The activities of herbaria are categorized as follows:

**1** Providing a standard reference collection for verifying the identification of newly collected plants. This is a major function of many small herbaria.

**2** Serving as a reference collection for plant taxonomy and other botany courses.

**3** Training graduate and undergraduate students in herbarium practices.

**4** Documenting the presence of a species at a particular location and providing data on its geographic range. It is often possible to go back to the exact spot where a plant was originally collected and to again find the plant material.

**5** Providing samples of the flora of an area. For example, it is possible for an ecologist to go to the herbarium and put together a series of specimens which represent the major vegetational components of a region. By studying these specimens, much time may be saved while in the field. Also, a taxonomist writing the flora of a region would go through the herbarium to ascertain which species are represented in the flora.

**6** Pointing out the existence of classification problems. A preliminary examination of herbarium specimens may indicate that a species contains plants which do not combine the characters normally listed in manuals, therefore suggesting the need for additional studies.

**7** Offering raw data of the plants themselves. Data are available in the form of vegetative and reproductive morphology; pollen samples; leaf samples for chemical analysis; anatomical samples; data for distribution maps; and ecological, economic, and ethnobotanical data from the labels.

**8** Preserving type specimens and serving as a repository of chromosome, chemosystematic, and experimental vouchers. Examination of type specimens allows a researcher to determine which plant was described by the original author and to find the exact specimen associated with the name. Modern taxonomic literature is documented by reference to individual plant specimens according to herbarium, collector, and collection number. If a mistake is made in the identification of a plant used for chromosome counts, the mistake may later be corrected if a voucher was prepared and deposited in a herbarium. It is now impossible to publish chromosome counts or chemosystematic data without reference to a deposited voucher.

## LABELS

Herbarium labels are an important and essential part of the permanent plant specimen (see Figure 8-2). The label provides pertinent collection data which

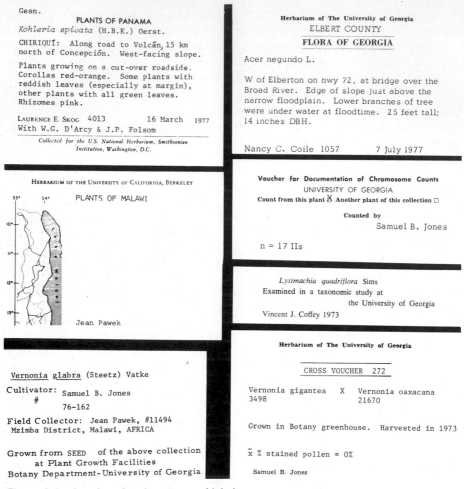

**Figure 8-2**   A series of various types of labels.

cannot be determined by examination of the specimen itself. Ideally, a herbarium label should be a miniature essay on the plant and its habitat. It is essential that labels be legible, neat, and permanent. The paper should be of high rag content, preferably 100 percent rag. Unless done by offset printing, labels should be typewritten. Xeroxed labels are acceptable only if reproduced on high rag content paper. Labels should be approximately $2^3/_4$ by $4^1/_4$ inches, which is adequate to accommodate the necessary data. Larger labels take up space needed for mounting the specimen. They should never be folded. Although often used by governmental agencies, printed labels which designate the data to be provided are cluttered and are seldom satisfactory.

Each herbarium label should contain the following information:

**1**   *Heading.*   State or province; county or parish; country (if necessary); name of the institution (or person) with which the specimen originated.

**2** *Scientific name.* Genus, specific epithet, author or authors. When the complete scientific name (including the author) is known, the reference to the original publication may be found in *Index Kewensis.*

**3** *Locality.* Be specific with localities. Refer to some town, latitude and longitude, or a permanent geographic feature. Indicate the place of collection in terms of distance in a particular direction from the nearest permanent and readily located point of reference. Some labels have a map with a spot or arrow indicating the location. Ideally, someone else should be able to read your label and by following your directions go back to the same location.

**4** *Habitat.* Describe the kind of place where the plant was growing— vegetation type, moisture, parent material, soil, elevation (if critical), direction of slope, side of mountain range (because of differences in moisture due to rain shadow), physiographic region, and so forth.

**5** *Date of collection.*

**6** *Name of collector.* The full name of the collector should be used. If several people are on the field trip, place their names on the label.

**7** *Collection number.* The literature of plant systematics identifies and refers to the specimen by the collector's name and collection number.

The following additional information may be placed on labels when appropriate: (1) associated plants, (2) flower color, (3) pollinators, (4) bark, (5) abundance, (6) height or diameter at breast height (DBH), (7) life form, (8) economic uses, (9) folklore, and so on.

Labels are glued by one edge to the lower right corner of the specimen sheets. In addition to the collection label, other labels may be found on herbarium sheets. Annotation labels are added when an expert examines the sheet and checks its identification. They are usually 2 by 11 centimeters in size. The annotation label will carry the full name of the person who examined the specimen along with the date examined. Other labels may indicate that pollen has been removed from the specimen or that it is a cytological or phytochemical voucher.

## MOUNTING

Mounting is the process by which a specimen is attached to a sheet of mounting paper and a label affixed at the lower right corner. Most herbaria in North America use a standard-size herbarium or mounting sheet of $11^1/_2$ by $16^1/_2$ inches. Various paper companies and biological supply houses can supply suitable mounting sheets. The quality of the paper used will vary according to the needs of the herbarium. For research herbaria, paper approaching 100 percent rag content is used for durability. Teaching collections use less expensive paper with lower rag content. The paper should be relatively stiff to prevent damage to the specimens when the sheets are handled.

A number of methods are used for attaching specimens to the mounting sheets (see Figure 8-3). A good quality paste or glue such as Swifts Z-5032, Elmer's Glue-All, Nicobon B, or Wilhold 128 is adequate. "Yes" brand paste is used by some to affix labels to sheets. Glue is applied to the back of the speci-

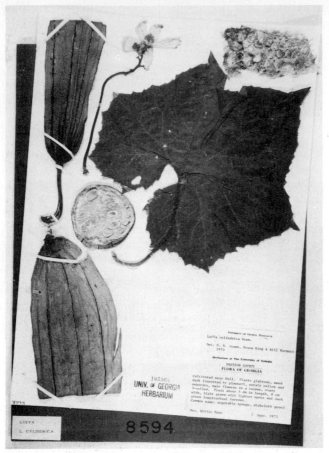

**Figure 8-3**  A plant specimen and genus folder. Note the fruit sections and the use of straps.

mens, and then the specimens are pressed onto the mounting sheets. Bulky parts may be sewn or fastened down with strips of linen tape. Archer's (1950) plastic formula is sometimes used for strapping bulky materials. Resyn 35-6262 is useful for specimens with leaves of hard textures, since it retains elasticity. Weights of scrap iron, heavy washers, large nails, and so on are used to hold down the specimens until the glue has set. Valuable loose parts, such as seeds, fruits, or flowers, are placed in fragment folders glued to the mounting paper.

## FILING

Plant specimens are filed in herbaria by some preselected arrangement. In small teaching herbaria, the filing system may be entirely alphabetical by family, genus, and species. In research herbaria, the specimens are arranged by family or genus according to one of the well-known systems of classification. When filed by families or genera, the folders are numbered, and an index is displayed in the herbarium. Many herbaria use family and genus numbers from Dalla Torre and Harms (1900–1907), in which each family and genus has a place in a numbered sequence based on the Engler system of classification.

Specimens are filed in manila genus folders $16^5/_8$ by 12 inches in size when folded. Names of included plant material are placed on the lower edge of the folder. Folders are often segregated geographically, and various color schemes are employed to indicate political and physiographic areas. Many large herbaria segregate specimens also by the continent from which the material is collected. Such segregation prevents unnecessary handling of specimens and facilitates locating the specimens. Species within a genus are filed alphabetically or by sections to show relatedness.

Special arrangements are required for filing bulky material such as palm leaves and flowers or gymnosperm cones. These are often tagged and placed in boxes designed to be stored in the pigeonholes of herbarium cases. Standard herbarium cases are insect- and dust-proof filing cases made of welded steel construction with two tiers of pigeonholes, each 19 inches deep, 13 inches wide, and 8 inches high.

Dummy sheets and folders may be used to refer the user to other folders when familiar names are changed in the course of taxonomic revisions or to specimens filed elsewhere in the herbarium. Most herbaria will have a folder at the end of each genus for specimens not determined to species.

## INSECT CONTROL

Precautions must be taken to safeguard herbarium collections from damage by insects. All the compounds used as repellants or fumigants are hazardous to some degree. Mandatory heating or fumigation upon entry of the specimens, containment and occasional use of fumigants in the collection, and a herbarium climate inhospitable to insects must be the cornerstones of any workable and safe (to humans) policy.

Particularly troublesome insect pests of dried material are the cigarette beetle, the drugstore beetle, and the black carpet beetle, often collectively called *dermestid beetles*. The most destructive pest is the cigarette beetle, which can complete its life cycle in 45 to 50 days with three to six generations per year. Any of these pests may ravage specimens during periods of unattended storage. Recommendations for control and precautionary measures vary, but they may be summarized and grouped under one of the three following categories: (1) heating, (2) repellents, and (3) fumigants.

Incoming specimens may be treated by *heating* in a specially constructed cabinet at 60°C for a period of six hours. This temperature will effectively kill eggs, larvae, pupae, and adults of the dermestids. If the herbarium is kept clean, at below 21°C, and with a relative humidity of 30 to 40 percent and if all incoming specimens are heated, chances of infestation may be greatly reduced.

*Repellents* are substances that keep insects away from the herbarium specimens because of their offensive odor or taste. Paradichlorobenzene (PDB) and naphthalene have been widely used as repellents in herbaria. A 2- to 3-ounce cloth bag refilled once a year is sufficient for a standard herbarium case. Contact with either repellent should be kept to a minimum. For people working in the herbarium eight hours a day, five days a week, air exposure should not exceed 75 parts per million (ppm) for PDB and 10 ppm for naphthalene.

*Fumigants* are chemicals used in the form of a gas and applied in an enclosure to kill insects. Fumigation of herbaria has been a necessary and common procedure for many years. Recent information indicates that many of the chemicals routinely used for this purpose may be extremely dangerous to humans, and in many instances, the traditional methods of application of these chemicals are also hazardous. The Federal Insecticide, Fungicide, and Rodenticide Act prohibits the sale of fumigants except for precisely those uses for which they are labeled. Herbaria are required to comply with the Occupational Safety and Health Act.

Dowfume-75 has been cleared by the Environmental Protection Agency as a fumigant for control of dermestid beetles and other pests in herbaria. Fumigation of the herbarium with Dowfume-75 should be conducted only by a trained fumigator wearing a full-face gas mask. Large herbaria use hospital-type sterilizers for a fumigation chamber. All incoming specimens are fumigated with either ethylene dioxide, methyl bromide, or ethylene dibromide at a temperature of 38°C for three to four hours. All such equipment must be carefully designed and operated.

Under controlled conditions, dichlorvos resin strips (Shell No-Pest Strips) are suitable for use as a fumigant in herbarium cases (Lellinger, 1972). One-third of a No-Pest resin strip should be applied in each case for seven to ten days twice a year. The cases must not be opened during fumigation. Work with the dichlorvos resin strips should only be assigned for short periods, perhaps around 30 to 60 minutes for cutting and placing and 15 minutes for removing each day. The work location should be well ventilated and excessive exposure to dichlorvos avoided. Disposable plastic gloves should be used while handling the material. Workers must be trained to wash carefully after handling dichlorvos or any toxic chemicals. If exposures are to be excessive, a full-face gas mask should be worn.

Three parts of ethylene dichloride (1,2 dichloroethane) mixed with one part of carbon tetrachloride ($CCl_4$) was once used for the fumigation of herbarium cases. The mixture was applied at the rate of 6 ounces in a beaker placed on the top shelf of each case. The case remained sealed for four or five days. Ethylene dichloride is explosive without $CCl_4$ and has the property of causing injury to the human liver and kidney from either excessive single or repeated exposures (Monro, 1969). Carbon tetrachloride is extremely toxic to humans, causing liver damage (Monro, 1969). These two chemicals should be considered extremely dangerous to the persons conducting the fumigation.

## TYPE SPECIMENS

Type specimens (or *types,* as they are usually called) are specimens upon which the name of a taxon is based (see Chapter 3). Type specimens are often housed separately from the general collection because of their inherent value. Every precaution should be taken to reduce unneccesary handling and to prevent damage or loss of type specimens. Some herbaria include types in the general collection, but they are filed separately in specially colored and marked genus folders.

## LOANS

Revisionary studies require the examination of as much relevant material as possible. This may necessitate obtaining material on loan. Travel to other herbaria to study their holdings often is too costly or not practical; for this reason, specimens may be borrowed for serious taxonomic studies. This service by herbaria to research workers in taxonomy also benefits the lending institution since the individual requesting the loan has the responsibility of annotating the specimens and revising identifications. Herbaria lend to other institutions and only rarely to individuals. Loans are requested by the curator of the herbarium for study by a specific researcher.

Specimens arriving from lending institutions should be carefully checked for transit damage and the packing list verified. Many institutions number and code the sheets lightly with pencil in the lower left corner on borrowed specimens. Consecutive numbers and the herbarium abbreviation facilitate later return of the loan. Borrowers must be meticulously careful to adhere to all provisions of the loan. Specimens are studied only on the institutional premises and are housed in standard metal herbarium cases when not in use. Dissections are made with care and all fragments placed in fragment folders. Special permission may be required to remove pollen or other samples. The borrower must annotate the specimens before returning the loan. Annotations for revisionary studies are made not directly on the collection labels or on the sheets, but on annotation labels which are glued above the herbarium label. Each annotation label should have the complete scientific name of the plant (genus, specific epithet, and author), full name of the person making the annotation, and the year in which the annotation was made. A convenient way to agree with an identification is by using an exclamation point (!), meaning "I have seen it."

Since specimens are irreplaceable, details of packing for return shipment are highly important. Ideally, the original boxes and packing materials are saved and used for the return shipment.


## EXCHANGES

Exchanges of duplicate specimens among herbaria and collectors are a means of augmenting a collection at minimal cost. Exchanges are generally made on a one-for-one basis. Specimens for exchange are normally unmounted and loose in the pressing paper and include adequate labels. Arrangements for an exchange should be made before shipping the specimens. Poor-quality specimens made for class purposes are not normally acceptable by most curators as exchange material.


## HERBARIUM ETHICS

Herbaria should provide visitors with policy statements including such information as: (1) hours, (2) filing arrangements, (3) reshelving, (4) loan procedures, (5) annotation labels, (6) use of library collections, (7) microscopes, and so on. Visitors have an obligation to adhere to the policy statements and to act accordingly.

Specimens are fragile. They are mounted on paper which is flexible and should not be bent. The sheets should be kept flat. Folders should not be leafed through by turning sheets like pages in a book. Instead, they should be lifted one at a time and put aside. When carrying specimens, support the stack on a sheet of cardboard (e.g., a press ventilator). One should never attempt to force an extra-large stack of specimens into a herbarium case pigeonhole. Heavy objects, elbows, books, and so forth must not be placed on specimens. Specimens should be studied only with a long-armed microscope to avoid bending the sheets. Loose material clearly identifiable with a particular specimen should be placed in fragment folders. Damaged specimens should be put aside for repair and called to the attention of curators. Flowers and fruits from herbarium specimens should be dissected sparingly. First soften by boiling or with a wetting agent; then examine the specimen; and then return the specimen to the fragment folder. Notes from the dissection may be recorded on the fragment folder. Reshelve specimens only with permission and always with extreme care.

Visits to other herbaria are often profitable in terms of examining critical specimens and collections, searching the undetermined folders, and exchanging ideas with other systematists. A note in advance will help the curator to prepare for your visit and to provide a place to work and a microscope. Many herbaria are closed in the evenings or on weekends. The privilege of visiting after normal hours may be offered but should not be requested.

## CHALLENGES FACING HERBARIA

Systematic collections of plants in herbaria provide a permanent record of the earth's flora and its diversity. The specialized literature and libraries associated with herbaria are the written record of the natural history of plants. Herbaria and the taxonomists associated with them have a variety of roles: education, service identifications, preparation of revisionary treatments, and development of classifications. Herbaria have an educational role in the training of under-graduates. At the graduate level, the student in systematics is an apprentice scientist engaged in original research. Many herbaria maintain programs of education in natural history for the general public and attempt to create an understanding of and appreciation for the living environment.

## SELECTED REFERENCES

Anderson, E.: "The Technique and Use of Mass Collections in Plant Taxonomy," *Ann. Mo. Bot. Gard.,* **28:**287–292, 1941.

Archer, W. A.: *Collecting Data and Specimens for Study of Economic Plants,* U.S. Department of Agriculture Miscellaneous Publications, no. 568, 1945.

———: "New Plastic Aid in Mounting Herbarium Sheets," *Rhodora,* **52:**298–299, 1950.

Bailey, L. H.: "The Palm Herbarium with Remarks on Certain Taxonomic Practices," *Gentes Herb.,* **7:**153–180, 1946.

Beaman, J. M.: "The Present Status and Operational Aspects of University Herbaria," *Taxon,* **14:**127–133, 1965.

Cronquist, A.: "The Relevance of the National Herbaria to Modern Taxonomic Research in the United States of America," in V. H. Heywood (ed.), *Modern Methods in Plant Taxonomy,* Linnean Society, Academic, London, 1968, pp. 15–21.

Dalla Torre, C. G. de, and H. Harms: *Genera Siphonogamarum ad Systema Englerianum Conscripta,* W. Englemann, Leipzig, 1900–1907.

Davis, P. H.: "Hints for Hard Pressed Collectors," *Watsonia,* **4:**283–289, 1961.

Fosberg, F. R., and M. H. Sachet: "Manual for Tropical Herbaria," *Regnum Veg.,* **39,** 1965.

Franks, J. W.: *A Guide to Herbarium Practice,* Handbook for Museum Curators, part E, sec. 3, Museums Association, London, 1965.

"Fumigants . . . Procedures, Precautions and Institutional Responsibility for Their Safe Use," *ASC Newsletter,* **4**(1):5–6, 1976.

Holmgren, P. K., and W. Keuken: "Index Herbariorum. Part I. The Herbaria of the World," 6th ed., *Regnum Veg.,* **92,** 1974.

Irwin, H. S., W. W. Payne, D. M. Bates, and P. S. Humphrey: *America's Systematic Collections: A National Plan,* Association of Systematics Collections, 1973.

Kobuski, C. E., C. V. Morton, M. Ownbey, and R. M. Tryon: "Report of the Committee for Recommendations on Desirable Procedures in Herbarium Practice and Ethics," *Brittonia,* **10:**93–95, 1958.

Lellinger, D. B.: "Dichlorvos and Lindane as Herbarium Insecticides," *Taxon,* **21:**91–95, 1972.

McNeill, J.: "Regional and Local Herbaria," in V. H. Heywood (ed.), *Modern Methods in Plant Taxonomy,* Linnean Society, Academic, London, 1968, pp. 33–44.

Monro, H. A. U.: *Manual of Fumigation for Insect Control,* 2d ed., FAO Agricultural Studies, no. 79, 1969.

Rollins, R. C.: "Deep-freezing Flowers for Laboratory Instruction in Systematic Botany," *Rhodora,* **52:**289–297, 1950.

————: "The Role of the University Herbarium in Research and Teaching," *Taxon,* **14:**115–120, 1965.

Savile, D. B. O.: *Collection and Care of Botanical Specimens,* Canada Department of Agriculture Publication, no. 113, 1962.

Shetler, S. G.: "The Herbarium: Past, Present, Future," *Proc. Biol. Soc. Wash.* **82:**687–758, 1969.

Smith, C. E.: *Preparing Herbarium Specimens of Vascular Plants,* Agricultural Information Bulletin, no. 348, A.R.S., U.S.D.A., 1971.

Stern, W. L., et al.: "Lindane and Dichlorvos for Protection of Herbarium Specimens Against Insects," *Taxon,* **17:**629–632, 1968.

Turrill, W. B.: "Plant Taxonomy, Phytogeography, and Plant Ecology," *Vistas Botany,* **4:**187–224, 1964.

# Chapter 9

# Methods of Identifying
# Vascular Plants

Identification is an integral part of all taxonomic work. Unidentified plants are identified by comparing them with identified plant specimens. This process, combined with the determination of the correct name applicable to that plant, is sometimes referred to as *specimen determination*. Aids to identification have been developed so that with some training and practice, one may readily identify most groups of plants in North America with no major problem. In order to make identifications, it is necessary to possess: (1) knowledge of taxonomic methods, characters and terms; (2) knowledge of manuals and other resources, such as herbaria; and (3) experience in the identification of plants.

Identification assumes that a classification scheme exists which has distinguished groups of plants and has applied names to them. Once specimens are identified and scientific names applied, the information stored in the classification system becomes available. The names of plants permit retrieval of information, such as chromosome numbers, natural products, distribution maps, and so on, from the system. Accurate identifications are a prerequisite to using classifications as information-retrieval mechanisms, and are fundamental to such research fields as biogeography, biochemistry, ecology, genetics, and physiology—i.e., the whole realm of science involving plants, including agricultural and biomedical disciplines. Names are a vital part of our language and are ul-

**157**

timately important to everyone. They help us to communicate facts concerning the world around us.

Most systematists spend at least some part of their time doing identification work for others or themselves. They use several methods to achieve correct identifications. Persons familiar with the flora of an area can usually recognize thousands of species and can apply a name at some rank to specimens. It may be necessary to aid and improve identification by comparing the specimen with previously identified plants in the herbarium.

Unknown specimens are often identified by means of *keys*. A key is a device for easily identifying an unknown plant by a sequence of choices between two (or more) statements. If a herbarium is available, direct comparison is used after an identification has been made by keying. If the identification cannot be checked locally, some collections may be sent to a person who has become an expert in the identification of that particular group of plants.

## OBSERVING THE PLANT BEFORE IDENTIFICATION

Examining a specimen carefully before beginning identification is a good habit to develop. The beginner should select a freshly collected plant with roots (on herbs), stems, leaves, flowers, fruits, and seeds. A specimen with all these parts will facilitate identification. For practice, make sure the plant selected has large flowers on which the parts can be easily seen. It is frustrating for a beginner to attempt to identify plants with small complex flowers. To observe the plant properly, you will need a good hand lens of 10X magnification, a pair of sharp-pointed forceps, a straight dissecting needle, and some single-edged razor blades.

Steps in observing the plant include the following (refer to Chapter 10 for diagrams and definitions of morphological terms):

1  Determine whether it is woody or herbaceous. If herbaceous, is it annual or perennial?
2  Observe the flower and name its parts.
3  Count the number of sepals and petals.
4  Determine whether the sepals and petals are connate or separate.
5  Count the number of stamens. Observe where they are attached. Note any fusion of the filaments or anthers. Observe the disposition of the anthers.
6  Count the number of pistils, styles, and stigmas of the gynoecium.
7  Remove the perianth and stamens. Make a cross section of the ovary with the razor blade. Count the number of locules. Observe the number of ovules and placentation.
8  Select another flower and make a longitudinal section of the entire flower through its center. Note the ovary position and the fusion of the perianth.
9  Note the leaf type, arrangement, and venation.
10  Note the distribution and kinds of surface coverings.

When these characters are determined in advance, the process of identification will be much easier and you will be confident of step-by-step decisions. After the plant has been carefully examined, the next step in identification is keying.

## IDENTIFICATION WITH KEYS

It is far more efficient to identify a specimen by the use of keys than to shuffle through a stack of previously named herbarium specimens until a likely comparison is found. In North America, unknown plants can usually be identified by means of keys. A key is fundamentally similar to an outline where the topics are arranged in order of descending importance. The use of a key provides the correct identity of a specimen by a process of elimination. Keys have long been used to identify plants, and most manuals contain them.

Plants were the subject of description and illustration in classical and medieval writing, especially in those works where accurate identification was a matter of practical importance. This eventually led to the development of bracketed diagrams in the seventeenth century which functioned both as a classification device or *conspectus* and as an identification tool. According to Voss (1952), the Latin *clavis*, or "key," was apparently not used in connection with such diagrams until Linnaeus used it in 1736 (and then with reference to a diagram in which he was classifying botanists, not plants). The use of modern keys for identification is usually credited to Lamarck (1778) in his *Flore française*.

Keys are constructed using contrasting characters to divide the plants in the key into smaller and smaller groups. Each time a choice is made, a number of taxa are eliminated. Statements in the keys are based on the characters of the plants. For example, a key might separate taxa using the following choices: (1) herbaceous versus woody, if herbaceous, the woody plants are eliminated, (2) the next choice, zygomorphic flowers versus actinomorphic; if zygomorphic, the plants with actinomorphic flowers are eliminated; and so forth. Each time a choice is made, the number of taxa which remain is reduced by the use of contrasting characters. If sufficient numbers of characters are contrasted, the number of possibilities is eventually reduced to one.

The use of a key is analogous to traveling a highway which forks repeatedly, each fork having roadside directions. If the traveler follows the proper direction, the destination will be reached. If the traveler lacks information or if the road signs are faulty, the destination may not be reached, except by trial and error.

In most manuals, the first step in identification of an unknown plant involves the use of keys to determine the family. Next, a key to genera will provide the generic name. After the genus is determined, the keying process is repeated within the genus for a determination of the species.

Descriptions are normally available in manuals for families, genera, and species. By consulting descriptions, the likelihood of errors due to mistakes in observations or selection of choices will be reduced. A few manuals use keys with many characters in each choice and do not use descriptions. In this case, additional references with illustrations or descriptions are useful.

## Suggestions for the Use of Keys

**1**   Obtain as much information as possible regarding the characters of the unknown plant before starting to key. Attempts to use the key are likely to fail if a specimen consists of a single leaf or just a flower.

**2**   Select appropriate keys for the plant material and for the geographic area where the plant was obtained.

**3**   Read the introduction to the keys for abbreviations and other details.

**4**   Always read both choices carefully, observing punctuation.

**5**   Be sure you understand all the terms found in each choice. Use a glossary.

**6**   If the specimen doesn't seem to fit the key and all choices are unlikely, you probably made a mistake. Retrace your steps.

**7**   If both choices seem possible, try going both ways.

**8**   Confirm your choices by reading descriptions.

**9**   Verify your results by comparing the specimen with an illustration or with a herbarium specimen.

## Types of Keys

Most keys used today are *dichotomous*. A dichotomous key presents two contrasting choices at each step. The pair of choices in a dichotomous key is called the *couplet*. The key is designed so that one part of the couplet will be accepted and the other rejected. The first contrasting characters in each couplet are referred to as the *leads*. These are usually the best contrasting characters. Characters following the lead are *secondary key characters*.

A few keys may not be dichotomous and may provide three or four choices, but pairs of choices are preferred.

Several different formats for arranging the couplets are used for plant identification. The *indented* or *yoked* key is the one most widely used in manuals for the identification of vascular plants. In the indented key, each of the couplets is indented a fixed distance from the left margin of the page. An example of the indented key is shown below (Jones, 1976):

**1**   Plants usually floating on the water of swamps and
ponds; stems hollow and inflated.                                    **1**   *Hottonia*
**1**   Plants terrestrial, not floating; stems solid, not en-
larged.
    **2**   Leaves all basal.                                              **3**   *Dodecatheon*
    **2**   Leaves cauline as well as basal.
        **3**   Leaves alternate.

    **4**  Plant usually less than 1 decimeter in height; flowers and fruits are sessile; mature fruit circumscissile.    **6**  *Centunculus*

    **4**  Plant usually more than 1 decimeter in height; flowers and fruits on stalks 1 to 2 centimeters long; mature fruit five-valved.    **2**  *Samolus*

  **3**  Leaves opposite.

    **5**  Flowers yellow; perennial; mature fruit valved.    **4**  *Lysimachia*

    **5**  Flowers scarlet, salmon, blue, or white; annual; mature fruit circumscissile.    **5**  *Anagallis*

In the second type, the *bracket* or *parallel* key, the two couplets are always next to each other in consecutive lines on the page. At the end of each line there is either a name or a number referring to a couplet later in the key.

**1**  Plants usually floating on the water of swamps and ponds; stems hollow and inflated.    **1**  *Hottonia*

**1**  Plants terrestrial, not floating; stems solid, not enlarged.    **2**

  **2**  Leaves all basal.    **3**  *Dodecatheon*

  **2**  Leaves cauline.    **3**

**3**  Leaves alternate.    **4**

**3**  Leaves opposite.    **5**

  **4**  Plant usually less than 1 decimeter in height; flowers and fruit sessile; mature fruit circumscissile.    **6**  *Centunculus*

  **4**  Plant usually more than 1 decimeter in height; flowers and fruits on stalks 1 to 2 centimeters long; mature fruit five-valved.    **2**  *Samolus*

**5**  Flowers yellow; perennial; mature fruit valved.    **4**  *Lysimachia*

**5**  Flowers scarlet, salmon, blue, or white; annual; mature fruit circumscissile.    **5**  *Anagallis*

Both the indented and bracket keys have advantages and disadvantages. When an indented key is used, the lines become more indented for each couplet. This can be uneconomical, since additional pages are required. In contrast, the bracket keys make good use of page space. Indented keys may have the second part, or leg, of the couplet on a page much later than the first part of the couplet. This may cause confusion when one tries to find the second part of the couplet. In bracket keys, both parts of the couplet are adjacent.

The indented key groups similar elements so that their morphology can be grasped visually, often making identification easier. In the example of the indented key, *Centunculus* and *Samolus* are grouped under leaves alternate. *Lysimachia* and *Anagallis* are grouped under leaves opposite. In the example of the bracket key, this grouping is not immediately apparent.

Modern dichotomous keys are conventionally published in either the indented or bracketed style. Although various other styles have been proposed, these traditional formats serve their purpose well and have gained general acceptance. The major considerations in key construction are ease, quickness, and accuracy.

When preparing a key, several techniques should be followed:

1  The key should be dichotomous.

2  The first word of each lead of the couplet should be identical. For example, if the first lead of a couplet begins with the word *stamens,* the second lead of the same couplet must begin with the word *stamens.*

3  The two parts of the couplet should be made up of contradictory statements so that one part will apply and the other part will not.

4  Avoid the use of overlapping ranges or vague generalities in the couplets, that is, 4 to 8 millimeters versus 6 to 10 millimeters long, or large versus small, or southern versus northern.

5  The couplets should be written to make positive statements. An example to avoid is "leaves narrow versus leaves not narrow."

6  Use features that are readily observable. Avoid using geographical location as a sole separation character.

7  The leads of consecutive couplets should not begin with the same word, since this may cause confusion when one looks away to observe the plant specimen.

8  It may be necessary to provide two sets of keys in some groups; flowering versus fruiting material, vegetative versus flowering, or staminate versus pistillate in dioecious plants.

9  Couplets of a key may be numbered or lettered, or may use some combination of lettering and numbering, or may be left blank in the case of indented keys.

## How to Locate Keys and Manuals

Appendix 2 (page 341) contains references to manuals for the identification of vascular plants found in North America north of Mexico. In addition, lists of books are provided for identification of special groups such as grasses, ferns, legumes, and others. For most references, the title alone provides the geographic area. If not, annotations are provided at the end of the reference.

Floras are constantly being written or updated, and new books for plant identification can be found in libraries and bookstores. Inquiries can also be made of herbarium curators or botany professors who teach plant taxonomy for suggestions of books to aid in plant identification.

## UNCONVENTIONAL IDENTIFICATION METHODS

Keys are the traditional method of identification in systematic botany. If keys are well written, if adequate specimens are in hand, and if the person doing the keying is careful, then the specimen can be successfully identified. Keys, however, have several major disadvantages. The use of certain characters is required even if the character is not available on the unknown specimen. Entry is usually available only at one point in the key. Attempts to improve upon the traditional identification process using keys have resorted to either polyclaves or computer techniques for identification.

## Polyclave Identification

A *polyclave* is a multientry, order-free key implemented in several different formats. One form is a diagnostic key which utilizes cards that are placed on top of one another to eliminate taxa which disagree with the specimen to be identified. Another form is a computer-stored multientry key. Still another polyclave is a printed table or matrix giving the status of various taxa and characters useful for separating the taxa.

Polyclaves have a tremendous advantage over dichotomous keys, since the user of the key rather than the author of the key selects the characters to be used. The user is free to select appropriate characters for each unknown specimen. The route taken to a particular taxon may differ considerably from one specimen to another. The logic of identification with a polyclave is the same as that in a key, but the user is free to choose any character, in any sequence, thus avoiding the rigid format of traditional keys. This freedom of choice is particularly valuable in attempts to identify fragmentary material.

A number of *field key polyclaves* are in existence: (1) cards with holes, commonly referred to as "peek-a-boo" or "window" cards; (2) edge-punched or "key sort" cards; (3) semitransparent overlays; and (4) standard computer cards.

The peek-a-boo or window card key to world angiosperm families uses a card for each character (Hansen and Rahn, 1969). Round holes are punched beside a family number for the families which have that character. The plant to be identified is examined and its characters noted. With a good specimen, it is possible to find 20 to 25 characters. One goes through the cards selecting those which have cards corresponding to the characters found on the plant. The cards are then put on top of each other in a stack. Families having all the observed characters will be indicated by the holes, or "windows," which are easily seen when the cards are held in front of a light. The logic is simple. Each time a card is added, families not perforated in this new card are excluded. With 20 to 25 cards, it is likely that the family may be determined or the search narrowed to several families. Until recently, field key polyclaves had to be manufactured specially, but now they may be punched by computer on standard computer cards (Pankhurst, 1974).

Two general kinds of *computer-based polyclaves* have been developed. One kind, developed for qualitative taxon-character data, employs elimination. The other, developed for taxon-character frequency tables, employs likelihood ratios or other probabilistic techniques. In the latter case, the computer, on the basis of the characters supplied, can indicate taxa that have been eliminated as well as the taxa that remain. The computer may also prompt the user by suggesting additional characters that might be observed, e.g., locule number. If locule number is unavailable, the user could input other characters that are available from the specimen.

## Computerized Identification

A computer functions from procedures derived from a computer program. The computer program is a machine-independent procedure or *algorithm,* which is a set of rules selected from some group of commonly understood rules or actions. An algorithm is evaluated by carrying out the action of a computer program in the specified order. This produces some described result, no less and no more. The result is produced in the last step of the evaluation of the algorithm.

A traditional taxonomic key is actually an example of an algorithm. The basic repertoire of a taxonomic key includes directions for continuing the sequence of evaluation on the basis of directed observations of a specimen. The result, the identification of a specimen, is given at the end point of the path taken through the key.

Computers are machines intended only for the evaluation, and not the writing, of algorithms. The computer can be programed to identify or to construct a key only if an initial algorithm has been successfully prepared by a taxonomist.

The major research efforts in computerized identification may be grouped into four major approaches: (1) computer-stored dichotomous keys, (2) computer-constructed keys, (3) simultaneous character-set methods or matching coefficients, and (4) automated pattern recognition.

*Computer-stored keys* may be programed to provide a dialog between the user and the computer. The computer asks a question, awaits an answer, then asks another question appropriate to that response. In general, computer-stored keys offer no advantage over printed keys, and have the disadvantage of requiring access to a computer for each identification. Until computers are cheaper and easier to use than books, this approach will attract little interest.

The second major approach, *computer-constructed keys,* shows promise as a technique. Traditional dichotomous keys may be constructed by computer from machine-readable files of taxonomic data. The challenging aspect of this topic lies more in the realm of data collecting and coding than in the computer algorithm itself. Character definition and coding are particularly troublesome in handling all taxonomic information. If these problems are overcome, computer aids to constructing keys and diagnostic descriptions could be helpful.

*Simultaneous character-set methods* are those in which values or states of a predefined set of characters are recorded for the unknown plant and then submitted to a computer program. The computer program suggests one or more possible identifications or concludes that the specimen differs from those considered as possibilities. Because of the time and expense involved, it is likely that character-set methods will prove useful only in very difficult situations when traditional identification methods are not adequate, e.g., hybrid or apomictic swarms, or with cultivated plants.

*Automated pattern-recognition systems* providing fully automated specimen identification may be possible for at least certain organisms. This technique combines automated character observation by optical scanning and pattern-recognition procedures. Such programs and techniques have been developed for analysis of chemical spectra and photomicrographs of chromosomes, recognition of abnormal cells in cytological samples, pattern recognition of remote sensing imagery in vegetation and agricultural surveys, and so on. Its successful use in plant identification will most likely combine analysis of the chemical constituents with pattern recognition.

Morse (1974) presented a system of computer programs to aid in the identification of plants and related aspects of plant taxonomy. His computer programs include routines for specimen identification, key construction, and description preparation. Also included are programs for comparing taxa, inverting data files, and producing punched-card field keys. He recognizes that routine application of computer-stored data matrices to specimen identification presents problems: (1) terminals must be located in herbaria; and (2) a network of accessible taxonomic data matrix files must be prepared and be available. Until a critical mass of such data files are accessible via a network of computer terminals, there will be insufficient use to justify the cost.

Identification aided by the computer is possible with present technology. In the future, libraries of computer programs and taxonomic data matrices may play a role in information processing and data retrieval in taxonomy. In the immediate future, identifications will be done with traditional keys and manuals.

## SELECTED REFERENCES

Bossert, W.: "Computer Techniques in Systematics," *Systematic Biology,* National Academy of Science Publication, 1692, Washington, D.C., pp. 595–614, 1969.

Duke, J. A.: "On Tropical Tree Seedlings. I. Seeds, Seedlings, Systems, and Systematics," *Ann. Mo. Bot. Gard.,* **56:**125–161, 1969.

Furlow, J. J., and J. H. Beaman: "Sample Taxonomic Data Matrices for Vascular Plants Prepared by Students at Michigan State University," in John J. Furlow and John H. Beaman (comps.), *Flora North America Report No. 56:1–118,* FNA Secretariat, Smithsonian Institution, Washington, D.C., 1971.

Hansen, B., and K. Rahn: "Determination of Angiosperm Families by Means of a Punched-card System," *Dan. Bot. Ark.,* **26:**1–46, plus 172 punched cards, 1969.

Jones, S. B.: "Mississippi Flora. VI. Miscellaneous Families," *Castanea,* **41:**189–212, 1976.

Lamarck, J. B. P.: *Flore Françoise,* Imprimerie Royale, Paris, 1778.

Morse, L. E.: "Specimen Identification and Key Construction with Time-sharing Computer," *Taxon,* **20:**269–282, 1971.

———: *Computer Programs for Specimen Identification, Key Construction and*

*Description Printing Using Taxonomic Data Matrices,* Publications of the Museum of Michigan State University Biological Series, East Lansing, 1974a.

————: "Computer-assisted Storage and Retrieval of the Data of Taxonomy and Systematics," *Taxon,* **23:**29–43, 1974b.

————, J. H. Beaman, and S. G. Shetler: "A Computer System for Editing Diagnostic Keys for Flora North America," *Taxon,* **17:**479–483, 1968.

Osborne, D. V.: "Some Aspects of the Theory of Dichotomous Keys," *New Phytol.,* **62:**144–160, 1963.

Pankhurst, R. J.: "A Computer Program for Generating Diagnostic Keys," *Computer J.,* **13:**145–151, 1970.

————: "Botanical Keys Generated by Computer," *Watsonia,* **8:**357–368, 1971.

————: "Automated Identification in Systematics," *Taxon,* **23:**45–51, 1974.

————(ed.): *Biological Identification with Computers,* Academic, London, 1975. (Articles are generally free of computer jargon.)

Shetler, S. G.: "Demythologizing Biological Data Banking," *Taxon,* **23:**71–100, 1974.

Voss, E. G.: "The History of Keys and Phylogenetic Trees in Systematic Biology," *J. Sci. Lab. Denison Univ.,* **43:**1–25, 1952. (An excellent paper with many references.)

# Chapter 10

# Terminology
# of Flowering Plants

Because the identification and classification of plants is based chiefly on the details of their external features, a knowledge of the terminology of plant morphology is essential. The description of plant structures is called *phytography,* and it includes the descriptive terminology of whole plants and their component parts.

The major components of a plant are referred to as *organs.* The *vegetative organs* are the roots, stems, and leaves. The *reproductive organs* are the flower, fruit, and seed. Adjectives referring to each of these organs are used to designate their various states or forms. The commonly encountered terminology for descriptive plant taxonomy is illustrated and defined in this chapter and is arranged under broad headings for each organ. Mastery of these selected terms will allow one to use most manuals. The best way to learn the terms is to study them concurrently with selected plant material. The learning experience is much more pleasant and rewarding if critical features are seen in actual plant material.

## VEGETATIVE MORPHOLOGY

### Duration

Duration is the length of time an individual plant exists. Note the following definitions having to do with duration:

*Annual.* A plant that lives for one growing season and then dies. A *winter annual* germinates in the fall, overwinters, produces seeds in the spring, and then dies.

*Biennial.* A plant that requires two growing seasons to complete the life cycle of two years' duration.

*Herbaceous perennial.* A nonwoody plant that lives for several years; the shoot system dies back each winter.

*Woody perennial.* A tree or shrub. The shoot system remains above ground during winter; the stems and the roots become woody and live for a number of years. Some woody perennials are *monocarpic,* living for several years then flowering and dying—e.g., *Agave,* the century plant.

### Roots

Unlike stems, roots (Figure 10-1) do not have regularly spaced buds and leaves. They branch irregularly and may produce adventitious buds. In general, roots provide relatively few features for identification and classification; however, they are often essential to determine whether the plant is an annual or a perennial. In some families, such as grasses, it is absolutely essential that roots be collected if the specimens are to be correctly identified. Terminology essential to the description of roots includes:

*Adventitious.* Developing from something other than the hypocotyl or another root.

*Aerial.* Growing in the air.

*Fibrous.* Thread-like and usually tough.

*Fleshy.* Relatively thick and soft, with storage tissue.

Fibrous                                Tuberous                                Tap

**Figure 10-1**   Forms of roots.

*Haustorial.* Specialized, penetrating other plants and absorbing nutrients from them.

*Primary.* Developing from the radicle of the embryo; the root that first appears from the seed.

*Secondary.* Developing from the primary root; branch roots.

*Tap.* A main primary root that more or less enlarges and grows downward.

*Tuberous.* Term applied to a relatively thick and soft root: enlarged and with storage tissue.

## Stems

Stems (Figure 10-2) are the main axes of plants. As opposed to roots, they have nodes and internodes and usually have leaves with a bud in the leaf axil. Stems may bear flowers. Many stems are highly specialized, particularly those underground stems which function as storage or overwintering mechanisms. Variation in stems is of importance in classification, and stems provide many useful taxonomic characters. Many genera of native trees may be easily recognized in their winter condition by examination of their stem features. Essential stem terminology includes the following:

*Acaulescent.* Stemless, or with the stem inconspicuous.

*Arborescent.* Becoming tree-like and woody, usually with a single main trunk.

*Bud.* An immature stem tip; an embryonic shoot (see also page 172).

*Bulb.* A short, basal, underground stem surrounded by thick fleshy leaves, e.g., an onion. *Bulbils* or *bulblets* are small bulbs.

*Caespitose.* Term applied to stems growing in a mat, tuft, or clump.

*Caudex.* The hard, overwintering base of a herbaceous perennial.

*Caulescent.* With a distinct stem.

*Cladophyll.* A flattened, leaf-like, green stem.

*Climbing.* Clinging to other objects, e.g., English ivy.

*Corm.* A short, upright, hard or fleshy bulb-like stem, usually covered with papery, thin, dry leaves, e.g., *Gladiolus.*

*Decumbent.* Term applied to stems lying flat on the ground, but turning up at the ends.

*Erect.* Growing upright.

*Fruticose.* Shrub-like and woody.

*Herbaceous.* Not woody; dying to the ground at the end of the growing season.

*Internode.* The part of the stem between two successive nodes.

*Lenticels.* Lens-shaped or wart-like patches of parenchymatous tissue on the surface of a stem.

*Node.* The area of a stem where a leaf and bud are normally borne.

*Pith* (Figure 10-3 ). The spongy tissue in the center of a stem. The pith may be (1) *continuous,* or solid throughout; (2) *diaphragmed,* or continuous, but with firmer cross plates at intervals; or (3) *chambered,* in which the tissue between plates has disappeared.

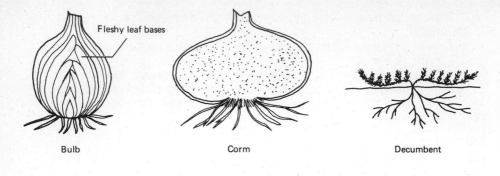

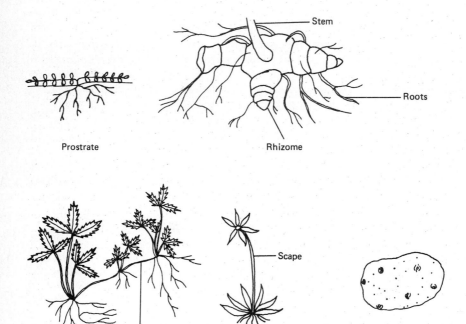

**Figure 10-2**  Specialized stems.

*Prickle.* A sharp pointed extension of the cortex and epidermis.

*Prostrate.* Growing flat on the ground.

*Rhizome.* A horizontal, prostrate, or underground stem with reduced scale-like leaves, as in German iris.

*Rootstock.* The underground portion of a plant; often referring to a caudex or a rhizome.

*Runner.* A horizontal aboveground stem, usually rooting and producing plants at the nodes, e.g., strawberry. See *Stolon.*

*Scandent.* Climbing.

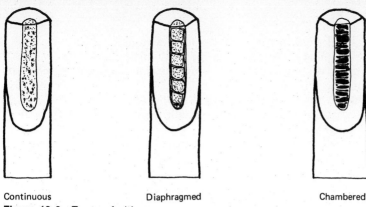

Continuous          Diaphragmed          Chambered

**Figure 10-3** Types of pith.

*Scape.* Leafless flowering stem, arising from the ground, sometimes with small, scale-like leaves.

*Scapose.* Having a scape.

*Scars* (Figure 10-4). The remains of a point of attachment. These include: (1) *leaf scars,* where a leaf was attached; (2) *stipule scars,* where stipules were attached (in some taxa, e.g., Magnoliaceae, they form a ring around the stem called a *stipular ring*); and (3) *scale scars,* where the bud scales are sloughed off. In the temperate zone, the age of twigs may be determined by counting each set of *terminal bud scale scars.*

*Shrub.* Plant having several main stems, more or less woody throughout, and usually less than 4 to 5 meters in height.

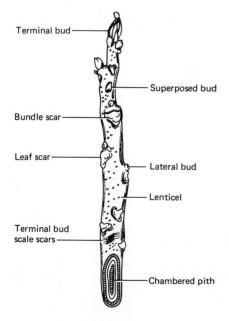

Terminal bud

Superposed bud

Bundle scar

Leaf scar

Lateral bud

Lenticel

Terminal bud
scale scars

Chambered pith

**Figure 10-4** A stem tip in the winter condition. The terminal bud scale scars indicate the location of last year's terminal bud scales.

*Stolon.* Horizontal stem rooting at the nodes; a runner.

*Suffrutescent.* Tending to be woody, especially at the base.

*Tendril.* A small, twisting appendage which attaches a scandent plant to other plants or objects.

*Thorn.* A reduced, sharp, pointed stem. See *Prickle.*

*Tree.* A woody plant, usually with one main trunk and normally 4 to 5 meters or more in height.

*Tuber.* An enlarged fleshy tip of an underground stem; e. g., Irish potato.

*Woody.* Hard in texture and containing secondary xylem.

## Buds

Buds are short, embryonic stem tips bearing leaves or flowers or both. They may be quite useful in the identification of woody plants in the winter condition. Some of the common descriptive terms for buds include the following:

*Accessory.* An extra bud produced to either side of an axillary bud.

*Adventitious.* Term applied to a bud developing from some place other than a node of a stem.

*Axillary.* Term applied to a bud located in the axil of a leaf.

*Dormant.* Inactive; may be owing to dry weather or winter condition.

*Flower bud.* A stem tip containing or producing embryonic flowers.

*Lateral.* Produced on the side of a stem, as opposed to *terminal.*

*Leaf bud.* A stem tip containing leaves.

*Mixed.* Term applied to a bud containing both embryonic leaves and flowers.

*Naked.* Not covered by bud scales.

*Pseudoterminal.* Term applied to a lateral bud which takes over the function of the terminal bud, e.g., persimmon.

*Reproductive.* Containing embryonic flowers.

*Scaly.* Covered by bud scales; a scaly bud is sometimes called a *covered* bud.

*Terminal.* Located at the stem tip.

*Vegetative.* Containing embryonic leaves.

## Leaves

Leaves are generally broad, flattened, and photosynthetic and are borne at the nodes of a stem. Just above the point of attachment of the leaf base or petiole, there is an axillary bud. Leaves are of considerable value in identification and classification. Some flowering plants have a particular form of leaf and may be recognized by their leaves alone. Some angiosperms are leafless and a few have leaves which are variable in form.

**Leaf Parts**   A complete leaf (see Figure 10-5) is composed of a blade, petiole, and stipules, any one of which may be lacking or highly modified. The ex-

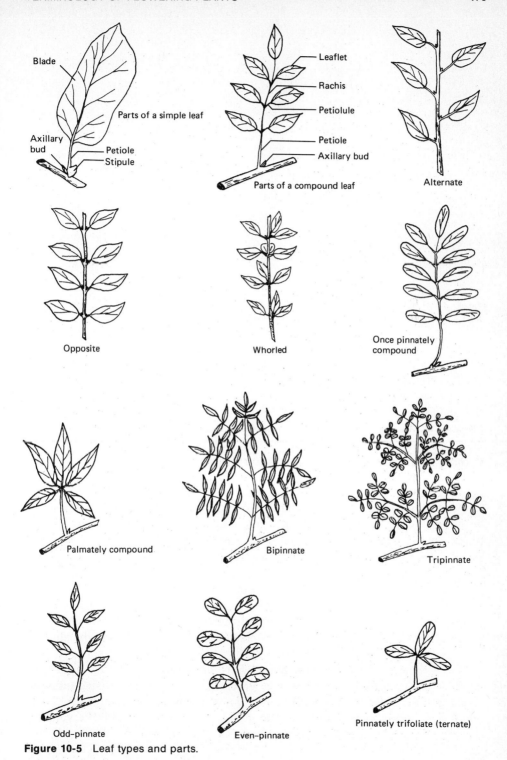

**Figure 10-5**  Leaf types and parts.

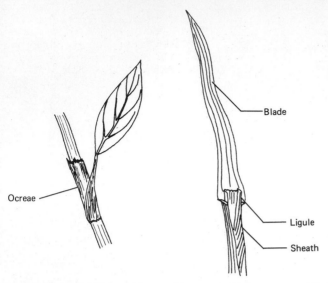

**Figure 10-6** Specialized parts of leaves. *Left:* An ocreae of *Polygonum*. *Right:* A grass leaf with a ligule and sheath.

panded, flattened part of the leaf is the *blade* (or *lamina*). The supporting stalk is called the *petiole*. The base of the petiole is variously shaped, or it may be rather enlarged and termed a *pulvinus*. In some taxa, the petiole is lacking and the leaf is said to be *sessile*. *Stipules* are a pair of appendages which, when present, either are located at the base of the petiole or are attached to the stem at the node on either side of the petiole. Stipules are usually green and leaf-like or scale-like, but they may be modified into tendrils (e.g., *Smilax*) or prickles (e.g., *Robinia*), or they may form a papery sheath around the stem called an *ocrea* (e.g., *Polygonum* and *Rumex*) (Figure 10-6). A plant lacking stipules is said to be *exstipulate*. In some plants, such as the grasses and the sedges, the basal part of the leaf surrounds the stem and is called the *sheath*. In grasses, there often is a *ligule,* which is a small portion of the sheath extending beyond the junction of the sheath and the blade (Figure 10-6).

**Venation**   The positioning of the veins (vascular bundles) in the leaf blade is called *venation* (Figure 10-7). Some leaves, especially those of monocots, have *parallel* venation, i.e., the veins occur side by side without intersecting. In contrast, most dicots have *net* or *reticulate* venation, with veins that branch and intersect. There are two main patterns of net venation: (1) *palmate,* with the main veins radiating from the point where they join the petiole; and (2) *pinnate,* with one central vein or *midrib* which has lateral veins arising along its length and at angles from it. In *dichotomous* venation, the veins fork twice into equal-sized branch veins. This is a common pattern in ferns and is the venation pattern of the leaves of *Ginkgo*.

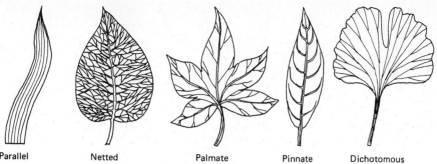

Parallel          Netted          Palmate          Pinnate          Dichotomous

**Figure 10-7**   Venation of leaves. Palmate and pinnate are also netted.

**Leaf Arrangement, or Phyllotaxy**   Stem, or *cauline,* leaves are generally arranged in one of three rather definite ways: (1) *alternate,* having one leaf at each node, usually arranged in spirals around the stem; (2) *opposite,* having leaves paired at each node on opposite sides of the stem; and (3) *verticillate,* or *whorled,* having three or more leaves at each node. See Figure 10-5.

The spiral patterns associated with the alternate arrangement are usually definite for a given genus. They may be determined by observing the number of nodes required to make a complete turn around the stem. When alternate leaves appear on just two sides of the stem, they are called *two-ranked.*

Leaf arrangement is generally a reliable character, but some genera (e.g., *Helianthus*) may have opposite leaves below and alternate leaves near the inflorescence. In a few instances, opposite or whorled may intergrade as a result of either extremely fast or slow growth. With woody plants, leaf arrangement should be determined by examining normal, terminal, long shoots rather than by observing lateral, dwarf, or spur shoots; accurate determination of the leaf arrangement may be difficult because of the very short internodes on spur shoots.

In *acaulescent* species (without conspicuous leafy stems), the leaves often form a cluster at ground level and are said to be *basal* or *radicle* (Figure 10-8). An acaulescent plant, such as dandelion, has a stem, but the internodes are extremely short, thereby producing a stacked spiral of leaves that forms a rosette on the ground. *Iris* has two-ranked leaves with overlapping leaf bases and the leaves folded together lengthwise; this arrangement is called *equitant* (Figure 10-8).

**Leaf Type**   See Figure 10-5. A leaf with the blade in a single part, although it may be variously divided, is a *simple* leaf, while one with the blade divided into smaller, blade-like parts is termed *compound.* The individual blade-like parts of a compound leaf are called *leaflets* or *pinnae.* The main axis of the leaf to which the leaflets are attached is called the *rachis.* The leaflets may be sessile or attached to the rachis by stalks called *petiolules.* To determine whether a leaf is simple or compound, follow it to the stem until an axillary bud is reached.

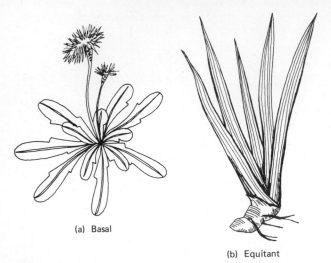

(a) Basal

(b) Equitant

**Figure 10-8**  Leaves with specialized disposition: *(a)* basal and *(b)* equitant.

Buds occur only in the axils of a complete leaf, not in the axils of the leaflets. In some plants, especially if immature, it may be difficult to find the axillary bud. It may be necessary to determine whether a leaf is a variously divided simple leaf or if it is a compound leaf. Generally, if the leaflets are well-developed and separate from the rachis, the leaf is regarded as compound.

If the leaflets are attached to both sides of one central rachis, the leaf is *once-pinnately compound.* If they diverge from a common point at the end of the petiole, much like fingers from the palm of the hand, the leaf is *palmately compound.* Pinnately compound leaves may have the leaflets divided. This type of leaf is said to be *twice-pinnately compound* or *bipinnate.* If the leaflets are divided twice, the leaf is *thrice-pinnately compound* or *tripinnate.* A once-pinnately compound leaf with paired leaflets and one leaflet at the tip is *odd-pinnate,* because it has an odd number of leaflets. If it does not have a leaflet at the tip, it is *even-pinnate.* A leaf with three leaflets is called *trifoliate* or *ternate.*

**Duration of Leaves**  Leaves are often temporary organs, and *duration* refers to the length of time they function. In the temperate zone, duration is usually related to the onset of lower temperatures, but in regions with seasonal rainfall, it may be regulated by drought. The following definitions are important:

*Deciduous.* Falling at the end of the growing season; not evergreen; referring to leaves falling in the autumn.
*Evergreen.* Persistent, not deciduous; often used to describe the plant.
*Fugacious.* Falling shortly after development; soon falling.
*Marcescent.* Withering at the end of the growing season, but not falling until the following spring.

*Persistent.* Evergreen, not deciduous; functioning over two or more growing seasons; often used to describe the leaves.

**Shape of Blade**   Although the overall size of the leaf blade may be environmentally influenced, the shape of a leaf or leaflet (Figure 10-9) is often characteristic for the species. Some common leaf shapes are listed below:

*Cordate.* Heart-shaped, with a basal notch.
*Cuneate.* Wedge-shaped, tapering toward point of attachment.
*Deltoid.* Triangular.
*Elliptical.* Having the shape of a flattened circle, usually more than twice as long as broad.
*Hastate.* Arrowhead-shaped.
*Lanceolate.* Lance-shaped, tapering from a broad base to the apex; much longer than wide.
*Linear.* Long and narrow with nearly parallel sides.
*Oblanceolate.* Lanceolate, but with the broadest part near the apex.
*Obovate.* Ovate, but with the broadest part near the apex.
*Ovate.* Egg-shaped, with the broadest part toward the base.
*Peltate.* With the petiole attached to the lower surface of the blade rather than on the margin; shield-shaped.
*Perfoliate.* Term used to describe opposite leaves having their bases united around the stem and the stem apparently passing through the leaves.
*Reniform.* Kidney-shaped.
*Subulate.* Tapering from a broad base to a sharp point.

**Apex of Leaf Blade**   The tip of a leaf farthest removed from the petiole is termed the *apex* (Figure 10-9). Some common types are listed below:

*Acuminate.* Tapering gradually to a sharp, prolonged point.
*Acute.* Ending in a point that is less than a right angle, but one that is not acuminate; distinct and sharp, but not drawn out.
*Aristate.* With a bristle at the tip.
*Cuspidate.* Tipped with a sharp and rigid point.
*Emarginate.* Shallowly notched and indented at the apex.
*Mucronate.* Abruptly tipped with a small point, projecting from the midrib.
*Obtuse.* With a blunt or rounded tip.
*Rounded.* With a broad arch at the apex.
*Truncate.* Cut squarely across at the apex.

**Base of Leaf Blades**   The part of the leaf blade where the petiole is attached is the *base* (Figure 10-9). Note the following definitions:

*Auriculate.* With ear-like appendages at the base.
*Cordate.* Heart-shaped, with a notch at the base.
*Cuneate.* Wedge-shaped, gradually narrowed toward point of attachment.
*Oblique.* Asymmetrical, having one side of the blade lower on the petiole than the other; inequilateral.
*Rounded.* With a broad arch at the base.

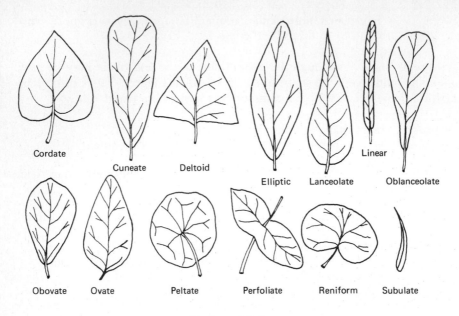

Cordate

Cuneate    Deltoid

Linear

Elliptic    Lanceolate    Oblanceolate

Obovate    Ovate    Peltate    Perfoliate    Reniform    Subulate

(a) Shape of blade

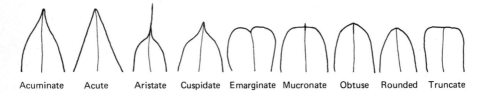

Acuminate    Acute    Aristate    Cuspidate    Emarginate    Mucronate    Obtuse    Rounded    Truncate

(b) Apex of blade

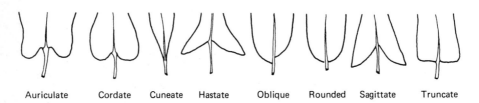

Auriculate    Cordate    Cuneate    Hastate    Oblique    Rounded    Sagittate    Truncate

(c) Base of blade

**Figure 10-9**    Leaf blades. *(a)* Shape of blade; *(b)* shape of apex; *(c)* shape of base.

*Sagittate.* Term describing basal lobes drawn into points on either side of the petiole, like the base of an arrowhead.

*Truncate.* Cut squarely across at the base.

**Margins of Leaf Blades**    The edge of a leaf blade is referred to as the *margin* (Figure 10-10). Margins fall into these categories:

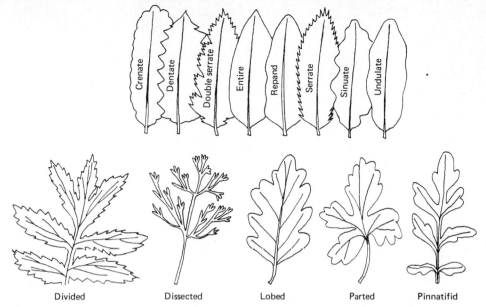

**Figure 10-10**  Leaf margins.

*Crenate.* With low rounded or blunt teeth.

*Dentate.* Having sharp, marginal teeth pointing outward.

*Dissected.* Cut into more or less fine divisions.

*Divided.* Cut into distinct sections, extending to the midrib or base.

*Doubly serrate.* With small serrations on larger serrations.

*Entire.* Smooth; devoid of any indentations, lobes, or teeth.

*Lobed.* Divided into parts separated by rounded sinuses extending one-third to one-half the distance between the margin and the midrib.

*Parted.* Cut or dissected almost to the midrib.

*Pinnatifid.* A pinnately parted leaf, divided almost to the midrib.

*Repand.* Slightly wavy.

*Revolute.* With the margin rolled inward toward the underside of the leaf.

*Serrate.* Having marginal teeth pointing toward the apex.

*Sinuate.* Having a deeply wavy margin.

*Undulate.* Having a slightly wavy margin.

**Modifications of Leaves**  Many plants produce leaves that are quite different from the customary foliage leaves. Some of these are not easily recognizable as leaves. Among the modifications are the following:

*Bract.* A greatly reduced or highly modified leaf often found in the inflorescence or subtending a flower; e.g., lemma and palea of grasses, or the brightly colored bracts of poinsettias.

*Bud scale.* A small scale surrounding the bud.

*Bulb scales.* Fleshy, succulent storage leaves; e.g., the bulb of an onion.

*Chaff.* A bract at the base of the achene in the family Compositae; i.e., a pale.

*Glume.* A bract, usually found in pairs at the base of the grass spikelet.

*Involucral bract.* Subtending the head of the family Compositae; i.e., a phyllary.

*Lemma.* The outer of the two bracts subtending the grass floret (see *Palea*).

*Pale.* See *Chaff.*

*Palea.* The inner bract subtending the grass floret. See *Lemma.*

*Phyllary.* A bract subtending the head in the Compositae family, sometimes called an involucral bract.

*Pitcher.* A tubular, insectivorous leaf able to hold water; e.g., *Sarracenia.*

*Spathe.* An enlarged bract enclosing an inflorescence; e.g., *Calla* and *Anthurium.*

**Surfaces, Surface Covering, and Texture**   The surface features (Figure 10-11) of angiosperms provide many valuable taxonomic characters. The terminology tends to be somewhat subjective and in practice is difficult to apply:

*Arachnoid.* With entangled hairs, giving a cobwebby appearance.
*Bloom.* The waxy coating on the parts of certain plants.
*Canescent.* Gray pubescent.
*Ciliate.* With a marginal fringe of hairs.
*Coriaceous.* A tough and leathery texture.
*Fimbriate.* Finely cut into fringes.

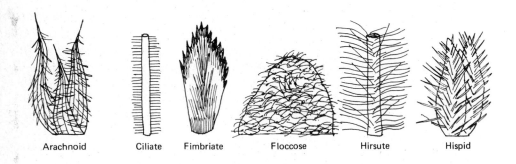

Arachnoid        Ciliate     Fimbriate         Floccose            Hirsute           Hispid

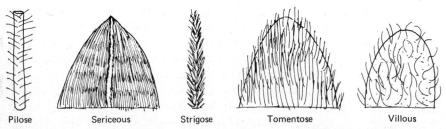

Pilose        Sericeous        Strigose         Tomentose               Villous

**Figure 10-11**   The more common types of pubescence or surface coverings.

*Floccose.* With irregular tufts of loosely tangled hairs; having a wooly appearance.

*Glabrate.* Nearly glabrous, or becoming glabrous with age.

*Glabrous.* Without pubescence of any kind; smooth.

*Glandular.* Having glands or small secretory structures.

*Glandular-punctate.* Dotted across the surface with glands.

*Glaucous.* Having a waxy appearance due to a bloom or powdery coating of wax, i.e., a waxy bloom.

*Hirsute.* With long, shaggy hairs, often stiff or bristly to the touch.

*Hispid.* With stiff, rough hairs.

*Lanate.* Wooly, with long, intertwined, coiled hairs.

*Pilose.* With scattered, long, slender, soft hairs.

*Pubescent.* Covered with short, soft hairs.

*Rugose.* Wrinkled.

*Scabrous.* Rough to the touch.

*Scarious.* Thin and membranous, usually dry.

*Scurfy.* Covered with scales.

*Sericeous.* With soft, silky hairs, usually all pointing in one direction.

*Stellate.* With star-shaped hairs.

*Stinging hairs.* Hairs which cause a stinging or burning sensation when they come into contact with the skin.

*Strigose.* Stiff hairs often appressed (i.e., pressed next to the stem) and pointing in one direction.

*Tomentose.* With densely matted soft hairs; wooly in appearance.

*Velutinous.* Velvety.

*Villous.* Covered with long, fine, soft hairs.


## REPRODUCTIVE MORPHOLOGY

Much of the classification of flowering plants is based upon reproductive structures, and a knowledge of the terminology of flowers, fruits, and seeds is essential to identify plants and to understand classifications.

A *flower* is a highly modified shoot with specialized appendages. Flowers may arise in the axil of a leaf or, more often, in the axil of a reduced leaf called a *bract.* Ripened ovaries in the flower develop into the *fruit.* Fruits may have other floral structures associated with them and normally contain *seeds,* which are ripened ovules. The seed germinates and produces a new plant.


### Parts of a Generalized Flower

The major components of the unmodified flower (Figure 10-12) are the *perianth,* the *androecium,* and the *gynoecium.* These structures are attached to the *floral axis,* which in its laterally extended form is the *hypanthium;* sometimes in simpler flowers the floral axis is called the *receptacle.* The *perianth,* or *floral envelope,* is subdivided into the *calyx* (composed of the sepals) and the *corolla* (made up of the petals). A group of *stamens,* where the pollen is

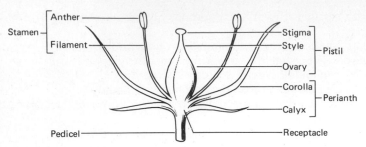

**Figure 10-12**   The parts of a generalized flower.

produced, is called the *androecium;* and the *gynoecium* is the collection of *carpels,* which may be separate or united into a *pistil.* Some terms associated with the generalized flower are defined below:

*Androecium.* The male part of the flower; a collective name for the stamens of a flower and the parts derived from stamens. The androecium occupies a position inward from the perianth.

*Anther.* That part of the stamen where the pollen is produced.

*Calyx.* The collective name for the sepals. The calyx occupies the outermost position in the flower. Together with the corolla, the calyx makes up the perianth. The calyx is usually green but is sometimes colored; e.g., *Fuchsia.* Sometimes the calyx intergrades into the corolla; e.g., water lily.

*Carpel.* A simple pistil; one part of a compound pistil. The carpel is the innermost part of the flower, bearing one or more ovules. It is thought to have evolved from ovule-bearing leaves, i.e., megasporophylls. (See Chapter 6.)

*Corolla.* The inner part of the perianth, composed of the petals. The corolla is usually larger than the calyx and is often brightly colored. The petals may be fused or separate.

*Filament.* The stalk-like part of the stamen which supports the anthers and which is attached to the receptacle and may be adnate to the petal(s) for part of its length.

*Fruit.* Mature ovary or ovaries containing seed.

*Gynoecium.* The female part of the flower; the collective term for the carpels or pistils. It is the innermost part of a flower and is composed of one or more carpels, which may or may not be united into a compound pistil.

*Hypanthium.* A floral cup or tube formed from the receptacle or from fusion of the bases of the sepals, petals, and stamens. The hypanthium appears to bear sepals, petals, and stamens on the rim of its cup or tube.

*Ovary.* The usually enlarged, basal portion of the pistil where the ovules are borne. The ovary, sometimes with associated floral structures, develops into the fruit.

*Ovule.* The structure in the ovary which develops into the seed.

*Pedicel.* The stalk of each individual flower.

*Peduncle.* The stalk bearing the entire inflorescence or a solitary flower.

*Perianth.* The collective term for the floral envelope, i.e., calyx and corolla.

*Petals.* The individual parts of the corolla occupying a position between the sepals and stamens.

*Pistil.* A structure composed of one or more carpels and usually having a stigma, style, and ovary. The pistils in a flower are collectively termed the *gynoecium.*

*Rachis.* The central axis of an inflorescence, e.g., grasses.

*Receptacle.* The portion of the stem which bears flower parts. It consists of several short nodes and internodes.

*Seed.* A matured ovule consisting of a seed coat (integument), an enclosed nucellus, embryo, endosperm, and the remains of the megagametophyte.

*Sepals.* The individual components of the calyx; the outermost whorl of the flower. The sepals are usually green, but may be petaloid and colored.

*Stamens.* The pollen-producing part of the flower, located just inside the corolla. Stamens usually have an anther and a filament. The stamens in a flower are collectively known as the *androecium.*

*Staminode.* A sterile, nonfunctional stamen.

*Stigma.* The portion of the style which is receptive to germination of pollen.

*Style.* The elongated stalk connecting the stigma and ovary.

## Inflorescences

The term *inflorescence* refers to the arrangement of the flowers on the plant (Figure 10-13). Some inflorescences are simple and readily distinguishable, but others are complicated aggregations that are difficult to characterize.

Several terms are associated with inflorescences and are essential to their understanding. *Determinate* inflorescences are those where the flowering sequence begins with the terminal flower at the tip of the stem or at the center of the flower cluster. In contrast, *indeterminate* inflorescences have a flowering sequence which starts at or near the base of the inflorescence, or the outside of the cluster, and proceeds upward or toward the center.

The following are some common types of inflorescences:

*Ament.* A deciduous, erect or lax, spike-like inflorescence, with scaly bracts and unisexual, apetalous flowers.

*Capitulum.* See *Head.*

*Catkin.* A soft spike or raceme of small unisexual flowers, the inflorescence usually falling as a unit. See *Ament.*

*Compound.* Composed of two or more simple inflorescences aggregated together, e.g., a panicle.

*Corymb.* A broad inflorescence in which the lower pedicels are successively elongate, giving the inflorescence a flat-topped appearance; indeterminate.

*Cyme.* A broad, more or less flat-topped inflorescence with the main axis terminating in a single flower which opens before the lateral flowers; determinate.

*Head.* A dense cluster of stalkless flowers; a capitulum.

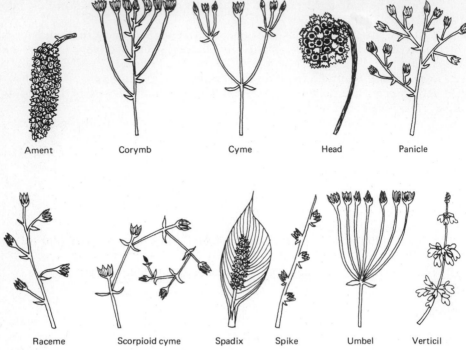

Ament          Corymb          Cyme          Head          Panicle

Raceme     Scorpioid cyme     Spadix     Spike     Umbel     Verticil

**Figure 10-13**   The major types of inflorescences.

*Panicle.* A compound inflorescence in which the main axis is branched one or more times and may support spikes, racemes, or corymbs.

*Raceme.* An inflorescence with a single axis and the flowers arranged along the main axis on pedicels; indeterminate.

*Scorpioid cyme.* A cyme which appears to coil, like a scorpion.

*Solitary.* With a single flower.

*Spadix.* A thick or fleshy spike-like inflorescence with very small flowers which are massed together and usually enclosed in a spathe, e.g. Araceae.

*Spike.* An inflorescence with a single axis and flowers without pedicels.

*Thyrse.* A compound, compact panicle with an indeterminate main axis and laterally determinate.

*Umbel.* An inflorescence of few to many flowers on pedicels of approximately equal length arising from the top of a peduncle; indeterminate; common in the Umbelliferae and Liliaceae.

*Verticil.* An inflorescence with the flowers arranged in whorls at the nodes.

## Numerical Plan

Most families of dicots have a numerical plan of 4 or 5, or multiples of 4 or 5. Monocots typically have 3-merous flowers or with parts in multiples of 3. For

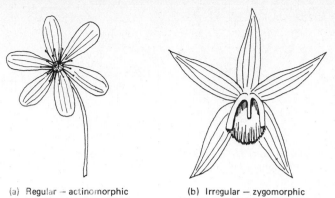

(a)  Regular – actinomorphic          (b)  Irregular – zygomorphic

**Figure 10-14**   The symmetry of flowers: *(a)* regular; *(b)* irregular.


example, the family Liliaceae has three sepals, three petals, six stamens, and a three-locular ovary. The ending -*merous* is used with a numerical prefix indicating the number of each of the parts. For example, a "5-merous calyx" implies that the calyx is made up of five sepals.


### Symmetry

The perianth in some flowers is so arranged that any line bisecting the flower through the central axis will produce symmetrical halves. Flowers having this radial symmetry (Figure 10-14) are called *regular* or *actinomorphic,* e.g., rose or petunia. Other plants have flowers which may be divided into symmetrical halves only on one line. These flowers have bilateral symmetry and are described as *irregular* or *zygomorphic;* e.g., snapdragon.


### Insertion of Floral Structures

The attachment and arrangement of floral parts is called *insertion.* The term *inserted* implies that one part is attached to or growing out of another part—e.g., stamens inserted on the petals. The parts of a flower may be fused together but eventually are attached to the receptacle. The petals normally alternate with the sepals in point of attachment, and the stamens alternate with the petals.

The *insertion of the ovary* relative to the position of the other floral parts is a widely used taxonomic character. If the ovary is attached to the receptacle above the attachment of the other floral parts, it is *superior.* The ovary is called *inferior* when it lies below the attachment of the perianth and androecium and is embedded in receptacle tissue.

*Flowers* may be classified into one of three groups: *hypogynous, perigynous,* or *epigynous,* depending upon the ovary, perianth, and androecium positions (Figure 10-15).

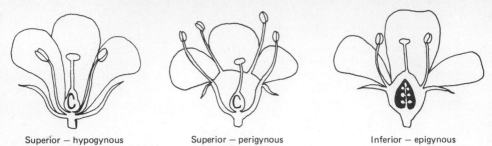

Superior — hypogynous          Superior — perigynous          Inferior — epigynous

**Figure 10-15**   Ovary position and insertion of flower parts. Superior and inferior describe the ovary; hypogynous, perigynous, and epigynous describe the flower.

In a *hypogynous* flower, the perianth and androecium are inserted around the base of the gynoecium. In some hypogynous flowers, the stamens may be adnate to the base of the petals, but the bases of the perianth are not fused together to form a floral cup. In addition, no floral structures are fused to the ovary. Hypogynous flowers have superior ovaries.

The *perigynous* flower has the perianth and stamens united by their bases to form a cup-like *hypanthium,* or else has a hypanthium developed from receptacle tissue. The hypanthium is free from the pistil, although it surrounds the ovary. The upper part of the sepals, petals, and stamens are inserted upon the rim of the hypanthium and appear to grow from it. The ovary is superior in a perigynous flower.

An *epigynous* flower has a hypanthium which is fused to the ovary, and therefore the sepals, petals, and stamens appear to arise from the top of the ovary. The ovary is obviously inferior in an epigynous flower, since it occurs below the point of attachment of the perianth and androecium. In some instances, the ovary is sunken in the receptacle. In some groups, the hypanthium extends only partially up the ovary, and the ovary is termed *half-inferior.* The hypanthium may extend beyond the top of the ovary, forming a cup around the style.

Many intergradations exist among the idealized flower types and ovary positions. To determine whether the flower is hypogynous, perigynous, or epigynous, view the flower after making a longitudinal cut.

**Union of Floral Parts**   Manuals use a number of terms to describe the connection (fusion or lack of fusion) among floral parts. Some of the more frequently encountered terminology include the following:

*Adhesion.* Union of unlike parts.
*Adnate.* United to a part of a different kind; e.g., stamens united to petals.
*Apo-.* A prefix meaning free, separate; e.g., apocarpous, which refers to separate carpels.
*Coalescence.* Union of similar parts; e.g., the sepals.
*Coherent.* Having like parts united, such as all the petals united.

*Cohesion.* Fused together.

*Connate.* United.

*Connivent.* Coming together, but not fused.

*Distinct.* Separate; not united.

*Free.* Not attached to a member of the same whorl nor to a member of another whorl of floral parts; not fused to another part.

*Gamo-.* A prefix meaning "united" or "fused"; e.g., *gamopetalous*, referring to united petals.

*Poly-.* A prefix indicating "many"; used in descriptions to imply many separate petals or sepals; e.g., *polypetalous* and *polysepalous*.

*Syn-.* A prefix meaning "united" or "fused"; e.g., *syncarpous*, "having fused carpels." The prefix becomes *sym-* before the letters b, m, and p, as in *sympetalous*, "having united petals."

**Presence or Absence of Floral Parts, and Sexuality**   The following are the essential terms in describing the sexuality and the presence or absence of floral parts:

*Apetalous.* Without petals, or with only a single perianth whorl.

*Bisexual.* Having both stamens and pistils in the same individual or flower.

*Complete.* Having calyx, corolla, stamens, and pistils.

*Dioecious.* Having staminate and pistillate flowers on different plants; e.g., American holly.

*Imperfect.* Refers to a flower having either stamens or pistils but not both; i.e., one sex but not both.

*Incomplete.* Used to describe a flower which lacks one of the four whorls of a typical flower, but usually with reference to the absence of petals or sepals or both.

*Monoecious.* Having stamens and pistils in different flowers on the same plant; e.g., corn.

*Naked.* Lacking a perianth or enclosing bracts.

*Obsolete.* Rudimentary; or not evident.

*Perfect.* Having both stamens and pistils in the same flower.

*Pistillate.* Term applied to plants or flowers with pistils, but no functional stamens.

*Polygamous.* Having perfect, pistillate, and staminate flowers all on the same plant. *Polygamo-dioecious:* polygamous, but mostly dioecious. *Polygamo-monoecious:* polygamous, but mostly monoecious.

*Staminate.* Term applied to plants or flowers with stamens, but no functional pistils.

*Unisexual.* Having only stamens or pistils, but not both, i.e., imperfect.

**Perianth**

See Figure 10-16. A calyx or a corolla of many separate sepals or petals is classified as *polysepalous* or *polypetalous*. If the sepals or petals are united, they are described as either *synsepalous* or *gamosepalous* and either *sympetalous* or *gamopetalous*. The corolla and sometimes the calyx may be formed into a basal cylindrical structure called the *tube*. The inside opening of the tube is referred

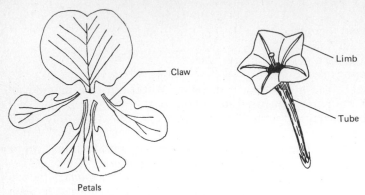

Petals

**Figure 10-16**   Corolla. Claw, limb, and tube.

to as the *throat*. The spreading part of a sympetalous corolla is the lobe or *limb*. The calyx often has prominent veins called *nerves*. The narrowed, petiole-like base of some petals or sepals is the *claw*. The *spur* is a hollow protrusion of the corolla or calyx; e.g., the corolla spur of *Linaria*. Several terms are used to refer to the shape of sympetalous corollas (Figure 10-17):

*Campanulate.* Bell-shaped.
*Funnelform.* Having the shape of an elongate funnel.
*Rotate.* Radiately spreading in one plane, with a short tube.
*Salverform.* A term used to describe a corolla with a cylindrical elongate tube and spreading rotate limbs which are abruptly flared.
*Tubular.* Cylindrical and hollow; having an elongated tube and short limb.
*Urceolate.* A term used to describe a corolla with united petals forming a tube that is expanded below the middle and narrow at the top, i.e., urn-shaped.

### Androecium

The collection of stamens in the flower is termed the *androecium* (Figure 10-18). Stamens are composed of the pollen-producing part, the *anther*, and the *filament*. Some taxonomic characters which are derived from the androecium include: number, fusion, and insertion. Terms used to describe the androecium include:

*Basifixed.* A term to describe an anther attached by its base to the filament.
*Connective.* Term describing the tissue between the two locules of an anther.
*Diadelphous.* The term applied to stamens which form two groups by the union of their filaments.
*Didynamous.* A term used to describe flowers with two long and two short stamens; common in the Labiatae.
*Epipetalous.* Having stamens inserted on the petals.
*Exserted.* Having stamens extending beyond the corolla.

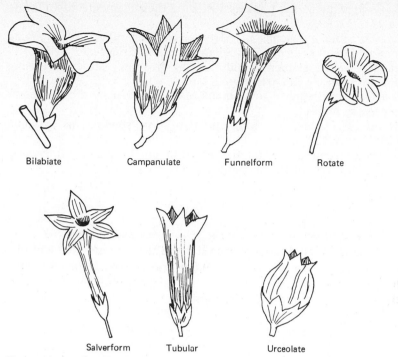

Bilabiate     Campanulate     Funnelform     Rotate

Salverform     Tubular     Urceolate

**Figure 10-17**   Corolla shapes.

*Extrorse.* Facing outward from the axis; e.g., anther sacs opening away from the center of the flower.

*Gynandrium or Gynostegium.* The result of fusion of stamens and pistil; e.g., milkweeds and orchids.

*Included.* A term used to describe stamens not protruding from corolla; not exserted.

*Introrse.* Facing inward toward the axis; e.g., anther sacs opening toward the center of the flower.

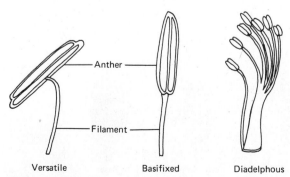

Anther

Filament

Versatile     Basifixed     Diadelphous

**Figure 10-18**   Anther types and diadelphous stamens.

*Monodelphous.* Having the stamens united by their filaments into a single structure; e.g., Malvaceae.

*Poricidal.* Opening by means of pores; e.g., the anther sacs opening through a pore at the apex, as in some Solanaceae and Ericaceae.

*Staminode.* A sterile stamen; a rudimentary or showy structure arising from the stamen whorl; e.g., *Penstemon.*

*Synantherous.* Term applied to stamens fused together by their anthers; e.g., the Compositae family.

*Tetradynamous.* Having four long and two short stamens, as in some members of the family Cruciferae.

*Versatile.* Attached at the middle; describing an anther attached crosswise at its midpoint to the apex of the filament; e.g. lily.

## Gynoecium

The collection of carpels in the flower is the *gynoecium* (see Figure 6-2). The *carpel* is generally considered to have been derived from an enrolled leaf-like structure bearing ovules. A gynoecium of one or more separate or individual carpels is described as *apocarpus,* and the individual carpel is a *simple pistil.* If the gynoecium is composed of two or more united carpels, it is *syncarpous* and is a *compound pistil.*

A typical gynoecium is composed of a pollen-receptive area at the apex called the *stigma,* a stalk-like *style,* and an enlarged basal portion bearing ovules called the *ovary.* A simple or *unicarpellate* ovary has one chamber within it where the ovules are borne. The chamber is called either a *"cell,"* or more properly, the *locule.* A *multicarpellate* ovary may have one or more locules. If an ovary has two or more locules, the interior wall separating the locules is the *septum* (plural, *septa*) (see Figures 10-19 and 10-20).

The stalk that attaches the ovule to the ovary wall is called the *funiculus.* The *placenta* is the point of attachment of an ovule to the ovary wall. It is sometimes difficult to determine the number of carpels in a gynoecium, but in many instances each group of placentae represents the location of a separate carpel. The arrangement of the placentae, or *placentation,* describes the way in which the ovules are arranged in the ovary.

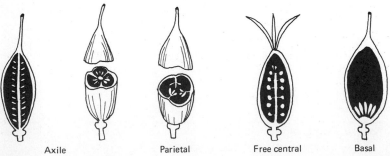

Axile      Parietal      Free central      Basal

**Figure 10-19** Placentation. The ovary with axile placentation has three locules; the others each have one locule.

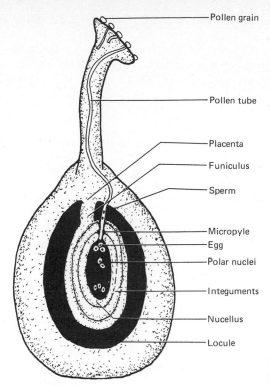

Pollen grain

Pollen tube

Placenta

Funiculus

Sperm

Micropyle

Egg

Polar nuclei

Integuments

Nucellus

Locule

**Figure 10-20** Pollen germination and fertilization in angiosperms.

The four main kinds of placentation are axile, parietal, free-central, and basal (see Figure 10-19). When there are separate locules for each carpel, with placentae at the center, the placentation is *axile*. If the septa between the carpels are absent, resulting in a one-chambered ovary with placentae on the walls, the placentation is *parietal*. A simple ovary of one carpel and several ovules usually has parietal placentation. A syncarpellate ovary in which the septa are absent and the placentae are on a central stalk arising from the base of the ovary is said to be *free central*. If the central stalk is absent and the placentation is directly on the floor of the locule, the placentation is *basal*.

## Pollination

The movement of pollen from the anther to the stigma is known as *pollination*. Some of the terms frequently related to the agents of pollination of flowers include:

*Anemophilous.* Wind-pollinated.
*Cantharophilous.* Beetle-pollinated.
*Chiropterophilous.* Bat-pollinated.
*Entomophilous.* Insect-pollinated.
*Hydrophilous.* Pollinated by water.

*Malacophilous*. Pollinated by snails and slugs.
*Ornithophilous*. Bird-pollinated.

The following are several additional, frequently encountered technical terms also relating to the pollination of plants:

*Allogamy*. Cross-pollination.
*Autogamy*. Self-pollination.
*Cleistogamous*. Characterized by self-pollination due to the flower buds' remaining closed.
*Dichogamy*. Term used of a plant which is unable to pollinate itself because stamens and carpels mature at different times.
*Protandrous*. Term applied when the pollen of a flower matures and is shed before the carpels mature.
*Progynous*. Term applied when the carpels mature before the stamens in the same flower.

## Fertilization and Embryo Development

See Figure 10-20. *Microsporogenesis* (production of microspores by meiosis) occurs in the anther, eventually giving rise to pollen grains or male gametophytes. The pollen grain of an angiosperm consists of two cells: the *tube cell* and the *generative cell*. When a pollen grain lands on a receptive stigma, the tube cell grows rapidly and forms a tube; then by digesting its way through the stigma, style, and ovary wall, the pollen grain enters the ovule through the micropyle. The generative cell produces two male gametes or sperms which follow the tube cell into the tube.

The outer wall of an angiosperm ovule is usually composed of two integuments. The *micropyle* is a minute opening between the integuments. The integuments later form the seed coat. The ovule contains a megasporangium called the *nucellus*. Megasporogenesis (production of megaspores by meiosis) occurs inside the nucellus, eventually giving rise to the *embryo sac* or female gametophyte. The embryo sac, which is embedded in the nucellus, has a group of eight nuclei. The three most important nuclei in the embryo sac are the *egg nucleus,* or female gamete, and the two *polar nuclei.*

When the pollen tube enters the embryo sac, the sperms are discharged into the embryo sac. One sperm fuses with the egg nucleus to form a *zygote*. The zygote rapidly develops into the *embryo* within the ovule. The second sperm unites with the polar nuclei to form a tissue called *endosperm*. The endosperm becomes a nutritive source for the embryo during embryo development and often during germination of the seed. Endosperm may or may not be present when the seed matures.

## Seed

A *seed* is a matured ovule containing an embryo and sometimes endosperm; it is surrounded by a seed coat. At maturation of the ovule and before germina-

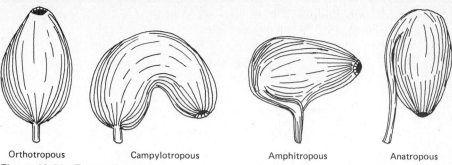

Orthotropous          Campylotropous          Amphitropous          Anatropous
**Figure 10-21**   Types of ovules.

tion of the seed, the embryo consists of an *epicotyle* (which will develop into the shoot), a *radicle* (which forms the primary root), a *hypocotyl* (which connects the epicotyl and radicle), and one or two *cotyledons,* or seed leaves. The seed coat, or *testa,* develops from the integuments and often has variations in its surface features. Among these are markings or patterns, wings, or tufts of hairs. Some seeds, e.g., litchi and nutmeg, have an *aril,* which is an outgrowth or covering of the seed developed generally from the *funiculus,* or seed stalk. The direction that the ovule or seed points, in relation to the funiculus that supports it, is sometimes an important taxonomic character (see Figure 10-21). The scar on the seed at the point where the funiculus was attached is the *hilum.* Some seeds, such as castor beans, have a *caruncle,* or appendage, near the hilum.

### Fruits

A *fruit* (Figure 10-22) is a matured ovary, including its contents. It may have other floral parts fused to it, including the receptacle, involucre, calyx, hypanthium, and so forth. Fruits may be diagnostic at the family level—e.g., the rose family—or used to delimit taxa within families—e.g., Compositae and Umbelliferae. There are numerous kinds of fruits, some easy to classify, and others more difficult.

**Definitions**   The following terms should be noted:

*Accessory structures.* Parts of fruits derived from nonovarian tissues.
*Beak.* Persistent style-base; the slender elongated ending of a fruit; e.g., *Brassica.*
*Circumscissile.* Dehiscent horizontally at or above the middle, the top part falling away as a lid.
*Dehiscent.* Opening and shedding contents in a regular and often distinctive manner.
*Endocarp.* The inner part of the pericarp, including the wall of the seed chamber.

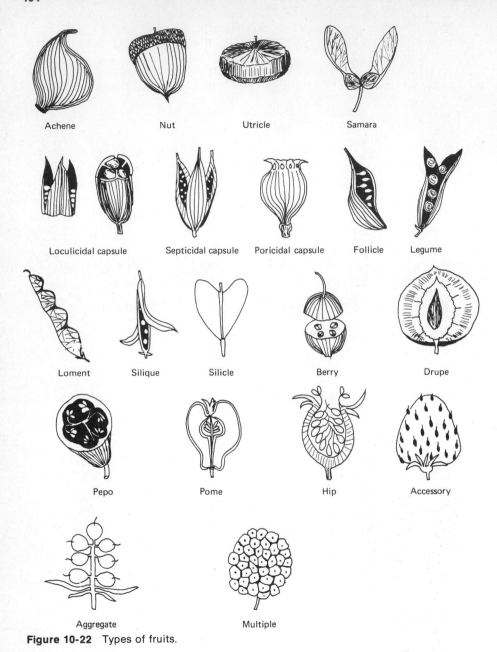

Achene          Nut          Utricle          Samara

Loculicidal capsule     Septicidal capsule     Poricidal capsule     Follicle     Legume

Loment     Silique     Silicle     Berry     Drupe

Pepo          Pome          Hip          Accessory

Aggregate          Multiple

**Figure 10-22** Types of fruits.

*Exocarp.* The outer part of the pericarp.

*Fleshy.* Soft and juicy.

*Indehiscent.* Without a specialized structure for opening; remaining closed and not shedding contents until opened by weathering, fire, or other external agents.

*Mesocarp*. The middle layer of the pericarp.

*Pericarp*. The entire wall of the fruit, from the skin to the lining of the seed chamber.

*Suture*. The line along which the fruit opens, or some other seam along the fruit. A suture usually represents the marginal fusions of the carpel or carpels.

*Valve*. A segment of a capsule.

**Fruit Types**   For ease of understanding, fruits may be divided into five major groups. Within each category, the various types of fruits are listed and described.

*1  Dry, indehiscent, and one-seeded*   This group contains the following types:

*Achene*. A dry, one-seeded fruit with a firm, close-fitting wall, but with the pericarp free from seed; examples include sunflower "seeds" and strawberry.

*Caryopsis*. A fruit in which the pericarp is fused with the seed; a grain; e.g., the fruit of the grasses.

*Grain*. See *Caryopsis*.

*Nut*. A fruit developed from a pistil with more than one carpel and having a hard woody coat; e.g., hickory, pecan, or oak.

*Nutlet*. A small nut.

*Samara*. An achene with a wing; e.g., maple.

*Utricle*. Usually a one-seeded fruit with a thin, bladdery, persistent, sometimes inflated wall; found in some amaranths.

*2  Dry, dehiscent, and having several to many seeds*   In this group are the following types:

*Capsule*. A fruit originating from two or more carpels. There are several types of capsules: *loculicidal*, dehiscent lengthwise along the back of the carpels (e.g., *Iris*); *poricidal*, dehiscent by pores at the top (e.g., poppy); *pyxis*, dehiscent in a circumscissile manner (e.g., *Plantago*); *septicidal*, dehiscent lengthwise at the junction of the carpels (e.g., *Yucca*).

*Follicle*. A fruit developing from a single pistil and dehiscing along one margin; e.g., *Delphinium*.

*Legume*. A one-locular fruit dehiscent on two sutures; e.g., beans and peas.

*Loment*. A flat legume that is constricted between seeds and falling apart into one-seeded joints; e.g., *Desmodium*.

*Silicle*. A short, two-locular fruit, composed of two valves which separate from the central partition, e.g., *Capsella* and other short-fruited members of the Cruciferae.

*Silique*. An elongated, two-locular fruit with two parietal placentae, and usually with two valves that separate from the partition on dehiscence; e.g., *Brassica* and other long-fruited members of the Cruciferae.

**3  Fleshy, derived from a syncarpous gynoecium of a single flower**  This group has the following types:

*Berry.* A fruit with a soft and fleshy pericarp; e.g., tomato, grape.

*Drupe.* A fruit with a fleshy mesocarp and a stony endocarp which forms a hard covering around the seed; e.g., peach, plum.

*Hesperidium.* A berry with a tough, leathery rind; e.g., a citrus fruit.

*Pepo.* A one-locular fruit developed from an inferior ovary with parietal placentae, having a hard outer covering; e.g., a gourd.

**4  Derived from a single or compound ovary and nonovarian tissue such as the receptacle or hypanthium; accessory fruit**  This group contains two types:

*Hip.* An aggregation of achenes surrounded by a cup-like receptacle and hypanthium; e.g., rose fruit.

*Pome.* A fruit developed from a compound inferior ovary with the seeds encased within a cartilaginous wall, as in an apple. The fleshy part of the pome is largely developed from the hypanthium.

**5  Resulting from the union of many separate carpels of a single flower, or the fusion of several fruits of separate flowers**  A fruit in this category can be aggregate or multiple:

*Aggregate.* A cluster of fruits clearly traceable to separate pistils of the same flower and inserted on a common receptacle; e.g., *Rubus* and *Ranunculus.*

*Multiple.* Term describing fruits on a common axis that are usually fused and derived from the ovaries of several flowers; e.g., mulberry. The *syconium* of the fig, which is a collection of achenes borne on the inside of a hollowed-out receptacle or peduncle, is a multiple fruit. The fruit of the pineapple results from a fusion of rachis, bracts, and ovaries.

## DESCRIPTIVE METHODS

### Floral Formulas

The floral formula method (Table 10-1) has been used for many years to describe graphically the floral morphology of the angiosperms. Although the formulas used by different authors may differ in a few details, they are all designed to compare graphically floral structure. Morphological features, such as the calyx (CA), corolla (CO), androecium (A), and the gynoecium (G), are represented by letters. The number of parts of each whorl is shown by numerals. $CA^3$ or $A^6$ convey the message that the plant has a calyx of three sepals and an androecium of six stamens. Symbols and other distinguishing marks, relatively few in number, are used to describe further the flowers; e.g., $CA^{⑤}$, a calyx of five fused sepals, or $\overline{G}$, an inferior ovary, versus $\underline{G}$, a superior ovary.

**Table 10-1   Floral Symbols, Formulas, and Examples**

| | | | |
|---|---|---|---|
| CA | Calyx, sepals | $CO^X$ | Petals variable in number |
| CAZ | Calyx zygomorphic | $CO^0$ | Apetalous |
| CO | Corolla, petals | $A^6$ | Stamens 6 |
| COZ | Corolla zygomorphic | $A^{4+2}$ | Stamens 6, 4 in one set, 2 in another |
| CO(Z) | Corolla sometimes zygomorphic | $A^{4-5}$ | Stamens either 4 or 5 |
| A | Androecium | $A^{4(5)}$ | Stamens 4 rarely 5 |
| G | Gynoecium | $G^{③}$ | Gynoecium of 3 fused carpels |
| $\underline{G}$ | Ovary superior | | |
| $\overline{G}$ | Ovary inferior | $A^{⑩}$ | Androecium of 10 stamens fused by their filaments |
| $\overline{\underline{G}}$ | Ovary either superior or inferior | $A\ G$ | Androecium and gynoecium united |
| X | Variable | | |
| ∞ | Many | $CA^3CO^3A^6\underline{G}^{③}$ | Sepals 3, petals 3, stamens 6, gynoecium of 3 united carpels, ovary superior |
| 4+2 | 4 in one set, 2 in another | | |
| 2—3 | 2 or 3 | $CA^{⑤}CO^0A^5\underline{G}^{⑤}$ | Calyx of 5 united sepals, corolla absent, stamens 5, gynoecium of 5 united carpels, ovary superior 10 |
| 0 | None | | |
| ( ) | Exceptional | | |
| ◯ | Fused | | |
| ⌢ | United above | | |
| ⌣ | United below | | |

Most procedures have advantages and disadvantages, and floral formulas are no exception. One problem is that there has been no attempt to standardize the symbols and systems. Formulas, when broadly generalized, may not convey to the user all the variation present within a family. Families may contain genera which are somewhat atypical of the majority of the genera. A floral formula of $A^{3-6}$ conveys the fact that the group has some plants with three stamens and others with six stamens, but it does not convey the relative frequency of each. Actually, six stamens may be very common and three is very rare. That could be shown by $A^{6(3)}$. On the other hand, if the formula system is made increasingly complicated by handling all special cases, the purpose of the system is defeated. When allowances are made for its weakness, the floral formula may be a valuable tool. Floral formulas are provided for each of the families treated in Chapter 14, Angiosperm Families.

## Floral Diagrams

Floral diagrams (Figure 10-23) visually depict the essential features of a flower in cross section. The floral parts are represented by semidiagrammatic symbols or ideographs. This illustrates both the number of whorls and floral parts and other features, such as fusion of structures and flower symmetry. Although

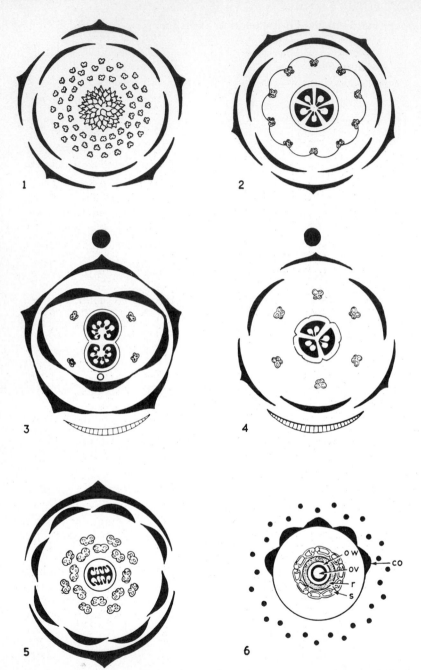

**Figure 10-23** Representative floral diagrams. *1, 2, 5,* and *6* are terminal flowers; *3* and *4* are axillary—the stem is represented by the black dot above, and the leaf in the axil of which the flower is borne is shown below. *1, Ranunculus. 2, Geranium. 3, Penstemon* (irregular). *4, Lilium. 5, Sanguinaria. 6, Taraxacum,* with calyx a pappus represented by dots (co, corolla; s, stamen; o w, ovary wall; ov, ovule; r, receptacle). *(Swingle, 1946; with permission).*

ovary position must be shown in a second drawing using a longitudinal section, this drawback does not diminish the overall value of such diagrams.

## SELECTED REFERENCES

Featherly, H. I.: *Taxonomic Terminology of the Higher Plants,* Iowa State College Press, Ames, 1954.

Gray, A.: *The Elements of Botany for Beginners and for Schools,* American Book, New York, 1887. (Reprinted by American Environmental Studies, 1970.)

Jackson, B. D.: *A Glossary of Botanic Terms, with Their Derivation and Accent,* 4th ed., Duckworth, London, 1928. (Reprinted by Hafner, 1971.)

Lawrence, G. H. M.: *Taxonomy of Vascular Plants,* Macmillan, New York, 1951.

Lindley, J.: *Excerpt From Illustrated Dictionary of Botanical Terms,* Stanford, Stanford, Calif., 1964. (This is a reprint of an excerpt from Lindley's *Illustrated Dictionary of Botanical Terms,* 1848.)

Radford, A. E., W. C. Dickison, J. R. Massey, and C. R. Bell: *Vascular Plant Systematics,* Harper and Row, New York, 1974.

Stearn, W. T.: *Botanical Latin,* 2d ed., David and Charles, Newton Abbot, England, 1973.

Swingle, D. B.: *A Textbook of Systematic Botany,* McGraw-Hill, New York, 1946.

Systematics Association Committee for Descriptive Biological Terminology: "I. Preliminary List of Works Relevant to Descriptive Biological Terminology," *Taxon,* **9:**245–257, 1960.

———: "II. Terminology of Simple Symmetrical Plane Shapes (Chart 1)," *Taxon* **11:**145–156, 1962.

Chapter 11

# Selected Literature
# of Systematic Botany

The literature of plant systematics is one of the oldest and most complicated literatures of the sciences. Plants have been named and described in innumerable books and periodicals in many languages. Taxonomists must have a thorough knowledge of the literature and bibliographic aids, because these tools are an essential part of identification, nomenclature, and revisionary studies.

*Indexes to plant names and illustrations* provide references to the original publication of names and the placement of a particular plant in a classification scheme. These indexes form the basis of a taxonomic library and are essential for nomenclatorial research. Taxonomists must not overlook the important *bibliographies, indexes, and guides* for location of the literature. There is no one book or series of volumes that treats all species of vascular plants known to occur on earth. There are, however, several important *comprehensive works* that present classifications of families, genera, or species. Stafleu (1967) and Stafleu and Cowan (1976) are excellent sources of information on these works, especially for dates of publication of works published over a long period of time.

The following annotated lists are not exhaustive, but are intended to summarize commonly used bibliographic resources.

## INDEXES TO PLANT NAMES AND ILLUSTRATIONS

Christensen, C. F. A.: *Index filicum,* 12 parts, 4 supplements, Hagerup, Hafniae, 1906–1965. (The fourth supplement was compiled by R. E. G. Pichi-Sermolli, *Regnum Veg.,* 37. This index provides the source of original publication of generic and specific names of ferns.)

Dalla Torre, C. G. de, and H. Harms: *Genera siphonogamarum ad systema Englerianum conscripta,* W. Engelmann, Leipzig, 1900–1907. (*Genera siphonogamarum* provides a numbered list of families and genera of seed plants arranged according to the Engler system. In addition, names are given for subfamilies, tribes, and so on, together with the names of genera assigned to them. Included is an extensive index. Many herbaria use the Dalla Torre-Harms numbers for filing and indexing their collections. *Genera siphonogamarum* has been updated with the issue of the *Register* [*Register zu de Dalla Torre et Harms,* J. Cramer, Weinheim, West Germany, 1958]).

*Gray Herbarium Index:* Gray Herbarium, Harvard University, Cambridge, Mass., 1896–   . (This card index is issued quarterly to subscribers and lists all new names and new combinations of names for vascular plants from the New World. Included are the names of genera, species, and infraspecific taxa. All Latin names published for Western Hemisphere plants since 1885 have been included and are arranged alphabetically by genus. In some ways it duplicates *Index kewensis,* but is invaluable as a source of names for infraspecific taxa in the New World. The index for 1896–1967 was reprinted in 1968 by G. K. Hall, Boston, in 10 volumes.)

*Index kewensis plantarum phanerogamarum:* 2 volumes, 16 supplements, Oxford, 1893–   . (Compiled at the Royal Botanic Gardens at Kew, *Index kewensis* is an alphabetical index to the generic names and binomials of seed plants. It is used to determine the source of the original publication of a generic name or binomial. It has worldwide coverage for names published since 1753. The basic volumes include names from 1753 to 1885, and supplements have been regularly published since that time. All known specific epithets published for a particular genus are in alphabetical order under each generic name. Also included are authors of the names, brief literature citations, and the geographical origin of the plants. Names considered to be synonyms appear in italics in the older volumes. Names of infraspecific taxa are excluded. In recent supplements, indication of synonymy has been discontinued but references to illustrations have been added. The original volumes were compiled by B. D. Jackson under the direction of J. D. Hooker. *Guide to Index kewensis and Its Supplements i-xiv,* compiled by E. Rouleau, Montreal, 1970, provides an index to genera and families.)

*Index Londinensis to Illustrations of Flowering Plants, Ferns and Fern Allies:* 6 volumes, with one 2-volume supplement, 1929–1931; supplement, 1941. (Compiled at Kew by O. Stapf, *Index Londinensis* is an index to the illustrations of vascular plants from 1753 to 1935. It is arranged alphabetically by genus and species. The names used in *Index Londinensis* are those accompanying the illustrations and may be obsolete. To locate illustrations of vascular plants published from 1753 to 1935, *Index Londinensis* should be used; after 1935, *Index kewensis* has this information.)

Pichi-Sermolli, R. E. G. See Christensen, C. F. A.

Rouleau, E. See *Index kewensis plantarum phanerogamarum.*

Willis, J. C.: *A Dictionary of the Flowering Plants and Ferns,* 8th ed., revised by H. K.
A. Shaw, Cambridge, London, 1973. (Willis's, *Dictionary* provides an alphabetical
list of all generic names published since 1753 and all family names published since
1789. It provides concise information on the author or authors, synonyms, family,
number of species, and distribution of genera. Descriptions are given for the families. This is a handy reference for checking spellings, synonyms, and number of
species in a genus.)

## BIBLIOGRAPHIES, INDEXES, AND GUIDES TO THE LITERATURE OF SYSTEMATIC BOTANY

*Biological Abstracts and BioResearch Index:* BioSciences Information Service,
Philadelphia. [Biological Abstracts (BA), 1929- , is a bi-monthly publication listing
abstracts of papers appearing in periodicals. *BioResearch Index* (BioI), 1965- ,
is a monthly publication indexing research reports from symposia, meetings, and
so on, which may or may not appear in periodicals. Each issue of both BA and
BioI contains author, biosystematic, and cross-subject indexes, as well as a
biosystematic index arranged by division, class, and family. Both publications are
available on computer tape, and it is possible to obtain printouts of literature citations for various taxonomic groups, i.e., genera, tribes, or families.]

Blake, S. F., and A. C. Atwood: "Geographical Guide to Floras of the World,"
U.S.D.A. Misc. Pub. No. 401 and 797, 1942–1961. (An index to older floras and
floristics studies.)

Chaudhri, M. M. See Lanjouw, J.

Federov, A. A. (ed.): *Chromosome Numbers of Flowering Plants,* Academy of Sciences of the USSR, V. L. Komarov Botanical Institute, 1969. (An index to chromosome numbers published for plants through 1966. Literature citations are
provided. See Moore, 1973, for chromosome numbers published since 1967.)

Henderson, D. M., and H. T. Prentice: "International Directory of Botanical Gardens," 3d ed., *Regnum Veg.,* **95,** 1977.

Holmgren, P. K., and W. Keuken (eds.): "Index Herbariorum. Part I. The Herbaria of
the World," 6th ed., *Regnum Veg.,* **92,** 1974. (This guide to the herbaria of the
world is arranged by the cities in which they are located. It provides addresses and
general information for each herbarium and also has a list of standard acronyms for
herbaria abbreviations.)

Hunt Botanical Library: *Botanico-Periodicum-Huntianum,* G. H. M. Lawrence, et al.
(eds.), Hunt Botanical Library, Pittsburgh, 1968. (B-P-H is a list of over 12,000 botanical or botanically related periodicals. It is often used to identify obscure journal
citations and to provide standard abbreviations for literature citations. It is especially helpful in locating references cited in *Index kewensis.*)

*Index to American Botanical Literature: Bull. Torrey Bot. Club,* 1886- . (This index
appears in each issue of the *Bulletin* and is later issued on cards. The listings are
arranged alphabetically by author under subject groupings in the journal. The
card file is alphabetical by author. The index for 1886–1966 was reprinted in
1969 by G. K. Hall, Boston, in 4 volumes, and a one-volume supplement for

1967–1976 was issued in 1977. The index permits a review of all the papers written by an author.)

*Index Herbariorum.* See Lanjouw, J.; Holmgren, P. K.

Jackson, B. D.: *Guide to the Literature of Botany,* Longmans, Green, London, 1881. (A guide to older botanical literature including about 6000 titles not listed in Pritzel's *Thesaurus.* Jackson and Pritzel should be used together. Facsimile reprint Otto Koeltz Science Publishers, Koenigstein, 1974.)

*Kew Record of Taxonomic Literature Relating to Vascular Plants:* Kew Royal Botanic Gardens, Kew, 1971–  . (An annual publication listing all taxonomic literature published in periodicals, books, and papers. Coverage is worldwide. Entries are arranged in systematic groups.)

Lanjouw, J., and F. A. Stafleu: "Index Herbariorum. Part 2. Index to Collectors," (1) A–D; (2) E–H; (3) I–L; (4) M; *Regnum Veg., 2, 9, 86,* and **93,** 1954, 1957, 1972, 1976. (This alphabetical list of collectors cites where their collections are deposited and is useful in locating types. M. M. Chaudhri, I. H. Vegter, and C. del Wall compiled I–L; I. H. Vegter compiled M.)

Lawrence, G. H. M. See Hunt Botanical Library.

Moore, R. J. (ed.): "Index to Plant Chromosome Numbers 1967–1971," *Regnum Veg.,* **90,** 1973 (See Federov, 1969, for an index to chromosome numbers published before 1967. Literature citations are provided. Indexes have also been published for 1972, 1973-1974, *Regnum Veg.,* **91** and **96,** 1974 and 1977.)

Pritzel, G. A.: *Thesaurus literaturae botanicae,* rev. ed., 1872–1877. (This is a basic reference and guide to the botanical literature up to 1851 and should be used with Jackson. Pritzel is extremely valuable for verifying old original descriptions and those not listed in Stafleu. Facsimile reprint, Otto Koeltz Antiquariat, Koenigstein, 1972.)

Solbrig, O. T., and T. W. J. Gadella: "Biosystematic Literature; Contributions to a Biosystematic Literature Index (1945–1964)," *Regnum Veg.,* **69,** 1970. (A handy source of selected references to literature dealing with biosystematics.)

Stafleu, F. A.: "Taxonomic Literature," *Regnum Veg.,* **52,** 1967. (A selective guide to older botanical publications. It provides dates of publication which are important in matters of priority. Included are concise bibliographies on the lives and works of important taxonomists.)

——— and R. S. Cowan: "Taxonomic Literature," 2d ed., *Regnum Veg.,* **94,** 1976–  . (An enlarged version of the 1967 edition. This will be a multivolume work when complete.)

Tralau, H. (ed.): *Index Holmensis, a World Phytogeographic Index,* vol. 1., *Equisetales—Gymnospermae;* vols. 2–3, *Monocotyledonae;* vols. 4–  , *Dicotyledonae,* Scientific Publishers, Ltd., Zurich, 1969–  . (An alphabetical index to plant distribution maps.)

Vegter, I. H. See Lanjouw, J.


## COMPREHENSIVE WORKS OF A BROAD SCOPE

Baillon, H.: *Histoire des plantes,* 13 volumes. Hachette, Paris, 1867–1895. (*The Natural History of Plants,* a translation of the first 8 volumes, was published

1871–1888. *Histoire des plantes* treats families and genera of vascular plants, is well illustrated, and has an extensive bibliography.)

Bentham, G., and J. D. Hooker: *Genera plantarum,* 3 volumes. Reeve, London, 1862–1883. (*Genera plantarum* provides generic descriptions for seed plants and was prepared from the plants themselves rather than from the literature. Written in Latin, it is a valuable reference for classifications between the sectional and family levels.)

Candolle, A. P. de, et al.: *Prodromus systematis naturalis regni vegetabilis,* 17 volumes and 4 index volumes, Paris, 1824–1873. (Intended to treat all species of seed plants, but only covers the dicots. The first seven volumes were prepared by A. P. de Candolle, the others by contributors, and edited by his son Alphonse. The title *Prodromus* indicates that de Candolle considered it only a preliminary treatment, but it remains the only comprehensive revision of some genera.)

Copeland, E. B.: "Genera Filicum, the Genera of Ferns," in *Annales cryptogamici et phytopatholgici,* vol. 5, Chronica Botanica, Waltham, Mass., 1947. (A treatment of the ferns arranged in a phylogenetic sequence. Descriptions, keys, and references to the literature are given.)

Cronquist, A.: *The Evolution and Classification of Flowering Plants,* Houghton Mifflin, Boston, 1968. (Provides a comprehensive system of classification for the angiosperms.)

Engler, A.: *Das Pflanzenreich,* Nos. 1–107, Berlin, 1900–1968. (*Das Pflanzenreich* is an incomplete series of monographs on the plant kingdom. An index to the series is provided in Stafleu, 1967, pp. 133–145.)

————: *Syllabus der Pflanzenfamilien,* 12th ed., 2 volumes (edited by H. Melchior and E. Werdermann), Gebrüder Borntraeger, Berlin, 1954–1964. (Provides the most recent synopsis of the Engler system of classification. Volume 1 treats the bacteria through the gymnosperms. Volume 2 treats the flowering plants and contains descriptions of the higher categories and has an extensive bibliography.)

———— and K. Prantl: *Die natürlichen Pflanzenfamilien,* 23 volumes, W. Engelmann, Leipzig, 1887–1915. (This monumental work contains keys and descriptions for all families and genera of plants from algae to Compositae. Written in German, it is well illustrated and contains extensive literature citations. The plants are arranged according to the Engler system of classification.)

———— and ————: *Die Natürlichen Pflanzenfamilien,* 2d ed., W. Engelmann, Leipzig, 1924– . (The 2d edition is not yet complete. Twenty-one volumes have thus far been published.)

———— and ————: *Syllabus der Pflanzenfamilien,* 2d ed., Berlin, 1898. (This treatise has been published in many editions; the first edition is cited as *Syllabus der Vorlesungen . . .* 1892.)

Hutchinson, J.: *Genera of Flowering Plants,* 2 volumes, Clarendon, Oxford, 1964–1967. (This is an attempt to update Bentham and Hooker's *Genera plantarum.* Hutchinson provides family and generic descriptions, as well as a classification scheme.)

————: *Key to the Families of Flowering Plants of the World,* Clarendon, Oxford, 1967. (This is perhaps the best key to families on a worldwide basis.)

————: *The Families of Flowering Plants,* vol. I, *Dicotyledons,* vol. II, *Monocotyledons,* 3d ed., Clarendon, Oxford, 1973. (Provides a synopsis of Hutchinson's scheme of classification. Treatments of families are accompanied by descriptions and illustrations.)

Melchior, H. See Engler, A., 1954–1964.

Rendle, A. B.: *The Classification of Flowering Plants,* vol. 1, *Gymnosperms and Monocotyledons;* vol. 2, *Dicotyledons,* 2d ed., University Press, Cambridge, 1925–1930. (A classification in the Englerian sense with excellent descriptions of families.)

Takhtajan, A.: *Die Evolution der Angiospermen,* Gustav Fischer Verlang, Jena, 1959.

———: *System et Phylogenia Magnoliophytorum,* Soviet Publishing Institution, Nauka, Moscow, and Leningrad, 1966. (In Russian.)

———: *Flowering Plants—Origin and Dispersal,* Smithsonian Institution Press, Washington, 1969. (This translation, by Charles Jeffrey, is quite readable and is recommended to the student.)

Wettstein, R.: *Handbuch der Systematischen Botanik,* 2 volumes, 4th ed., Franz Deuticke, Leipzig, 1933–1935. (Presents an improved Englerian classification.)

# Chapter 12

# Pteridophytes

The pteridophytes, which were much more dominant in past geologic periods, are represented today by over 10,000 living species: the wisk ferns (about 10), club mosses (about 300), spike mosses (about 700), quillworts (about 65), horsetails (about 25), and ferns (about 9280). Most pteridophytes are terrestrial, but many are epiphytes, a few are true aquatics, and some are xerophytes.

Most pteridophytes have roots, stems, and leaves, although in many temperate species of ferns, the stems consist of creeping rhizomes bearing adventitious roots (Figures 12-1 and 12-2). The leaves, or *fronds,* of ferns are quite distinctive and often variously dissected. The fronds usually show a characteristic unrolling during their development called *circinate vernation,* which produces fern fiddleheads. The leaves of ferns are large; the stems have a complex vascular system and a nodal leaf gap.

The life cycle of pteridophytes alternates between two distinct generations, represented by separate and unlike plants. For example, the familiar leafy fern plant represents the *sporophyte* phase of growth. It produces spores on the back of its leaves in small hollow cases called *sporangia.* These sporangia are borne on the underside of ordinary leaves or on very specialized fertile fronds or segments of a fertile frond. Sporangia usually occur in clusters called *sori.* The spore germinates to form the independent alternate generation

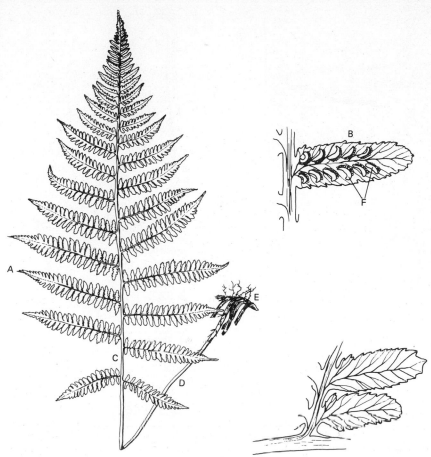

**Figure 12-1** Diagram of a frond of the lady fern. *A*, pinna; *B*, pinnule; *C*, rachis; *D*, stipe: *E*, rootstock; and *F*, sorus with a curved or straight indusium. *(From McVaugh and Pyron, 1951; with permission.)*

known as the sexual or *gametophytic* plant. Pteridophytes do not produce seeds.

The gametophytic fern plant is a small, haploid, green, heart-shaped *prothallus* rarely more than 1 centimeter in diameter. The prothallus bears sexual structures on its underside. Near its base are the male *antheridia*, which produce swimming sperms. When bathed in water from dew or rain, the sperms swim toward the female *archegonia* (containing eggs) located near the indented apex of the prothallus. The zygote formed by the union of the haploid male and female gametes gives rise to the sporophyte. The sporophyte is diploid. Later the cells in the sporangium undergo meiosis to form the haploid spores.

A few living pteridophytes are *heterosporous,* producing large *megaspores* and small *microspores* in separate sporangia, but most are *homosporous,* with all spores being alike. In heterosporous pteridophytes, the gametophytes are

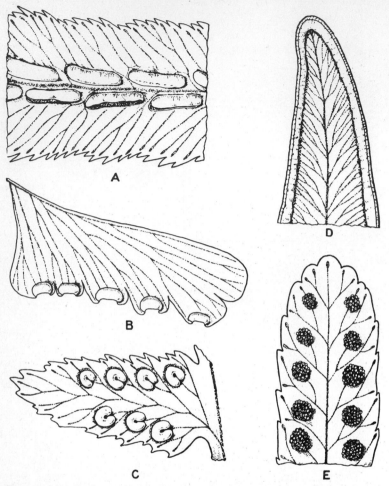

**Figure 12-2** Portion of the leaflets of five common ferns, illustrating differences in sori and indusia, 4x. *(a) Woodwardia; (b) Adiantum; (c) Dryopteris; (d) Pteridium; (e) Polypodium.* In *a* the indusium is elongate. In *b* and *d* a false indusium is seen, while in *e* there is no indusium. In *c* the indusium is reniform. *(From Haupt, 1953; with permission.)*

sexually differentiated and reduced in structure. The megaspores develop into megagametophytes which form archegonia, and the microspores develop into microgametophytes which produce antheridia.

## PTERIDOPHYTE CLASSIFICATION

Few groups of plants, especially the ferns, have undergone so many changes in taxonomy and nomenclature (Pichi-Sermolli, 1973). There is difficulty in understanding the interrelationships between the main groups of pteridophytes, as well as in delimiting and arranging the taxa at lower ranks (Pichi-Sermolli, 1973). Many classifications of the pteridophytes exist, derived in part from the different philosophies of the various pteridologists and complicated by the peculiarities of the pteridophytes.

The main groups of pteridophytes were extant by the Devonian Period. Today they consist of a complex of heterogeneous groups reflecting their antiquity and diverse evolutionary pathways. The fern families in particular are not always easily characterized and classified. Differences within and between the fern families are usually based on a series of small characters rather than on one or two sharp distinctions.

At the rank of division, Cronquist, Takhtajan, and Zimmermann (1966) recognize four groups of pteridophytes (see Figure 12-3): Psilotophyta (whisk ferns), Lycopodiophyta (club mosses, spike mosses, and quillworts), Equisetophyta (horsetails), and Polypodiophyta (ferns and water ferns). These groups represent distinct evolutionary lines; they are lumped together for convenience, not in an attempt to reflect relationships. The distinctions between the first three divisions are generally clear; however, the classification problem is not yet fully resolved for the ferns to the satisfaction of all pteridologists.

Crabbe, Jermy, and Mickel (1975) presented a useful generic sequence for arranging pteridophytes in the herbarium, along with a valuable working list of generic synonyms.

## EXPLANATION OF SOME SPECIAL TERMS ASSOCIATED WITH THE PTERIDOPHYTES

*Annulus.* A thick-walled ring of cells surrounding the sporangium which at maturity breaks the sporangium, releasing the spores.

*Antheridium* (plural, *antheridia*). The sperm-producing structure on the prothallus.

*Archegonium* (plural, *archegonia*). The egg-producing structure on the prothallus.

*Areole.* The small spaces in a leaf surrounded by veins.

*Auricle.* The basal lobe of a pinna.

*Circinate vernation.* The coiled arrangement of a young fern frond, a fiddlehead.

*Compound.* Made up of multiple parts.

*Dimorphic.* In ferns, fronds or parts of fronds differentiated into fertile and sterile portions.

*Fertile frond.* A frond or part of frond that bears sporangia.

*Free venation.* A term used to describe veins that do not form a network but pass directly to the margin of the leaf.

*Frond.* The entire leaf of a fern, including the petiole.

*Gametophyte.* A gamete-producing plant.

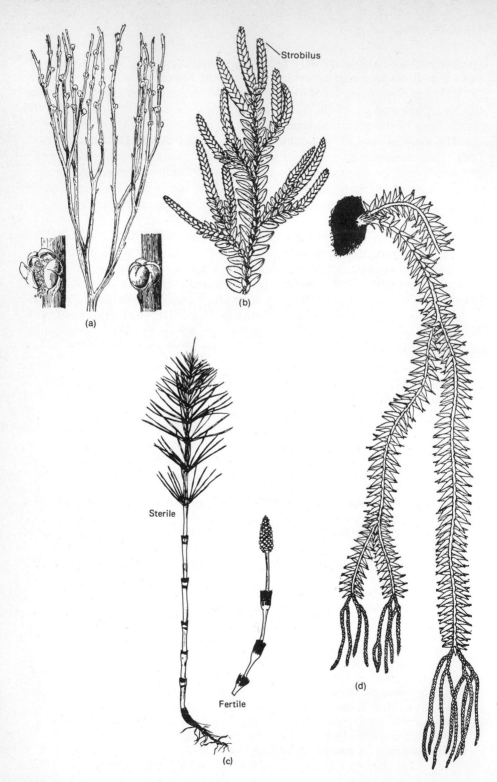

(a)

(b)

Strobilus

(c)

Sterile

Fertile

(d)

**Figure 12-3** *Opposite page: Psilotum, Selaginella, Equisetum,* and *Lycopodium. (a)* Upper portion of a shoot of *Psilotum nudum* with sporangia *(from Haupt, 1953; with permission). (b) Selaginella,* branch with leaves and strobili *(from Haupt, 1953; with permission). (c)* Sterile and fertile shoots of *Equisetum (from McVaugh and Pyron, 1953; with permission). (d) Lycopodium* with strobili *(from Smith, 1955; with permission).*

*Heterosporous.* Having two different kinds of spores, microspores and megaspores.

*Homosporous.* Having spores of only one kind.

*Indusium* (plural, *indusia*). The tissue covering a fern sorus. A false indusium is present in some with the leaf margin folded over the sori.

*Megaspore.* The larger of the two kinds of spores produced by heterosporous pteridophytes, developing into gametophytes which produce eggs.

*Microspore.* The smaller of the spores produced by heterosporous pteridophytes, developing into gametophytes which produce sperm.

*Midrib.* The central vein of a frond or the central vein of a division of a frond.

*Paraphyses.* Hair-like structures associated with the sori.

*Pinna* (plural, *pinnae*). A blade division spaced on rachis, as in a feather; the openings between pinnae extend to the midrib.

*Pinnate.* Divided into pinnae; pinnately compound.

*Pinnule* (plural, *pinnules*). The division of a pinna.

*Prothallus* (plural, *prothalli*). The gametophyte stage of a pteridophyte, developing from a spore.

*Rachis.* The main axis (or midrib) of a compound fern frond extending from the base of the blade upward.

*Rhizoids.* Hair-like structures on the gametophytes which absorb moisture.

*Rhizome.* An underground stem, usually horizontal.

*Rootstock.* The entire rhizome-root system.

*Scale.* In ferns, the thick, dry, straw-colored chaff on the rootstocks and stipes.

*Sorus* (plural, *sori*). A cluster of sporangia in ferns (see Figure 12-2).

*Spike.* In pteridophytes, an elongated cluster of sporophylls and sporangia; a strobilus; a special reproductive structure of the Ophioglossaceae.

*Sporophyte.* A spore-producing plant.

*Sporangiophore.* A special stalk bearing sporangia; e.g., *Equisetum.*

*Sporangium* (plural, *sporangia*). A spore-producing structure, site of meiosis.

*Spore.* A one-celled microscopic structure produced in sporangia, which germinates to form the gametophyte (Figures 12-4 and 12-5, page 212).

*Sporocarp.* A bean-shaped structure containing sporangia and spores.

*Sterile.* In ferns, a frond lacking sporangia.

*Stipe.* The petiole or stalk of a fern frond.

*Strobilus* (plural, *strobili*). The cone-like cluster of sporophylls.

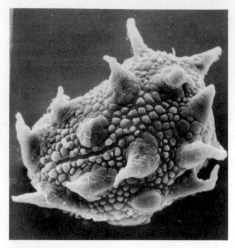

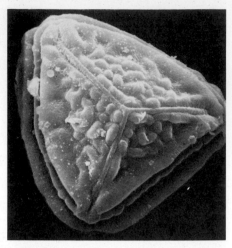

**Figure 12-4** SEM of a fern spore.
*Pteyopsis piloselloides*, 1500X. *(Courtesy
of Dr. W. H. Wagner, Jr.).*

**Figure 12-5** SEM of a fern spore. *Gym-
nogramma chrysophylla*, 2000X. *(Courtesy
of Dr. W. H. Wagner, Jr.)*

## MAJOR FAMILIES OF PTERIDOPHYTES
## NATIVE TO NORTH AMERICA

The families follow the sequence of Crabbe, Jermy, and Mickel (1975). The
Gleicheniaceae, with one species found only sporadically along certain Gulf of
Mexico coastal islands, is excluded, along with the Grammitidaceae, which is
known from one locality in North Carolina where it exists primarily as a gametophyte.

Listed below are the family names, common names, a brief diagnosis,
plus generalized distribution and habitat data.

### Eusporangiate Pteridophytes

These pteridophytes have large sporangia called *eusporangia,* with numerous
spores (several hundred to thousands in each sporangium). The sporangium develops from several initial cells on the shoot.

**1  Division Psilotophyta  *Psilotaceae: Whisk ferns*  Epiphytic or terrestri-
al; perennials; rhizome with rhizoids, rootless; leaves or enations small or
minute, one-veined; sporangia fused into groups of two or three; homosporous;
gametophytes subterranean; distribution largely tropical or subtropical (Figure
12-3). *Psilotum* is grown as a greenhouse plant and is found naturally in the
warmer parts of the Southeastern states (Figure 12-6). *Tmesipteris* is the other
genus in the family and it occurs in Australia, New Zealand, the Philippines,
and some islands of Polynesia.

**Figure 12-6** Photograph of *Psilotum.*

**2 Division Lycopodiophyta** *Lycopodiaceae: Club mosses, ground pine*
Epiphytic or terrestrial; perennials; stems dichotomously branching; leaves
simple with a single vein; sporangia in the axils of leaves; homosporous; gametophytes green when on the soil surface or nongreen when subterranean; distribution tropical and temperate (Figure 12-3). *Lycopodium* is common in some
parts of North America. Certain species were once gathered in large quantities
for making Christmas wreaths, and the dry spores were once used as flash
powder and condom lubricants. The lycopods were literally the forests of the
Carboniferous Period, which formed the present-day coal deposits so critical to
our energy needs. In addition, they have the longest continuous fossil record of
any group of vascular plants.

*Selaginellaceae: Spike mosses* Terrestrial or epiphytic; perennials; stems
dichotomously branching; leaves small with a single vein; heterosporous;
gametophytes form within the spore walls; largely tropical, but also temperate
in distribution (Figure 12-3). Several species of *Selanginella* are cultivated in
greenhouses as ornamentals. In nature, some inhabit xeric places, others
mesophytic to moist habitats; most species live in the wetter tropics.

*Isoetaceae: Quillworts* Terrestrial; perennials; stems short, erect, corm-
like, having secondary growth; leaves long, quill-like, with a single vein;
sporangia axillary, large, sunken in base of leaves; heterosporous; gametophytes forming within spore walls; distribution is tropical and temperate.
*Isoetes* is an emergent plant of shallow ponds or vernal pools and is easily
overlooked because of its grass-like appearance.

**3 Division Equisetophyta** *Equisetaceae: Horsetails, scouring rushes* Terrestrial; annuals or perennials; stems conspicuously jointed, ribbed, un-

branched, or with whorls of branches easily mistaken for leaves; leaves greatly reduced to a whorl of scales; strobili terminal; sporangia attached to special holders or sporangiophores; homosporous; gametophytes green and found on the soil surface; distribution both tropical and temperate (Figure 12-3). One genus only, *Equisetum:* several species are native to North America. The rough silica-containing stems were supposedly used by the early settlers to scour pots and pans, leading to the common name of scouring rush. Some species of *Equisetum* cause poisoning of livestock.

**4   Division Polypodiophyta   *Ophioglossaceae: Adder's tongue and grape ferns*** Terrestrial (two species epiphytic); perennials; leaves simple or much divided, differentiated into a sterile and a fertile portion; sporangia aggregated into an upright fertile region or spike; homosporous; gametophytes subterranean; distribution both tropical and temperate. *Ophioglossum* (adder's tongue) and *Botrychium* (grape fern) are both represented in North America. Some species of this family are rather small and easily overlooked.

### Leptosporangiate Pteridophytes

These pteridophytes have small sporangia called *leptosporangia,* which produce a relatively small number of spores, usually 128 spores or fewer in each sporangium. The sporangium develops from one cell or a small group of cells.

***Osmundaceae: Royal, cinnamon, and interrupted ferns*** Terrestrial; perennials; plants various, large rosette ferns (ours) or small "filmy" ferns; leaves pinnately compound with the fertile pinnae differentiated from the sterile; sporangia large with many spores, the family is intermediate between the typical eusporangiate and typical leptosporangiate condition; gametophytes occur on the surface of the soil and are green; distribution is both tropical and temperate. The Osmundaceae has a long fossil record. *Osmunda* is common in parts of North America: *O. regalis* (royal fern), *O. cinnamomea* (cinnamon fern), and *O. claytoniana* (interrupted fern).

***Schizaeaceae: Curly grass fern, climbing fern*** Terrestrial; perennials; morphology diverse, including grass-like tufted plants and vines; sporangia have an apical annulus, a common feature of the family; gametophytes occur either on the surface or below ground; distribution is both tropical and temperate. *Lygodium palmatum* (climbing fern) is widespread in the Eastern United States. One introduced species, *L. japonicum,* has escaped cultivation and is weedy in the warmer parts of the South. *Schizaea* is the curly grass fern. *Anemia* is native to the extreme Southern United States.

***Parkeriaceae: Water fern family*** Aquatic or semiaquatic on mud; leaves large, fleshy, divided; distribution tropical to warm temperate. Contains only the genus *Ceratopteris,* which occurs as a weed and is used in aquaria as an ornamental.

   ***Adiantaceae: Maidenhair, cloak, and lip ferns*** Terrestrial or epiphytic; perennials; small plants often of specialized habitats, such as basic rocks, or xerophytic habits, also in rich woods; pinnae dichotomously veined; sori along veins, in grooves, or along reflexed margins of pinnae; distribution tropical and temperate. *Adiantum pedatum* (northern maidenhair) is a beautiful fern which ranges over much of the Eastern United States. Largely tropical, southern maidenhair *(A. capillus-veneris)* grows on limestone rock in the Southern states. Some of the genera include: *Cheilanthes* (lip fern), *Pellaea* (cliffbrake fern), *Vittaria* (shoestring fern), *Pteris* (wall fern), *Acrostichum,* and *Notholaena.*

   ***Hymenophyllaceae: Filmy fern*** Terrestrial or epiphytic; perennials; occurring on moist rocks, trees, stream banks; leaves very thin, only one cell thick (except at the midvein); sori marginal, indusium tubular or bivalved; distribution mostly tropical, a few temperate. *Trichomanes* (bristle fern) and *Hymenophyllum* (tunbridge fern) are found in the Southeastern states.

   ***Polypodiaceae: Polypody, rockcap, resurrection ferns*** Typically epiphytic; perennials; rhizomes creeping and covered with scales; leaves often simple though usually lobed; sporangia forming a mound-like rounded sorus, indusia absent; distribution chiefly tropical, a few temperate. Some species of *Polypodium* are often cultivated in greenhouses as ornamentals. The staghorn fern (*Platycerium*) is commonly cultivated. The resurrection fern (*P. polypodioides*) is a common epiphytic fern in the Eastern states.

   ***Dennstaedtiaceae: Hayscented fern, bracken*** Terrestrial or rarely climbing; perennials; rhizome creeping; leaves large and divided; sori along leaf margin or on back of leaf, indusium cup-like, two-lipped, or lateral along the margin; distribution mostly tropical, but with a few temperate species. *Pteridium aquilinum* (bracken fern), ranging over much of the world, is one of the most widespread species of vascular plant. The common name of *Dennstaedtia punctilobula* (hayscented fern) is well-suited because of the strong fragrance of the dried leaves, which have an odor resembling alfalfa.

   ***Thelypteridaceae: Beech, marsh, New York ferns*** Terrestrial; perennials; rhizomes short and erect to creeping; petioles with two vascular strands; sori usually round, indusium kidney-shaped; spores bilateral (as opposed to tetrahedral), perispore or membranous spore covering present; distribution mostly tropical. *Thelypteris* is common in parts of North America.

   ***Aspleniaceae: Wood ferns, spleenworts, sensitive fern, cliff fern*** Primarily terrestrial, often growing on rocks; perennials; sori round or elongate, indusia kidney-shaped or linear; spores bilateral (as opposed to tetrahederal), surrounded by a perispore or membranous covering; distribution both temperate and tropical, one of the most abundant groups of ferns. Important genera in North America include *Asplenium* (spleenwort), *Onoclea* (sensitive fern), *Athyrium* (lady fern), *Woodsia* (cliff fern), *Cystopteris* (brittle fern), *Polystichum* (Christmas fern), and *Dryopteris* (wood fern). Walking fern (*Asplenium rhizophyllus*) is unusual because of its habit of "walking" across rocks vegetatively by plantlets produced by adventitious buds at the tip of the leaves. *Matteuccia* (ostrich fern) is a well-known edible species.

*Davalliaceae: Boston and rabbit's foot fern*   Mostly epiphytes; perennials; indusia kidney-shaped or somewhat tubular; spores bilateral with a perispore; distribution largely tropical, but also warm temperate. A popular house fern, *Nephrolepis exaltata* (Boston fern) is the source of numerous vegetatively propagated cultivars. *Davallia fejeensis* (rabbit's foot fern), with handsome rhizomes and divided leaves, is often used to form hanging fern balls.

*Blechnaceae: Chain fern*   Mostly terrestrial; perennials; indusia linear, opening toward midvein; spores bilateral, covered with a perispore; distribution mostly tropical, but with a few temperate species. Genera include *Blechnum* and *Woodwardia* (chain fern).

## Water Ferns

All these ferns are heterosporous with megaspores and microspores. The water ferns are definitely not fern-like in appearance. Two of the families are floating aquatics, and the third is emergent or rooted in mud. Crabbe, Jermy, and Mickel (1975) state that they place the water ferns here, " . . . not because we feel they are related to each other, but out of convenience. They are quite distinct and there are no firm data to justify placing them elsewhere."

*Marsileaceae: Water clovers, pillwort*   Emergent aquatic or rooted on mud; leaves resemble a small bunch of grass or clover with four or two leaflets. Reproductive structures are hard, modified basal leaflets called *sporocarps;* distribution tropical and temperate. *Marsilea* has clover-like leaves, whereas *Pilularia* has grass-like leaves. *Marsilea* is amphibious, growing in shallow water, and it is also able to survive on mud when the water level falls (Figure 12-7).

*Salviniaceae: Water spangles*   Floating aquatics; plants small; leaves in whorls of three, two floating and simple, and one very divided and dangling resembling a root; reproductive structures (sporocarps) occur among root-like leaves; distribution tropical and temperate. Salviniaceae has one genus, *Salvinia,* frequently used as an aquarium ornamental. Most species of *Salvinia* are native to Africa, but one species occurs in the United States.

**Figure 12-7**   Photograph of *Marsilea.*

*Azollaceae: Mosquito fern*   Floating aquatic; plants very small; leaves in two rows, scale-like and overlapping, harboring blue-green algae; distribution tropical and temperate. Azollaceae has one genus, *Azolla,* often grown as an ornamental in aquaria. Two species of *Azolla* occur in the United States, often coloring the water of swamps red when the plants are in full sun.

## SELECTED REFERENCES

Banks, H. P.: "The Early History of Land Plants," in E. T. Drake (ed.), *Evolution and Environment,* Yale, New Haven, 1968, pp. 73–107.

Bierhorst, D. W.: "The Systematic Position of *Psilotum* and *Tmesipteris,*" *Brittonia,* **29:**3–13, 1977.

Bower, F. O.: *The Ferns (Filicales),* 3 vols., Cambridge, London, 1923-1928.

Crabbe, J. A., A. C. Jermy, and J. T. Mickel: "A New Generic Sequence for the Pteridophyte Herbarium," *Fern Gaz.,* **11:**141–162, 1975.

Cronquist, A., A. Takhtajan, and W. Zimmermann: "On the Higher Taxa of Embryobionta," *Taxon,* **15:**129–134, 1966.

Evans, A. M.: "A Review of Systematic Studies of the Pteridophytes of the Southern Appalachians," in P. C. Holt (ed.), *The Distributional History of the Biota of the Southern Appalachians,* part II. *Flora,* Research Division Monograph 2, Virginia Polytechnic Institute and State University, Blacksburg, 1970, pp. 117–146.

Haupt, A. W.: *Plant Morphology,* McGraw-Hill, New York, 1953.

Holttum, R. E.: "The Family Names of Ferns," *Taxon,* **20:**527–531, 1971.

———: "Posing the Problems," in A. C. Jermy et al. (eds.), "The Phylogeny and Classification of the Ferns," *Bot. J. Linn. Soc.,* **67** (Suppl. 1): 1–10, 1973.

Klekowski, E. J.: "Genetical Features of Ferns as Contrasted to Seed Plants," *Ann. Mo. Bot. Gard.,* **59:**138–151, 1972.

McVaugh, R., and J. H. Pyron: *Ferns of Georgia,* University of Georgia Press, Athens, 1951.

Mickel, J. T.: "Phyletic Lines in the Modern Ferns," *Ann. Mo. Bot. Gard.,* **61:**474–482, 1974.

Pichi-Sermolli, R. E. G.: "Historical Review of the Higher Classification of the Filicopsida," in A. C. Jermy et al. (eds.), "The Phylogeny and Classification of the Ferns," *Bot. J. Linn. Soc.,* **67** (Suppl. 1):11–40, 1973.

Smith, G. M.: *Cryptogamic Botany,* vol. II, *Bryophytes and Pteridophytes,* 2d ed., McGraw-Hill, New York, 1955.

Taylor, T. N., and J. T. Mickel (eds.): "Evolution of Systematic Characters in the Ferns," *Ann. Mo. Bot. Gard.,* **61:**307–482, 1974. (Papers presented at the 1973 Amherst Symposium.)

Wagner, W. H.: "Some Future Challenges of Fern Systematics and Phylogeny," in A. C. Jermy et al. (eds.), "The Phylogeny and Classification of the Ferns," *Bot. J. Linn. Soc.,* **67** (Suppl. 1):245–256, 1973.

———: "Systematic Implications of the Psilotaceae," *Brittonia,* **29:**54–63, 1977.

# Chapter 13

# Gymnosperms

The gymnosperms are seed plants, but unlike the angiosperms, they are not a cohesive group with a monophyletic origin. The term *gymnosperm* refers to the mode of seed production and includes unrelated evolutionary lines with divergent forms. Two conspicuous and very distinct assemblages are the cycad line and the conifer-ginkgo line. Unusual gymnosperm forms are *Ephedra*, *Welwitschia*, and *Gnetum*. There are many fossil gymnosperms having numerous affinities.

Gymnosperms are differentiated from angiosperms by having ovules exposed on megasporophylls. The ovules are not enclosed by a carpel or carpels, and pollen germinates directly upon the ovule wall. Another distinguishing feature is that the endosperm of gymnosperms is derived directly from the haploid gametophytic tissue and is formed before fertilization.

The gymnosperms, consisting of 9 to 11 families, approximately 61 genera, and over 500 species, are widely distributed throughout all climates. In North America north of Mexico, the conifers are most common. They include the families Pinaceae, Taxodiaceae, Cupressaceae, and Taxaceae. Some other gymnosperms found in North America are *Zamia* (Cycadaceae), with one species native to Florida, and *Ephedra* (Ephedraceae), a genus of bushy shrubs found in Southwestern deserts.

Gymnosperms are more important than indicated by the numerical size of the group. Conifers occupy vast areas in almost pure stands and are the dominant vegetation in many places. They are of primary importance for forest products, providing lumber, paper pulp, and chemicals (turpentine and rosin). They also control erosion, protect watersheds, enhance recreational uses of forests, and are often the first trees to appear in secondary succession.

## CLASSIFICATION

Many of the modern species and genera have a markedly disjunct geographic distribution, and there are many species with a limited distribution. Such a distribution pattern reflects their great antiquity. The geological record of the gymnosperms extends from the late Devonian (see Table 6-1). Each of the families of conifers has a fossil record extending from the Mesozoic Era. Their fossil evidence shows a much wider geographic distribution than that of the living conifers. Leaves identical with *Ginkgo biloba* occurred as far back as the Permian. The group of gymnosperms represented today only by *Ginkgo biloba* was once almost worldwide in distribution, but is restricted now to a small and relatively inaccessible area in South China.

The gymnosperms have had controversial taxonomic interpretations. In some classifications, they are assigned to four independent divisions: the Cycadophyta, Ginkgophyta, Coniferophyta, and Gnetophyta. This is based upon the argument that the divergences among such extant genera as *Zamia, Cycas, Ginkgo, Pinus, Ephedra*, and so on overshadow their common gymnospermous features. Also, the Paleobotanical evidence suggests several long-distinct evolutionary lines and that the seed-producing habit (i.e., their gymnospermous features) arose several times during the history of land plants. Other interpretations suggest that the differences observed in the living gymnospermous genera should not be allowed to obscure their similarities. The major lines of gymnosperms are grouped into one division, the Pinophyta, and the subdivisions Cycadophytina, Pinophytina, and Gnetophytina.

## CLASSIFICATION OF GYMNOSPERM FAMILIES*

Division Pinophyta
  Subdivision Cycadophytina
    Class Lyginopteridopsida (fossil seed ferns)
    Class Cycadopsida
      Family Cycadaceae (native to North America)
    Class Bennettitopsida (fossils only)

*Following Cronquist, Takhtajan, and Zimmermann, 1966.

Subdivision Pinophytina
  Class Ginkgoöpsida (fossils and *Ginkgo biloba*)
    Family Ginkgoaceae (not native to North America)
  Class Pinopsida
    Subclass Cordaitidae (fossils only)
    Subclass Pinidae
      Family Pinaceae (native to North America)
      Family Cupressaceae (native to North America)
      Family Taxodiaceae (native to North America)
      Family Taxaceae (native to North America)
      Family Podocarpaceae (not native to North America)
      Family Cephalotaxaceae (not native to North America)
      Family Araucariaceae (not native to North America)
Subdivision Gnetophytina
  Class Gnetopsida
    Subclass Ephedridae
      Family Ephedraceae (native to North America)
    Subclass Welwitschiidae
      Family Welwitschiaceae (*Welwitschia*—native to West Africa)
    Subclass Gnetidae
      Family Gnetaceae (*Gnetum*—native to South America, Asia, and Africa)

Since the living gymnosperms belong to clearly distinct groups, the problem of classification is not one of intergradation, but an evaluation of their relationships and a determination of the rank to be accorded to each group. Some morphologists regard gymnosperms as a particular level of evolution rather than a taxonomic group. If true, this should be reflected in their classification. The fossil record suggests that the major groups of gymnosperms were derived from several ancestral groups.

Cycads (class Cycadopsida) appeared in the upper Triassic. They exhibit a number of primitive features, such as pinnately compound leaves, sparingly branched low habit, circinate vernation, and motile sperm. The cycads probably evolved from the pteridosperms or seed ferns (class Lyginopteridopsida), which were still abundant in the Triassic.

The group of gymnosperms represented today only by *Ginkgo biloba* (Class Ginkgoöpsida) probably originated in the Permian Period of the late Paleozoic Era and became widespread and moderately abundant during mid-Mesozoic times. The ancestor of the ginkgo lineage is not known. Ginkgoöpsida has some characters of both the conifers (freely branching stems) and the cycads (motile sperm). However, its relations are probably closer to the conifer line (Sporne, 1965).

When compared with other gymnosperms, the conifers (subclass Pinidae) are relatively advanced in some features, but primitive in others. Traces of early conifers have been identified from the late Carboniferous. The earliest co-

**Table 13-1  Conifer Families in North America**

| Families | Ovule, bract, and scale | Cones and seeds | Leaves |
|---|---|---|---|
| Pinaceae | Bract and scale distinct, ovules two on each scale | Woody cone; seeds terminally winged or wingless | Persistent or deciduous; needle-like or linear |
| Taxodiaceae | Bract and scale partially fused, ovules two to nine per scale | Woody cone; seeds laterally winged | Persistent or deciduous; linear or ovate |
| Cupressaceae | Bract and scale fused, ovules two to many on each scale | Woody, leathery, or fleshy cone; seeds laterally winged or wingless | Persistent; scale-like, or awl-shaped |
| Taxaceae | Ovulate structure comprising a single erect ovule | Aril or drupe-like; seed single, enclosed or partially enclosed by fleshy tissue | Persistent; linear |

*Source:* Adapted from Harlow and Harrar, 1968, with permission.

nifers may have been similar to *Araucaria heterophylla* (Norfolk Island pine). The entire group is fertilized by nonmotile sperm conveyed to the egg by a pollen tube.

The Pinaceae is essentially a family of the Northern Hemisphere and is the largest family of conifers (see Table 13-1). Half of the Cupressaceae occur in the Northern Hemisphere and the other half in the Southern Hemisphere. The Taxodiaceae had an important place in the paleoforests of the Northern Hemisphere, but today it shows only a relict distribution. Some workers have separated the Taxaceae from the conifer group because of the uniform presence of a single ovule covered or partially covered by a fleshy structure called an *aril*. The Taxaceae, however, share many vegetative features with the conifers. They are largely confined to the Northern Hemisphere. The most important family of conifers in the Southern Hemisphere is the Podocarpaceae. It has a wide range of morphological characters. In some species, the ovules are borne in cones; in others, there is a single terminal fleshy receptacle. Cephalotaxaceae is represented today by one genus, *Cephalotaxus,* restricted to eastern Asia, although in past geological history there were representatives in Europe and western North America. The fossil record of the Araucariaceae extends back to the Triassic and includes both hemispheres. Today it is found only in the Southern Hemisphere.

Subdivision Gnetophytina (*Ephedra, Gnetum,* and *Welwitschia*) is an unusual group with practically no fossil record. Its noteworthy features are the following: (1) the presence of xylem vessels (unique among gymnosperms), (2) compound strobili, (3) a perianth-like structure in the male flowers, (4) several envelopes around the ovary, and (5) a micropyle of a long, bristle-like tube. *Ephedra* pollen has been found in the Eocene. *Welwitschia*-like pollen has been found in the Permian, and possible *Gnetum* pollen is known from the Tertiary. Some have suggested that this group was the forerunner of the angiosperms; however, the prevalent view is that the group is a strange variation of the gymnosperms.

Information on fossil groups may be obtained from several of the selected references listed at the end of the chapter.

## MORPHOLOGY ESSENTIAL TO IDENTIFICATION OF GYMNOSPERMS

*Pinus* is used as a primary example, since it is the most familiar and widespread genus of gymnosperms in North America (see Figure 13-1).

### Vegetative Morphology

**Shoots**   The *spur shoots* of *Pinus* are lateral branches of determinate growth which abcise. The *long shoots* are lateral or terminal branches of indeterminate growth. Indeterminate short shoots forming a spur branch are found

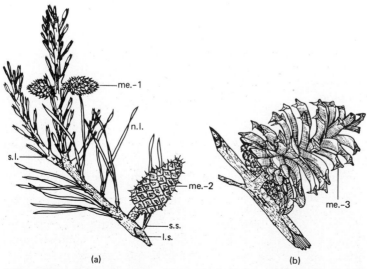

**Figure 13-1**   *Pinus virginiana:* s.l., scale leaf; n.l., needle leaf; l.s., long shoot; s.s., spur shoot; me-1, megastrobilus first year; me-2, megastrobilus second year; and me-3, megastrobilus three years old. *(From Bold, 1973; with permission.)*

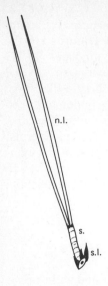

**Figure 13-2** Spur shoot or fascicle of *Pinus;* n.l., needle leaf; s.l., scale leaf; s., sheath.

in several genera including *Larix, Ginkgo,* and *Cedrus,* giving an unusual branching pattern.

**Needle Leaves** In *Pinus,* the needle leaves are produced on short branches (or *spur shoots*) in clusters (or *fascicles*) of one to eight needle leaves depending upon the species (Figure 13-2). A *sheath* of scale leaves surrounds the base of the fascicle.

**Scale Leaves** In addition to the needle leaves, inconspicuous scale leaves are produced in *Pinus* on the main branches or *long shoots* and at the bases of the spur shoots. The scale leaves are brownish, small, and triangular. With the exception of *Pinus,* most conifers have needle leaves or scale-like leaves arranged directly on the long shoots without basal sheaths.

### Reproductive Morphology

**Cones** The microsporangia and megasporangia of *Pinus* occur in separate cones, or *strobili* (singular, *strobilus*). In *Pinus,* both types of strobili occur on the same individual; therefore, *Pinus* is *monoecious.* This contrasts with the *dioecious* condition in which a single plant bears either microsporangia or megasporangia (e.g., *Juniperus*).

*Polliniferous cones or microstrobili* In *Pinus,* the microstrobili develop in clusters around the base of the terminal buds. When mature, these pollen-producing cones range in size from 1 to 4 centimeters long. A microstrobilus consists of a central axis bearing spirally arranged microsporophylls. Each microsporophyll bears on its lower side two elongate pollen sacs, or micro-

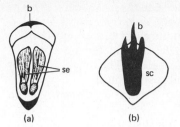

**Figure 13-3** Ovuliferous scales. *(a)* Ovuliferous scale of *Pinus:* se, winged seed (on upper side of scale); b, bract extending from lower side of scale. *(b) Pseudotsuga* ovuliferous scale: lower side with three-pronged bract exserted beyond the scale: sc, scale; b, bract.

sporangia. Each pollen mother cell gives rise to four characteristic winged pollen grains. Pines are wind-pollinated and produce large quantities of pollen.

*Ovuliferous cones or megastrobili* The megastrobili are borne on spur shoots near the tip of some of the long shoots of the current season's growth (see Figure 13-3). They are small and inconspicuous at the time of pollination. The ovules develop on *ovuliferous scales* or *cone scales* which in turn are borne in the axils of *bracts* spirally arranged on the strobilus axis. The ovuliferous scales each bear two ovules. In *Pinus,* the bract is short, but in *Pseudotsuga menziesii* (Douglas fir), the three-pronged bracts are exserted beyond the scales. In some conifers, such as the Taxodiaceae and Cupressaceae, the bract is partially or entirely fused with the scale.

Additional details of vegetative and reproductive morphology of the various groups of gymnosperms can be obtained by consulting Foster and Gifford (1974), Bold (1973), or Bierhorst (1971).

## ADDITIONAL TERMINOLOGY FOR IDENTIFICATION OF GYMNOSPERMS

### Vegetative Features

*Acicular leaf.* Needle-like.

*Awl-shaped leaf.* Narrow and sharp-pointed, gradually tapering from base to stiff point, relatively short, as in some leaves of *Juniperus.*

*Fascicled.* In clusters or bundles, as in needle leaves in *Pinus.*

*Linear leaf.* Long and narrow with parallel sides, as in *Picea, Taxus.*

*Persistent leaves.* Evergreen.

*Scale leaf.* Small, flat, appressed leaves, as in *Cupressus.*

*Sterigmata.* Small, woody, peg-like outgrowth of the twig bearing leaves or needles, as in *Picea.*

*Subulate.* Awl-shaped.

### Reproductive Features

*Apophysis.* The part of the cone scale which is exposed when the cone is closed.

*Aril.* Tissue that forms a fleshy covering of the seed, as in *Taxus, Torreya;* or rudimentary at the base of the fleshy seed, as in *Cephalotaxus.*

*Cone.* Aggregation of sporangia-bearing structures of either sporophylls and scales, or just of sporophylls forming a *strobilus.* The female cone (or megasporangiate strobilus) bears ovules and later seed. The male cone (or microsporangiate strobilus) produces pollen in microsporangia.

*Megasporophyll.* Modified leaf which bears ovules, as in *Cycas, Zamia* (and in conifers too).

*Microsporophyll.* Modified leaf which bears microsporangia.

*Serotinous.* Cones that remain closed long after the seeds are mature.


## FAMILIES OF GYMNOSPERMS NATIVE TO NORTH AMERICA


### Cycadaceae: Cycad Family

**Field Recognition**   *Stems palm-like, sparingly branched; leaves compound and fern-like; plants dioecious.*

**Description**   *Habit*   Stems palm-like in appearance, tuberous, usually armored, sparingly branched.

*Leaves*   Spirally inserted, pinnately compound, terminating the stem; young pinnae circinate, leaf bases remaining after the leaves drop.

**Reproduction**   Plants dioecious. Microsporangia and megasporangia produced on strobili (except in *Cycas* which has loosely grouped megasporophylls with ovules borne on petiole margins of sporophyll); cones variable in size. Seed usually drupe-like and often brightly colored.

**Systematics**   The Cycadaceae are sometimes treated as three families: Cycadaceae (*Cycas*), Stangeriaceae (*Stangeria*), and Zamiaceae (*Zamia*). The division is based largely on leaf characters.

**Size, Distribution, and General Information**   A family of about 10 genera and 100 species confined to tropical and subtropical regions, especially in Mexico, the West Indies, South America, Australia, and South Africa. One genus, *Zamia,* occurs in North America north of Mexico.

Some of the genera are *Cycas* (20);* *Zamia* (30–40) (coonties, Florida arrowroot); *Dioon* (3–5); *Ceratozamia* (4); *Microcycas* (1); and *Encephalartos* (30).

*Cycas* ranges from Africa and Madagascar to Southeast Asia and Australia. Often grown as an ornamental in the warmer parts of the United States, *C. revoluta* occurs in tropical China and Japan. Sago starch is obtained from the pith of *C. circinalis* of Indonesia and the Philippines.

---

*The number in parentheses is the estimated number of species in the genus.

*Zamia* was known to the Seminole Indians as "conti hateka," which means "white root" or "white bread plant." The starchy underground stem of *Z. floridana* was used by the Seminoles and later by the early settlers as a source of flour.

Indigenous to Mexico, *Dioon edule* has starchy seeds which are ground into a meal used for food.

**Economic Importance**   The Cycadaceae are of limited economic value. Some are grown as ornamentals in warmer regions. Several species have starchy stems or seeds which can be used as food. Leaves of several Cycadaceae are used in the florist trade. Some are poisonous.

## Pinaceae: Pine Family

**Field Recognition**   *Cone scales flattened and distinct from the subtending bract; bract usually shorter than the scale* (except in *Pseudotsuga*).

**Description**   *Habit*   Trees, rarely prostrate or creeping shrubs; branches whorled or opposite.
*Leaves*   Mostly evergreen (deciduous in a few genera), needle-like or linear, arranged in spirals or in fascicles.

**Reproduction**   Plants monoecious; microsporangia in small herbaceous strobili; each microsporophyll bearing two to six sporangia on the lower side. Ovulate strobili with scales in spirals, with two ovules borne on the upper surface; the scale subtended by bracts, closed until seeds are ripe; cones woody. Seeds winged or wingless.

**Size, Distribution, and General Information**   A family of about 10 genera and 250 species of the Northern Hemisphere, south to Sumatra, Java, Mexico, Central America, and the West Indies. Pinaceae is the largest and most important family of the gymnosperms.

The chief genera are *Pinus* (70–100) (pine); *Abies* (50) (fir); *Pseudotsuga* (7) (Douglas fir); *Tsuga* (15) (hemlock); *Picea* (50) (spruce); *Larix* (10–12) (larch, tamarack); *Pseudolarix* (2) (larch, golden larch); and *Cedrus* (4) (true cedar).

*Pinus,* the largest genus in the family, is characterized by the needles which are borne in fascicles on short shoots. The cones vary in length from around 2 or 3 centimeters to over 45 centimeters. Some important species in North America are *P. strobus* (white pine), *P. monticola* (Western white pine), *P. lambertiana* (sugar pine), *P. edulis* (piñon pine), *P. monophylla* (singleleaf piñon), *P. palustris* (longleaf pine), *P. taeda* (loblolly pine), *P. echinata* (shortleaf pine), *P. elliottii* (slash pine), *P. ponderosa* (ponderosa pine), *P. coulteri* (Coulter pine), and *P. radiata* (Monterey pine). The bristle cone pines (*P. aristata*), from high elevations in the American Southwest, are the oldest known living things, achieving 4000 to 4900 years of age. They grow only 2.5 centimeters per 100 years.

The firs (*Abies*) are evergreen trees. *Abies balsamea* (balsam fir) is an important tree of the boreal forest formation of Canada. Other species of North America are *A. fraseri* (balsam fir) of the southern Appalachians; and in western North America, *A. concolor* (white fir), *A. lasiocarpa* (alpine fir), *A. grandis* (grand fir), and *A. nobilis* (noble fir). *Pseudotsuga menziesii* (Douglas fir) is a valuable timber tree of western North America. *Tsuga* (hemlock) has flat needles with two white lines beneath. *Tsuga canadensis* is a large tree of eastern North America, and its western counterpart is *T. heterophylla*.

The spruces (*Picea*) are important timber, pulp, and ornamental trees. Some important species are *P. glauca* (white spruce) and *P. mariana* (black spruce), both important elements of the Canadian boreal forest; *P. abies* (Norway spruce), *P. rubens* (red spruce), *P. sitchensis* (Sitka spruce), and *P. engelmannii* (Engelmann spruce) of subalpine forests; and *P. pungens* (Colorado blue spruce), a widely planted ornamental.

The larches (*Larix*) are deciduous, differing from most other conifers which are evergreen. *Larix laricina* (tamarack) is characteristic of the boreal forest of Canada; *L. occidentalis* (western larch) is a large tree of the northern Rockies. Venetian turpentine is obtained from the European larch (*L. decidua*).

The true cedar (*Cedrus*) ranges from the Mediterranean to the Himalayas. *Cedrus libani* (cedar of Lebanon) is the cedar mentioned in the Bible, and it appears on the Lebanon flag. *Cedrus deodara* and *C. atlantica* are often grown as ornamentals.

**Economic Importance**   The Pinaceae are of considerable direct economic importance for lumber, pulpwood, chemicals (turpentine, rosin, essential oils, and so on), and ornamentals.

### Cupressaceae: Cypress Family

**Field Recognition**   *Leaves small and scale-like, opposite or whorled; cones generally smaller than other families of gymnosperms, cone scales opposite or whorled.*

**Description**   *Habit*   Trees or shrubs.
*Leaves*   Evergreen, opposite or whorled, small and scale-like, often closely appressed to branches.

**Reproduction**   Plants monoecious or dioecious. Male strobili small, terminal or axillary. Ovulate strobili terminal or lateral, on short branches; the mature cones small, dry and woody, or fleshy and berry-like at maturity. Seeds often winged.

**Size, Distribution, and General Information**   A family of about 19 genera and 130 species of worldwide distribution.

The chief genera are *Cupressus* (15–20) (true cypress); *Juniperus* (60) (red cedar, juniper); *Chamaecyparis* (7) (Port Orford cedar, white cedar); *Thuja* (5) (arborvitae); and *Callitris* (16) (cypress pine).

*Thuja* occurs in North America and eastern Asia. *Thuja plicata* (western red cedar) is a valuable timber species which yields a rot-resistant wood. It occurs along the northwestern coast of North America, and it reaches a height of about 75 meters. *Thuja orientalis* of China and Korea is often grown as an ornamental. *Thuja occidentalis* is found in eastern North America.

The cones of *Juniperus* become fleshy at maturity and are known as "juniper berries." A volatile oil extracted from the cones of *J. communis* is used in flavoring gin. A number of species and cultivars of juniper are used as ornamentals. The wood of *J. virginiana* (red cedar) is used in making cedar chests. It is the alternate host for the cedar-apple rust and should not be permitted to grow near apple orchards.

*Chamaecyparis* occurs in North America and eastern Asia. White cedar (*C. thyoides*) occurs along the east coast of North America; *C. lawsoniana* (Port Orford cedar), in Oregon and California; and *C. nutkatenis* (Alaska yellow cedar), from Alaska to Oregon.

*Cupressus,* the true cypress, has several species often grown as ornamentals. The Monterey cypress is *C. macrocarpa. Cupressus sempervirens* (Mediterranean cypress) occurs along the Mediterranean.

**Economic Importance**   Several of the genera yield valuable wood; others are cultivated as ornamentals. Crushed ovulate cones of Juniper are used for flavoring gin, and oil of cedar is obtained from *Thuja occidentalis.*

## Taxodiaceae: Taxodium or Bald Cypress Family

**Field Recognition**   *Trees; twigs and leaves deciduous or persistent; leaves scale-like or needle-like, sometimes dimorphic; cone scales flat or peltate, lacking distinct bracts, each scale producing two to nine seeds.*

**Description**   *Habit*   Trees, rarely shrubs, lateral branches deciduous or persistent.
*Leaves*   Scale-like to needle-like, sometimes dimorphic, persistent or deciduous.

**Reproduction**   Plants monoecious. Staminate strobili small, catkin-like clusters. Ovulate strobili woody, globose, terminal; sporophylls with two to nine ovules; the bracts and scales partially or completely fused; cone scales flat or peltate.

**Size, Distribution, and General Information**    A family of about 10 genera and 16 species from eastern Asia, Tasmania, and North America. Three genera are native to North America north of Mexico.

Some of the genera are *Taxodium* (3) (bald cypress, pond cypress); *Sequoia* (1) (coast redwood, redwood); *Sequoiadendron* (1) (big tree); *Metasequoia* (1) (dawn redwood); *Cryptomeria* (1) (Japanese cedar); *Cunninghamia* (3) (China fir); and *Sciadopitys* (1) (parasol pine, umbrella fir).

*Taxodium distichum* (bald cypress) is an important timber tree of river swamps of the Southeastern United States. Like the redwood, its wood is resistant to termites and wood-decaying fungi. The "knees" are unique features projecting upwards out of the soil or water. *Taxodium mucronatum,* the big cypress tree of Tule in the state of Oaxaca, Mexico, is about 12 meters in diameter and is estimated to be from 2000 to 4000 years old. It may have resulted from natural grafting of several trees planted close together.

*Sequoia sempervirens* (redwood) reaches about 102 meters in height and *Sequoiadendron giganteum* (big tree or giant sequoia) about 96 meters. Giant sequoia does not attain the height of the redwood but greatly surpasses it in diameter. The redwood is valued for its decay-resistant timber.

*Metasequoia glyptostrobioides* (dawn redwood), from central China, is often called a living fossil since it was found living after it was named and described originally from fossil material.

*Cryptomeria, Sciadopitys,* and *Cunninghamia* are often grown as ornamentals in parts of North America.

**Economic Importance**    The family is of considerable importance as a source of decay-resistant lumber. Several species are used as ornamental trees.

### Taxaceae: Yew or Taxus Family

**Field Recognition**    *Microsporophylls peltate; ovules solitary and terminal; seed surrounded by a fleshy aril.*

**Description**    *Habit*    Much-branched trees or shrubs.
*Leaves*    Evergreen, alternate, often two-ranked, with three light-green and two dark-green bands beneath.

**Reproduction**    Plants dioecious. Microsporophylls shield-shaped with six to eight microsporangia, occurring in strobili. Ovulate structure a solitary, terminal ovule with a fleshy aril at the base. Seed dry, nut-like, surrounded or enveloped by a fleshy, often brightly colored aril.

**Size, Distribution, and General Information**    A family of about 5 genera and 20 species of the Northern Hemisphere, south to Mexico, the Celebes, and New Caledonia.

The two genera occurring in North America are *Taxus* (10) (yew, taxus, Florida yew) and *Torreya* (6) (California nutmeg, stinking cedar, torreya).

Taxus is the largest genus in the family. Among its better-known species are *T. baccata* (common or English yew) of Europe, *T. canadensis* (Canada yew), and *T. brevifolia* (Pacific yew). *Taxus floridana* is endemic along the Apalachicola River in Florida. The wood of the yew is valuable. During the Middle Ages, *T. baccata* was the chief wood used in making bows. The leaves of *Taxus* are poisonous, but the aril is harmless.

*Torreya* has two species in North America: *T. californica* (California nutmeg) and *T. taxifolia* (stinking cedar) from Florida. The wood of stinking cedar is durable. This species is now rare, since many plants were destroyed for making fence posts.

**Economic Importance**   Several species and cultivars of *Taxus* are grown as ornamentals, and the wood is used in cabinet making. The family is of relatively little economic importance.

### Ephedraceae: Ephedra Family

**Field Recognition**   *Stems jointed and photosynthetic; horsetail like; with shrubby habit; leaves reduced to scales; cones small, deciduous, borne in stem axils.*

**Description**   *Habit*   Shrubs, much-branched; stem photosynthetic.
*Leaves*   Opposite, united, reduced to scales.

**Reproduction**   Plants usually dioecious. Staminate strobili compact, subglobose to oblong, compound, usually axillary, opposite or in whorls of three or four at nodes. Ovulate strobili elongated, opposite or in whorls of three to four at nodes, with several pairs of bracts along the axis; a single terminal ovule with two integuments, the outer of four bracts, the inner of two bracts which elongate at time of pollination. Seed has leathery covering, globose to cylindric, membranous or winged or forming a berry-like red structure.

**Size, Distribution, and General Information**   A family of one genus, *Ephedra,* and about 40 species of America and Eurasia. In the southwestern United States, *Ephedra* is called "joint fir" or "Mormon tea" from its use in making a beverage by Mormon pioneers. The Asiatic species, *E. sinica* and *E. equisetina,* are the source of the drug ephedrine.

**Economic Importance**   The family is important as the source of ephedrine and for species grown as ornamentals in desert areas.

### SELECTED REFERENCES

Arnold, C. A.: "Classification of the Gymnosperms from the Viewpoint of Paleobotany," *Bot. Gaz.,* **110:**2–12, 1948.

————: "Origin and Relationship of the Cycads," *Phytomorphology,* **3:**51–65, 1953.

Bailey, L. H.: *The Cultivated Conifers in North America,* Macmillan, New York, 1933.

Bierhorst, D. W.: *Morphology of Vascular Plants,* Macmillan, New York, 1971.

Bold, H. C.: *Morphology of Plants,* 3d ed., Harper and Row, New York, 1973.

Chamberlain, C. J.: *Gymnosperms: Structure and Evolution,* The University of Chicago Press, Chicago, 1935. (Reprinted by Dover, New York, 1966.)

Coulter, J. M., and C. J. Chamberlain: *Morphology of Gymnosperms,* The University of Chicago Press, Chicago, 1917.

Cronquist, A., A. Takhtajan, and W. Zimmermann: "On the Higher Taxa of Embryobionta," *Taxon,* **15:**129–134, 1966.

Dallimore, W., and A. B. Jackson: *A Handbook of Coniferae and Ginkgoaceae,* 4th ed., S. G. Harrison (ed.), St. Martin's, New York, 1967. (A standard reference work on the world's conifers, illustrated.)

Foster, A. S., and E. M. Gifford: *Comparative Morphology of Vascular Plants,* 2d ed., Freeman, San Francisco, 1974.

Harlow, W. M., and E. S. Harrar: *Textbook of Dendrology,* 5th ed., McGraw-Hill, New York, 1968.

Hutchinson, J.: "Contributions Towards a Phylogenetic Classification of Flowering Plants. III. The Genera of Gymnosperms." *Kew Bull.,* 49–66, 1924.

Ouden, P. den, and B. K. Boom: *Manual of Cultivated Conifers Hardy in the Cold- and Warm-temperate Zone,* Martinus Nijhoff, The Hague, Netherlands, 1965. (Illustrated, useful for ornamentals.)

Rehder, A.: *Manual of Cultivated Trees and Shrubs,* 2d ed., Macmillan, New York, 1940. (Comprehensive, useful for ornamentals.)

Sporne, K. R.: *The Morphology of Gymnosperms,* Hutchinson University Library, London, 1965.

# Angiosperm Families

In addition to nomenclature, history, literature, and theoretical considerations of systematics, students of taxonomy are expected to become familiar with the more important families of flowering plants. It is beyond the scope of this book to deal with all the families of angiosperms. The 90 families which have been selected for discussion occur in North America north of Mexico. Some were chosen because of their large number of species, others because of their economic importance or unusual features. The arrangement of categories follows that of Cronquist's latest version (personal communication).

Treatments of the families have been made as uniform as the individual situations permit. The characters of each family have been compiled from the literature, including Hutchinson (1973), Lawrence (1951), and Willis (1973). The floral formulas follow the method outlined in Chapter 10. Variation in number of floral parts is indicated in the following manner: "sepals (3)5(10)" indicates "normally 5 but sometimes 3 or 10 sepals;" ∞ means "variably numerous."

The number of taxa in the families was obtained from Willis (1973). Scientific names generally follow Hutchinson (1973) and Willis (1973). The number of species in a genus is indicated in parentheses, e.g., *Magnolia* (80). The distribution of each family is generalized and serves to indicate the region of natural occurrence. Illustrations of selected representative families have been provided to portray the more significant structural features. Several families that are likely to be confused are compared in Table 14-1.

**Table 14-1  A Comparison of Some Families That Are Likely to Be Confused***

| A. Families | Leaves | Stems | Perianth | Bracts† | Fruit |
|---|---|---|---|---|---|
| Juncaceae | Basal forming tufts, linear | Round, solid | 6-parted | Several | Capsule |
| Gramineae | Ligulate, sheaths open (mostly) | Round, hollow at internodes | Lodicule(s) | 2 (palea, lemma) | Grain or caryopsis (achene, berry) |
| Cyperaceae | Sheaths closed | 3-sided, solid | Bristles, hairs, scales | 1 | Achene, nutlet |

| B. Families | Leaves | Leaf bases | Gynoecium | Fruit |
|---|---|---|---|---|
| Alismataceae | Basal | Sheathing | Many separate carpels | Achene |
| Ranunculaceae | Compound, alternate, (opposite) | Mostly sheathing | 3-many separate carpels | Follicle, achene, berry, capsule |
| Rosaceae | Compound (simple), alternate | Stipulate | Highly variable, ovary free or adnate to hypanthium | Achene, aggregate, drupe, pome, follicle |
| Saxifragaceae | Alternate | Exstipulate | 2 (–5) united carpels | Capsule |
| Crassulaceae | Simple, alternate or opposite, usually succulent | Exstipulate | 3 (+) separate carpels subtended by glandular scale | Follicle |

| C. Families | Leaves | Symmetry | Stamens | Gynoecium | Fruit | Inflorescences | Other |
|---|---|---|---|---|---|---|---|
| Scrophulariaceae | Alternate (opposite) | Zygomorphic | (2) 4 (+ staminode) | 2 carpellate | Capsule, berry | Various | Some parasitic |
| Boraginaceae | Alternate | Actinomorphic | 5, epipetalous | 4-lobed ovary (bicarpellary) | 4 nutlets | Coiled cymes | Rough texture |
| Verbenaceae | Opposite or whorled | Zygomorphic-salverform | (2) 4 (5) didynamous | Single, terminal style | Drupe or nutlet | Cymes | Often woody in tropics |
| Lamiaceae | Opposite or whorled | Zygomorphic-bilabiate | (2) 4 didynamous | Gynobasic style, 4-lobed ovary (bicarpellary) | 4 nutlets | In whorls or axillary | Glandular, volatile oils |
| Solanaceae | Alternate | Actinomorphic-rotate to tubular (zygomorphic) | 5 (2, 2+2, 4) | 2 carpels | Berry | Cymose | Enlarged calyx in fruit (some), rank smell |

*Note that *not* all features are present in all members of the family and that there are *few* families distinguished by a single, uniform feature.

†Bracts subtending the inflorescence are not considered.

# CLASS MAGNOLIOPSIDA: Dicots

## SUBCLASS I, MAGNOLIIDAE; ORDER 1, MAGNOLIALES

### Magnoliaceae: Magnolia Family

**Field Recognition**   Trees or shrubs; leaves simple, alternate; stipules deciduous and leaving a stipular ring at each node; flowers conspicuous, actinomorphic; stamens and carpels numerous; $CA^3 CO^{6-\infty} A^\infty \underline{G}^\infty$.

**Description**   *Habit*   Shrubs or trees.
*Leaves*   Alternate, simple, pinnately veined, petioled; stipules present and enclosing the bud, early deciduous and leaving a stipular scar or ring at each node.
*Inflorescences*   Usually solitary.
*Flowers*   Bisexual, rarely unisexual, actinomorphic. Calyx often of 3 sepals, sometimes petal-like. Corolla of 6 to many petals. Androecium of numerous stamens, spirally arranged on the basal part of the floral axis. Gynoecium of numerous (rarely few), separate, simple pistils, each with a unilocular carpel, spirally arranged on the upper part of the elongated floral axis, placentation parietal, ovary superior. (See Figure 6-1, page 88.)
*Fruit*   A follicle, a samara, or (rarely) a berry, sometimes aggregated into a woody cone-like structure.
*Seed*   Often suspended within the follicle by a thread-like funiculus.
*Embryo*   Minute, in a large endosperm.

**Systematics**   Present-day members of the Magnoliaceae are relatively specialized in certain vegetative, floral, and cytological features. The family is characterized by oil passages in its parenchyma tissue. Other families in the order Magnoliales, such as the Winteraceae, are more primitive. For this reason, the flower of *Magnolia* should not be cited as an example of the most primitive type of flower. Magnoliaceae is divided into two tribes, Magnolieae and Liriodendreae. *Liriodendron* is the only genus in the Liriodendreae.

**Size, Distribution, and General Information**   A family of about 12 genera and 230 species occurring primarily in the warm temperate regions of the Northern Hemisphere, with aggregates of species in eastern North America and eastern Asia. It ranges southward into Malaysia, the West Indies, and Brazil. The majority of the species are Asiatic.
Some of the important genera are *Magnolia* (80) (bullbay, cucumber tree); *Liriodendron* (2) (tulip tree, yellow poplar); and *Michelia* (50) (banana shrub).
The evergreen *Magnolia grandiflora,* with its large shining leaves and handsome cream-white flowers, has become a symbol of the Deep South of the

United States. When its petals open wide, they form a soft cup 20 centimeters across. The tulip tree or yellow poplar (*Liriodendron tulipifera*) of the eastern United States often grows rapidly to the height of 30 meters or more, and its diameter at breast height (DBH) may be nearly 2 meters. Its flowers appear tuliplike, with three recurved sepals and six stiff petals. As the carpels ripen into samaras, the winged fruits break off and are carried away by the wind.

**Economic Importance** The Magnoliaceae are important as ornamentals and as a source of wood. *Magnolia soulangiana, M. stellata,* and other cultivars of Asiatic origin are widely grown as ornamentals in North America; *M. grandiflora* is a well-known ornamental tree. *Liriodendron tulipifera* is an important timber species in the southern Appalachians. *Michelia fuscata* is planted as an ornamental shrub in the warmer parts of the United States.

## SUBCLASS I, MAGNOLIIDAE; ORDER 2, LAURALES

### Lauraceae: Laurel Family

**Field Recognition** Aromatic trees or shrubs; leaves usually alternate; perianth small and undifferentiated; stamens 3-merous in several whorls, anthers with valvate dehiscence; fruit a single seeded drupe or berry; $CA^6 CO^0_{|}$ $A^{3+3+3+3} \underline{G}^1$.

**Description** *Habit* Aromatic trees or shrubs; one genus is a parasitic vine.

*Leaves* Alternate, rarely opposite, simple, pinnately veined, usually entire, deciduous in temperate regions, evergreen in tropics; without stipules.

*Inflorescences* Usually axillary, panicles, spikes, racemes, or umbels.

*Flowers* Usually bisexual, sometimes unisexual, actinomorphic. Calyx of 6 sepals, in 2 series united into a tube at the base. Corolla lacking. Androecium of 4 whorls of 3 stamens each, or 1 or more whorls reduced to staminodes or absent, adnate to perianth tube. Gynoecium a simple pistil with 1 carpel, 1 locule, and 1 ovule, placentation parietal, ovary superior, style 1, stigma 1.

*Fruit* A drupe or berry often surrounded at the base by a cupule derived from the persistent calyx tube.

*Embryo* Large, straight; endosperm absent.

**Systematics** The family is divided into two subfamilies: Lauroideae, which is arborescent; and Cassythoideae, with the parasitic twiner *Cassytha*. Lauroideae is classified into five tribes: (1) Perseeae (*Persea*); (2) Cinnamomeae (*Cinnamomum*); (3) Laureae (*Lindera*); (4) Cryptocaryeae (*Cryptocarya*); and (5) Hypodaphnideae (*Hypodaphnis*).

**Size, Distribution, and General Information**   A family of about 32 genera
and 2000 to 2500 species distributed in tropical and subtropical regions of both
hemispheres, with the greatest diversity in Southeast Asia. *Umbellularia
californica* (myrtle, bay tree, or pepperwood) occurs along the Pacific Coast in
the western United States. The family is common in the southeastern United
States. *Sassafras albidum* and *Lindera* (spicebush) extend into Canada.

Some of the important genera are *Cinnamomum* (250); *Persea* (150); *Sassafras* (3); *Lindera* (100) (spicebush); *Nectandra* (100); *Cassytha* (20); and
*Laurus* (2) (laurel).

According to Greek legend, the large evergreen shrub *Laurus nobilis*
(laurel) was sacred to Apollo, and branches were woven into crowns for victors
in the ancient Olympic games. Later, laurel became the symbol of triumph in
Rome. Today the shrub is cultivated in the warmer parts of the United States,
and laurel wreaths are still used ceremonially in some parts of the world.

**Economic Importance**   Economically, the family is of importance for its
aromatic oils. *Cinnamomum camphora* of China and Japan yields camphor; *C.
zeylanicum* is cultivated for cinnamon in Ceylon and elsewhere. The avocado is
*Persea americana. Sassafras albidum* yields oil of sassafras, and the bark from
the roots is boiled for tea. The leaves of *Sassafras* are used in Creole cooking
along with those of *Persea borbonia.* The woods of some species are used in
cabinetmaking or marine shipbuilding. *Ocotea bullata* (black stinkwood) from
southeastern Africa is used in making heavy, handsome furniture.

## SUBCLASS I, MAGNOLIIDAE; ORDER 6, NYMPHAEALES

### Nymphaeaceae: Water Lily Family

**Field Recognition**   Perennial aquatic herbs; leaves long-petioled, peltate;
flowers large, long-peduncled, floral parts often numerous; $CA^{3-\infty}$ $CO^{3-\infty}$ $A^{\infty}$
$\underline{G}^{\infty}$.

**Description**   *Habit*   Perennial aquatic herbs with large rhizomes.
*Leaves*   Alternate; floating, submerged, or emersed; pinnately veined, petiolate, peltate.
*Inflorescences*   Axillary and solitary.
*Flowers*   Bisexual, actinomorphic. Calyx of 3 to many sepals. Corolla of
numerous petals frequently grading into the stamens. Androecium of
numerous stamens. Gynoecium of numerous simple pistils each having 1 carpel, with 1 locule, ovules numerous, placentation parietal, ovary superior.
*Fruit*   A spongy berry.
*Seed*   With endosperm and perisperm and sometimes arillate.

**Size, Distribution, and General Information** A family of 3 genera and 76 species of cosmopolitan distribution. The three genera are *Nymphaea* (50) (water lily); *Nuphar* (25) (cow lily, spadderdock, yellow pond lily); and *Ondinea* (1).

Members of the water lily family are favorite plants for garden pools. The variously colored flowers of *Nymphaea* open just above the surface of the water. The golden-yellow flowers of *Nuphar advena* extend up and out of the water. *Nymphaea lotus* is considered to have been the original sacred lotus of Egypt. (*Nelumbo* [Nelumbonaceae], often called *sacred lotus,* was not introduced into Egypt until about 500 B.C.)

**Economic Importance** Horticultural hybrids of *Nymphaea* are numerous. *Nymphaea* sometimes becomes weedy in lakes and ponds.

## SUBCLASS I, MAGNOLIIDAE; ORDER 7, RANUNCULALES

### Ranunculaceae: Buttercup, Crowfoot Family

**Field Recognition** Mostly herbs; leaves with sheathing leaf bases, blades often divided; flowers mostly bisexual with spirally arranged, numerous stamens and carpels; $CA^{3-\infty} CO^{5-\infty} A^{\infty} \underline{G}^{3-\infty(1-3)}$.

**Description** *Habit* Annual or perennial herbs, sometimes shrubs or vines.

*Leaves* Mostly alternate, palmately compound, petioles sheathing at base; lacking stipules.

*Inflorescences* Solitary to paniculate, or in racemes or cymes.

*Flowers* Typically bisexual, actinomorphic or zygomorphic, parts spirally arranged. Calyx of 3 to many sepals. Corolla of 5 to many petals, rarely absent; nectiferous glands often present. Calyx and corolla often not differentiated. Androecium of 5 to many stamens, free, spirally arranged. Gynoecium of 3 to many simple pistils having 1 carpel, with 1 locule, ovules 1 to many, placentation parietal, ovary superior, style 1, and stigma 1. See Figure 14-1.

*Fruit* A follicle, achene, or berry, or rarely a capsule.

*Seed* With minute embryo and oily endosperm.

**Systematics** The perianth varies greatly and is of importance in delimiting genera. Generic concepts have varied widely over the past 70 years in North America, as illustrated by the different treatments in various floras.

**Size, Distribution, and General Information** A family of about 50 genera and 1900 species found primarily in the temperate regions of the Northern Hemisphere, with concentrations in eastern Asia and eastern North America.

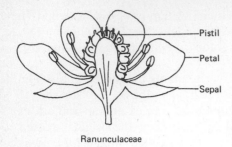

Ranunculaceae

**Figure 14-1** Longitudinal section of a generalized flower of *Ranunculus* (Ranunculaceae).

Some of the important genera are *Ranunculus* (400) (buttercup); *Clematis* (250) (Virgin's bower); *Thalictrum* (150); *Aquilegia* (100) (columbine); *Delphinium* (250) (larkspur); *Anemone* (150) (windflower); *Helleborus* (20) (Christmas rose, hellebore); and *Aconitum* (300) (monkshood).

Many people recognize the widespread, often weedy genus *Ranunculus*. *Hepatica* (liverleaf) grows across Eurasia and in North America, and its three-lobed leaves resembling the liver were formerly used to treat liver ailments. Cattle raisers in some parts of the West refer to *Delphinium* as "locoweed" because cattle that eat *Delphinium* may develop a staggering walk and other symptoms.

**Economic Importance**  The Ranunculaceae are important as ornamental, drug, and poisonous plants. They are best known for their flowers, e.g., columbine, larkspur, buttercup, windflower, and Christmas rose. Alkaloids are obtained from *Aconitum; A. ferox* of Nepal is the source of bikh poison. Wolfbane of Europe, *A. napellus,* yields aconite. Roots of the golden seal (*Hydrastis*), used to treat stomach ailments, are sold in the crude drug trade.

## SUBCLASS I, MAGNOLIIDAE; ORDER 7, RANUNCULALES

### Berberidaceae: Barberry Family

**Field Recognition**  Herbs or shrubs; stamens biseriate, the outer opposite the petals, anthers dehiscing by flaps; pistil 1; $Ca^{3+3} CO^{3+3} A^6 \underline{G}^1$.

**Description**  *Habit*  Perennial herbs and shrubs.
*Leaves*  Alternate or basal, simple or compound, deciduous or evergreen, often spiny; stipules absent.
*Inflorescences*  Solitary, cymes, or racemes.
*Flowers*  Bisexual, actinomorphic. Calyx of 3 to 6 sepals, free and distinct.

Corolla of 3 to 6 petals, free and distinct; petal-like nectaries sometimes present. Androecium of 6 stamens, distinct, anthers opening by flaps. Gynoecium of a simple pistil having 1 carpel with 1 locule, ovules few and basal, placentation basal or parietal, ovary superior.

*Fruit*  A berry or follicle.

*Seed*  With copious endosperm.

*Embryo*  Small (elongate in *Berberis*).

**Size, Distribution, and General Information**  A family of about 14 genera and 650 species widely distributed throughout much of the Northern Hemisphere and South America.

Some important genera are *Berberis* (450) (barberry); *Mahonia* (70) (Oregon grape); *Nandina* (1) (heavenly bamboo); and *Podophyllum* (11) (mayapple).

The Berberidaceae has several genera with disjunct eastern North American–eastern Asian ranges. *Podophyllum* has one species in eastern North America, others in eastern Asia. *Diphylleia* has one species in the southeastern United States and one in Japan. *Jeffersonia* has one species in eastern United States and one in Manchuria. *Caulophyllum* has one species in eastern United States and another in eastern Asia.

**Economic Importance**  The family is of economic importance as a source of many beautiful garden shrubs: *Berberis thunbergii* (Japanese barberry); *Mahonia aquifolium* (Oregon grape); and *Nandina domestica* (heavenly bamboo). The common barberry (*B. vulgaris*) is the host for the sexual phase of the wheat stem rust. A drug prepared from the rhizome of *Podophyllum peltatum* (mayapple) promotes the discharge of bile from the liver.

## SUBCLASS I, MAGNOLIIDAE; ORDER 8, PAPAVERALES

### Papaveraceae: Poppy Family

**Field Recognition**  Herbs; flowers bisexual; sepals deciduous; petals crumpled in the bud; stamens numerous in several whorls; ovary unilocular; fruit usually a capsule; sap sometimes milky or colored; $CA^{2-3} CO^{4-\infty} A^{\infty} \underline{G}^{2-\infty}$.

**Description**  *Habit*  Annual to perennial herbs, rarely woody, usually with milky or colored sap.

*Leaves*  Alternate, basal or cauline, often much divided.

*Inflorescences*  Solitary showy flowers.

*Flowers*  Bisexual, actinomorphic. Calyx of 2 to 3 distinct, early deciduous sepals. Corolla of 4 to 6 or more petals, in two series, often crumpled in the bud. Androecium of numerous stamens, free. Gynoecium a compound pistil of 2 or more united carpels, with 1 locule, placentation of 2 to many parietal placentas, ovary superior.

*Fruit*   A capsule opening by valves or pores.

*Seed*   Numerous and small.

*Embryo*   Minute, in copious fleshy or oily endosperm.

**Size, Distribution, and General Information**   A family of about 26 genera and 200 species mostly of Northern Hemisphere distribution.

Some important genera are *Papaver* (100) (true poppy); *Eschscholzia* (10) (California poppy); *Sanguinaria* (1) (bloodroot); *Argemone* (10) (prickly poppy); and *Chelidonium* (1) (celandine).

The true poppies (*Papaver*) have a milky white juice, and the capsule resembles a pepper shaker opening by a series of small holes around the rim of the flat top. Here the seeds are protected from the rain and can only escape when the capsule is shaken. *Papaver somniferum* is the opium poppy. The drug is obtained by cutting notches in the half-ripened capsules, from which the latex exudes. The seeds of this species are used to add a slight tang to breads and cookies. When pressed, the seeds yield a cooking oil.

One of the first spring wild flowers is *Sanguinaria canadensis,* or bloodroot. It is named for its bright orange-red juice which oozes quickly from its freshly cut rhizome. Its rhizome yields an alkaloid used in medicine.

**Economic Importance**   The family is of little economic importance except for opium and its narcotic alkaloids. It is illegal to grow the opium poppy in the United States. Several other species of *Papaver* are grown as ornamentals, including *P. orientale* (oriental poppy) and the Iceland poppy (*P. nudicaule*). The bright red flowers of the European field poppy are associated with some of the battlefields of World War I. The California poppy (*Eschscholzia california*), which has brilliant golden-orange flowers, is common in parts of the southwestern United States.

**SUBCLASS II, HAMAMELIDAE; ORDER 6, URTICALES**

**Ulmaceae: Elm Family**

**Field Recognition**   Trees or shrubs with watery sap; leaf bases oblique; fruit an evenly winged samara or drupe; $CA^{4-9} CO^0 A^{4-8} \underline{G}^{②}$.

**Description**   *Habit*   Trees or shrubs.

*Leaves*   Alternate, simple, deciduous, blades often oblique at base; stipules paired.

*Inflorescences*   Solitary, cymose, or fasciculate.

*Flowers* Bisexual or unisexual, reduced. Calyx of 4 to 8 united sepals. Corolla lacking. Androecium of 4 to 8 stamens, inserted at bottom of calyx. Gynoecium a compound pistil of 2 carpels, with 1(2) locules, ovules solitary in each locule, ovary superior, pistil 1, styles 2.

*Fruit* A samara or drupe.

*Seed* Without endosperm.

**Systematics** The family is classified into two well-marked subfamilies: Ulmoideae and Celtidoideae.

**Size, Distribution, and General Information** A family of about 15 genera and 200 species mostly of the north temperate zone.

Some important genera are *Ulmus* (45) (elm); *Zelkova* (6 to 7); *Planera* (1) (water elm); and *Celtis* (80) (sugarberry, hackberry).

Elms, such as *Ulmus americana* (American elm) and *U. laevis* (European white elm), were once important street trees in Western Europe and in eastern North America, but Dutch elm disease is gradually destroying these magnificent shade trees. Horticulturists are planting species of *Zelkova* as a possible replacement.

**Economic Importance** The family is of importance for timber and for shade trees. Elms and hackberry trees, especially the latter, are of prime importance as horticultural plants throughout the western half of the United States and Canada. *Ulmus pumila* is widely planted, as are other Oriental elms. Hackberry trees are the standard avenue tree in the Rocky Mountain and Great Basin states.

## SUBCLASS II, HAMAMELIDAE; ORDER 6, URTICALES

### Moraceae: Mulberry Family

**Field Recognition** Trees and shrubs, rarely herbs, with milky sap; flowers unisexual; stigmas usually 2; (CA$^4$ CO$^0$ A$^4$ G$^0$)(CA$^4$ CO$^0$ A$^0$ $\overline{G}^2$).

**Description** *Habit* Trees, shrubs, or rarely herbs, with milky sap.

*Leaves* Alternate, simple, pinnately or palmately veined, evergreen or deciduous, often lobed; stipules 2.

*Inflorescences* Plants monoecious or dioecious; flowers in racemes, spikes, umbels, or heads or on the inside of a hollow receptacle.

*Flowers* Unisexual, much reduced. Calyx of 4 sepals, free or united, sometimes absent. Apetalous. Androecium of 4 stamens, opposite the sepals. Gynoecium a compound pistil of 2 united carpels, with 1 locule, ovary superior to inferior, pistil 1, styles usually 2.

*Fruit* Basically drupes, often aggregated or united into a multiple fruit, or achenes within or upon a fleshy receptacle.

*Seed* With or without endosperm, embryo usually curved.

**Size, Distribution, and General Information** A family of about 53 genera and 1400 species mostly of tropical and subtropical distribution; a few are temperate.

Some important genera are *Morus* (10) (mulberry); *Maclura* (12) (hedge apple, osage orange); *Broussonetia* (7 to 8) (paper mulberry); *Ficus* (800) (fig, rubber plant, Indian rubber tree); and *Artocarpus* (47) (breadfruit, jackfruit).

The leaves of *Morus alba* (white mulberry) are used to feed silkworms. Mulberry fruits are edible, and *M. nigra* (black mulberry) is sometimes cultivated for its large and juicy fruits.

Breadfruit (*Artocarpus altilis*) became the center of historical adventure associated with the voyage of Captain William Bligh and his ship *Bounty* to Tahiti. Captain Bligh was to collect breadfruit plants for introduction to the West Indies, when Fletcher Christian (one of his officers) and many of the crew mutinied. Bligh was placed adrift in a small boat, but he managed to survive and later introduced breadfruit into Jamaica.

*Ficus elastica* (Indian rubber tree or rubber plant) is grown as a houseplant. Rubber was made from its sap before the use of the latex from Brazilian rubber trees of the family Euphorbiaceae. The commercial fig is *F. carica*. Some species of *Ficus* are known as strangler figs. They germinate high on the trunk of a tree and send roots down around the trunk of the supporting plant. The roots interlace, and they appear to be strangling the tree.

**Economic Importance** The Moraceae is perhaps most important as the source of several food plants—e.g., fig and breadfruit—and a few ornamentals such as the rubber plant and osage orange.

## SUBCLASS II, HAMAMELIDAE; ORDER 6, URTICALES

### Urticaceae: Nettle Family

**Field Recognition** Mostly herbs, many with stinging hairs; inflorescences cymose on short axillary shoots; ovary unilocular, style 1; (CA$^{4-5}$ CO$^0$ A$^{4-5}$ G$^0$) (CA$^{4-5}$ CO$^0$ A$^0$ $\underline{G}^1$).

**Description** *Habit* Fibrous herbs, or rarely subshrubs or small, soft-wooded trees.

*Leaves* Alternate or opposite, simple, sometimes with stinging hairs; stipulate.

*Inflorescences* Plants monoecious or dioecious; flowers are in cymes or heads, or reduced to a solitary flower.

*Flowers*  Unisexual, actinomorphic, greatly reduced. Calyx of 4 to 5 sepals, free or united, or calyx absent. Apetalous. Androecium of mostly 4 stamens, exploding when ripe. Gynoecium a simple pistil of 1 carpel, with 1 locule, ovary superior.

*Fruit*  An achene or fleshy drupe.

*Seed*  Mostly with endosperm.

**Size, Distribution, and General Information**  A family of about 45 genera and 550 species of tropical and temperate distribution.

Some important genera are *Urtica* (50) (nettle); *Laportea* (23) (wood nettle); *Pilea* (400) (clear weed); *Boehmeria* (100) (ramie); and *Parietaria* (30) (pellitory).

*Urtica* (stinging nettle) has sharp, stiff, glandular hairs. When people or animals brush against the plants, the hairs penetrate the skin. A material is released which causes a sharp burning pain or stinging rash that may last a few minutes or several hours.

**Economic Importance**  The family is of little economic importance. Species of *Pilea* are grown as houseplants. *Boehmeria nivea* is cultivated in China for the fiber, China grass or rhea, obtained from the inner bark; it is perhaps the longest and most silky of all vegetable fibers. In the tropics, *B. nivea* var. *tenacissima* yields the fiber ramie. The young tops of *Urtica* (stinging nettle) may be cooked and eaten like spinach.

## SUBCLASS II, HAMAMELIDAE; ORDER 8, JUGLANDALES

### Juglandaceae: Walnut Family

**Field Recognition**  Deciduous trees; leaves pinnately compound, alternate, aromatic; flowers unisexual with staminate flowers usually in catkins; $(CA^{3-6} CO^0 A^{3-\infty} G^0) (CA^4 CO^0 A^0 \overline{G} \oplus)$.

**Description**  *Habit*  Deciduous trees.

*Leaves*  Alternate, pinnately compound, aromatic; without stipules.

*Inflorescences*  Plants monoecious; staminate flowers in catkins.

*Flowers*  Apetalous, unisexual. Staminate flowers with a variable number of bracts and a calyx of 3 to 6 sepals; androecium of 3 to 100 stamens. Pistillate flowers with a number of bracts and a calyx of 4 sepals. Gynoecium a compound pistil of 2 to 3 united carpels, with 1 locule, ovule 1, ovary inferior, stigmas 2, often plumose.

*Fruit*  A drupe-like nut with a dehiscent or indehiscent leathery or fibrous husk, or sometimes a winged nutlet.

*Seed*  With 2 to 4 lobes, solitary.

*Embryo*   Very large, cotyledons often much contorted.

**Size, Distribution, and General Information**   A family of about 7 genera and 50 species occurring in the north temperate and subtropics south to India and Indochina and in the New World into the Andes of South America.

Some important genera are *Juglans* (15) (walnut) and *Carya* (25) (hickory, pecan).

**Economic Importance**   The Juglandaceae are valuable for the wood used in furniture making and for walnuts, pecans, and hickory nuts. *Juglans regia* is the English walnut; *J. nigra* is the black walnut of eastern North America. Black walnut trees bring the highest price of any timber tree in eastern North America. *Carya illinoensis* (pecan) is extensively cultivated in the southeastern United States. Some cultivars of pecan contain up to 70 percent vegetable oil.

## SUBCLASS II, HAMAMELIDAE; ORDER 10, FAGALES

### Fagaceae: Beech or Oak Family

**Field Recognition**   Trees or shrubs; leaves alternate (in oak, leaves and buds clustered at the end of the stems, pith 5-angled); fruit a nut, at least partially covered by a cupule of hardened bracts; ($CA^{4-7} CO^0 A^{4-40} G^0$) ($CA^{4-6} CO^0 A^0 \overline{G} ⊛$).

**Description**   *Habit*   Deciduous or evergreen trees and shrubs.
*Leaves*   Alternate, simple, pinnately veined; with stipules.
*Inflorescences*   Plants usually monoecious; staminate flowers solitary, in catkin-like racemes or in heads; pistillate flowers solitary or few in a cluster.
*Flowers*   Unisexual. Corolla lacking. Staminate flowers with 4 to 7 calyx lobes and 4 to 40 stamens. Pistillate flowers solitary or in 2 to 3 flowered cymules; calyx lobes 4 to 6; gynoecium a compound pistil of 3 to 6 carpels, with 3 to 6 locules, placentation axile, ovary inferior.
*Fruit*   A 1-seeded nut.
*Seed*   Solitary by abortion.

**Systematics**   The family is classified in three subfamilies: (1) Fagoideae (*Fagus* and *Nothofagus*); (2) Castaneoideae (*Lithocarpus, Castanopsis, Castanea,* and *Chrysolepis*); and (3) Quercoideae, (*Quercus* and *Trigonobalanus*).

The important genus *Quercus* is divided in our area into two subgenera: Leucobalanus (white oak) and Erythrobalanus (red oak). The red oaks have sharp points on each lobe around the margin of the leaf, and the white oaks have rounded lobes. The white oaks have small, thin-walled summer wood pores, whereas the red oaks have large, thick-walled summer wood pores.

**Size, Distribution, and General Information** A family of about 8 genera and 900 species of cosmopolitan distribution but absent from tropical Africa southwards.

Some important genera are *Quercus* (450) (oak); *Fagus* (10) (beech); and *Castanea* (12) (chestnut, chinquapin).

*Castanea dentata* (American chestnut) was one of the dominant species of the forests of the Appalachian Mountains, but almost none remain today. It supplied nutritious nuts and a decay-resistant wood. The chestnut was doomed by a devastating fungus that was introduced into New York in 1904. The few American chestnuts remaining are generally sprouts that arise from the roots of a former tree. They are short-lived since they are infected by fungus spores carried on the feet of birds. Most chestnuts used for food today come from the European chestnut (*C. sativa*). The fruit of the Oriental chestnut (*C. mollissima*), widely planted in the southeastern states, is undesirable.

Oaks provide a hard, heavy, durable lumber used in construction, in hardwood floors, and for other purposes. White oak (*Q. alba*) is used in making watertight barrels which are charred and used to age bourbon whiskey.

**Economic Importance** The family is of considerable economic importance for the lumber produced by its members. The bark and wood of *Castanea* is used in tanning. Cork is obtained from the bark of *Quercus suber*. Many species of oak and beech are cultivated as shade or ornamental trees.

## SUBCLASS II, HAMAMELIDAE; ORDER 10, FAGALES

### Betulaceae: Birch Family

**Field Recognition** Deciduous trees or shrubs; leaves simple, serrate; staminate flowers in catkins; (CA$^{0 \, or \, 4}$ CO$^0$ A$^{2-20}$ G$^0$) (CA$^0$ CO$^0$ A$^0$ $\overline{\text{G}}$②).

**Description** *Habit* Deciduous trees and shrubs.

*Leaves* Alternate, simple, pinnately veined; with stipules.

*Inflorescences* Plants monoecious, both staminate and pistillate flowers in catkins.

*Flowers* Unisexual, in condensed cymules. Perianth lacking. Staminate flowers in catkins, naked or minutely 4-parted, stamens 1 to 4 in each flower (2 to 20 per cymule). Pistillate flowers 2 or 3 in each cymule subtended by a 3-lobed bract; gynoecium a compound pistil of 2 united carpels, ovary inferior.

*Fruit* A small, 1-seeded samara or nut.

**Systematics** The family is sometimes divided and treated as two families, the Betulaceae and Corylaceae.

**Size, Distribution, and General Information** A family of about 5 genera

and 142 species, Betulaceae occurs particularly in the north temperate zone, but in the New World extends into tropical mountains and along the Andes to Argentina.

Some of the genera are *Alnus* (35) (alder); *Betula* (60) (birch); *Carpinus* (35) (hornbeam, ironwood); *Corylus* (15) (hazelnut, filbert); and *Ostrya* (10) (hop hornbeam).

American Indians once used the bark of paper birch (*Betula papyrifera*) for covering the frames of canoes. Many alders grow in bogs or swamps and add nitrogen to the soil from large nodules of nitrogen-fixing bacteria on their roots.

**Economic Importance**   The family is of some importance as a source of ornamental trees; hardwood timber is obtained from *Betula* (birches); hazelnuts and filberts are produced from *Corylus;* and a wintergreen-like oil is extracted from the twigs of *Betula*. Some birch is used for veneer, and *Alnus rubra* (red alder) of western North America provides a wood that is a good imitation of mahogany.

## SUBCLASS III, CARYOPHYLLIDAE; ORDER 1, CARYOPHYLLALES

### Cactaceae: Cactus Family

**Field Recognition**   Succulent, fleshy habit, usually spiny herbs, with the spines arranged in areoles; flowers solitary and showy with numerous perianth parts; stamens numerous; ovary inferior; $CA^X CO^\infty A^\infty \bar{G}^{(2-\infty)}$.

**Description**   *Habit*   Fleshy, herbaceous or woody, sometimes branched or tree-like, with spines or bristles or both at areoles.
*Leaves*   Usually scale-like, or flat and fleshy, much reduced.
*Inflorescences*   Flowers solitary and usually showy.
*Flowers*   Bisexual, actinomorphic or nearly zygomorphic. Perianth parts numerous, showing a gradual transition from sepals to petals. Calyx often petaloid. Petals epigynous in several series. Androecium of numerous stamens, epipetalous or inserted at base of petals. Gynoecium a compound pistil of 4 united carpels, with 1 locule, ovules numerous, placentation parietal, ovary inferior.
*Fruit*   A berry, often spiny or bristly.
*Seed*   Little or no endosperm.

**Systematics**   Over the years, the family affinities have been somewhat uncertain; but discovery of the presence of the betalain pigments rather than anthocyanins has finalized the placement of Cactaceae in subclass Caryophyllidae. This supports evidence obtained from embryology, floral morphology, and anatomy. The genera are not well defined.

**Size, Distribution, and General Information** A family of about 50 to 150 genera and 2000 species chiefly localized in the drier regions of tropical America, but extending from British Columbia to Patagonia, and reaching altitudes of over 3600 meters in the Andes. Several are epiphytes. The only Cactacteae native to the Old World are species of *Rhipsalis,* perhaps introduced, in Africa, Madagascar, Mauritius, Seychelles, and Ceylon. Several species of *Opuntia* have become naturalized and are often troublesome weeds in South Africa, India, and Australia.

Some genera are *Opuntia* (250) (prickly pear, beaver tail, cholla); *Cereus* (50) (hedge cactus); *Echinocactus* (10) (barrel cactus); *Mammillaria* (200 to 300) (pincushion cactus); and *Lemaireocereus* (25) (organ pipe cactus).

The symbol of the deserts of the American Southwest is the giant saguaro cactus (*Carnegiea gigantea*). It may reach 15 meters in height with a diameter of 50 centimeters. Saguaro seedlings grow very slowly, and because of human activity, the saguaros may be doomed to extinction.

The short days and long nights of December bring the colorful Christmas cactus (*Zygocactus truncatus*) of Brazil into flower. In Mexico, cacti are set in rows to form living fences. The jumping chollas (*Opuntia*) of the American Southwest obtained their name from the case by which their terminal segments become hooked on animals passing the plant. The cholla segment, once fallen or removed from the animal, will usually take root.

**Economic Importance** The Cactaceae are important as ornamentals. The fruits of the prickly pear are edible, and the stems of some are used as emergency fodder. *Lophophora williamsii* (peyote), which contains an alkaloid called *mescaline,* is hallucinogenic and is eaten by certain groups of American Indians for religious purposes. The dried bodies of mealybugs which live on *Opuntia* yield the red dye cochineal which is used in cosmetics.

## SUBCLASS III, CARYOPHYLLIDAE; ORDER 1, CARYOPHYLLALES

### Chenopodiaceae: Goosefoot Family

**Field Recognition** Herbs or shrubs, often succulent; flowers small and greenish, scarious bracts absent; $CA^{\widehat{2-5}} CO^0 A^{2-5} \widehat{\underline{G}^{2-3}}$.

**Description** *Habit* Often xerophytic or halophytic annual or perennial herbs, or sometimes shrubs; sometimes succulent with nearly leafless stems, sometimes jointed.

*Leaves*   Usually alternate and simple; sometimes fleshy or reduced to scales; stipules absent.

*Inflorescences*   Cymose, often bracteate, plants dioecious.

*Flowers*   Bisexual or unisexual, actinomorphic. Calyx of 2 to 5 sepals, small, united, persistent in fruit. Corolla absent. Androecium of 2 to 5 stamens, opposite the calyx lobes. Gynoecium a compound pistil of 2 to 3 carpels, with 1 locule, ovule solitary, ovary superior or inferior, styles 1 to 3.

*Fruit*   A nutlet, indehiscent.

*Seed*   Embryo usually surrounding the endosperm.

**Size, Distribution, and General Information**   A family of about 102 genera and 1400 species of worldwide distribution, but centered in xeric and saline habitats.

Some of the important genera are *Chenopodium* (100 to 150) (lamb's quarter, goosefoot); *Beta* (6) (beet, Swiss chard); *Atriplex* (200) (saltbush); *Salicornia* (35) (glasswort); and *Salsola* (150) (glasswort, Russian thistle).

The common weed lamb's quarter (*Chenopodium album*) has plagued farmers in many agricultural regions but is avidly sought as a pot herb. Russian thistle (*Salsola iberica*) is a noxious prickly weed which has become established across North America. The stubby glassworts (*Salicornia*), which grow in salt marshes, lack leaves. Their stems will noisily break, hence the name *glasswort*.

**Economic Importance**   The family is important as a source of beet sugar, the garden vegetable beet, and Swiss chard, all from *Beta vulgaris*. Spinach is *Spinacia oleracea*. *Chenopodium ambrosioides* is the source of wormseed or Mexican tea, used as a vermifuge, and *C. quinoa* is a food plant in South America. Its seeds, boiled like rice, were a staple food in the Inca Empire.

## SUBCLASS III, CARYOPHYLLIDAE; ORDER 1, CARYOPHYLLALES

### Amaranthaceae: Pigweed Family

**Field Recognition**   Herbs, rarely woody; inflorescence dense or congested; flowers minute, green, subtended by dry, scarious, papery bracts; stamens united by their filaments; $CA^{3-5} CO^0 A^5 \underline{G^{(2-3)}}$.

**Description**   *Habit*   Annual or perennial herbs, rarely undershrubs or climbers.

*Leaves*   Alternate or opposite, simple, usually entire or nearly so; without stipules.

*Inflorescences* Spikes, heads, or racemes; dense and congested; plants dioecious or polygamodioecious; often spiny because of the firm bristle tips on the inflorescence bracts.

*Flowers* Bisexual or unisexual, actinomorphic, with scarious bracts. Calyx of 3 to 5 sepals, free or united, more or less dry and membranous. Corolla lacking. Androecium usually of 5 stamens, opposite the sepals, filaments united at the base into a short tube. Gynoecium a compound pistil of 2 to 3 carpels, with 1 locule, ovules solitary or rarely several, ovary superior, styles 1 to 3.

*Fruit* A circumscissile capsule or a utricle or nutlet, dehiscing by a lid or indehiscent.

*Seed* Globose, embryo annular, surrounding the copious endosperm.

**Size, Distribution, and General Information** A family of about 65 genera and 850 species, mostly of tropical America and Africa but extending into the temperate zones.

Some important genera are *Celosia* (60) (celosia, cockscomb); *Amaranthus* (60) (pigweed, careless weed); *Iresine* (80) (gizzard plant, blood leaf); *Alternanthera* (200) (alligator weed); and *Gomphrena* (100) (globe amaranth).

Best known of the tumble weeds of western North America is *Amaranthus graecizans*. It grows in disturbed soil or waste places. It drys into a ball, breaking loose from the soil and spreading seeds as the wind rolls it around.

**Economic Importance** The family is of little economic importance except for several widespread weeds in *Amaranthus, Iresine,* and *Acnida,* as well as ornamentals from *Amaranthus, Celosia, Gomphrena,* and *Iresine.*

### SUBCLASS HI, CARYOPHYLLIDAE; ORDER 1, CARYOPHYLLALES

### Portulaceae: Purslane Family

**Field Recognition** Annual or perennial herbs, leaves usually fleshy; sepals 2; placentation basal; pistil with 2 to 5 styles and a unilocular ovary; $CA^2 CO^{4-6} A^{8-10} \underline{G^{(2-3)}}$.

**Description** *Habit* Annual or perennial herbs or undershrubs.

*Leaves* Alternate or opposite, often fleshy, simple; usually with scarious or setose stipules.

*Inflorescences* Flowers solitary, or cymose or racemose.

*Flowers* Bisexual, actinomorphic, showy. Calyx of 2 sepals, free or united at base. Corolla of 4 to 6 petals, imbricate, free or united at base, soon falling. Androecium of 4 to many, free stamens. Gynoecium a compound pistil of 2 to 3 carpels, with 1 locule, placentation basal, ovules 1 to many, ovary superior or half inferior, styles 2 to 5.

*Fruit*    A capsule dehiscing by valves or circumscissile; rarely a nut and indehiscent.

*Seed*    Embryo curved around the perisperm.

**Systematics**    Recent anatomical evidence suggests that the 2 sepals are bracts and the "petals" are actually the sepals.

**Size, Distribution, and General Information**    A family of about 19 genera and 580 species, cosmopolitan in distribution but occurring particularly in western North America and southern South America.

Some important genera are *Portulaca* (200) (purslane, moss rose); *Talinum* (50); *Claytonia* (35) (spring beauty); and *Calandrinia* (150) (red maids).

The flowers of many Portulaceae remain closed in wet weather and open only when the sun is shining. This can be observed in the showy *Portulaca grandiflora* from Argentina. *Claytonia* (spring beauty) signals the departure of winter and the arrival of spring in parts of North America.

**Economic Importance**    The family is of little economic importance except for several ornamentals, notably *Portulaca grandiflora* and several species of *Calandrinia, Lewisia,* and *Talinum.* The weedy *P. oleracea* is sometimes used as a pot herb and in salads.

## SUBCLASS III, CARYOPHYLLIDAE; ORDER 1, CARYOPHYLLALES

### Caryophyllaceae: Pink Family

**Field Recognition**    Herbs with swollen nodes; leaves opposite, connected at the base with a transverse line; pistil with free central placentation; fruit a many-seeded capsule opening by teeth or valves; $CA^5$ $CO^{5(0)}$ $A^{5-10}$ $\underline{G}$.

**Description**    *Habit*    Annual or perennial herbs with swollen nodes.

*Leaves*    Opposite, simple, connected at the base by a transverse line, entire; stipules absent or scarious if present.

*Inflorescences*    Cymose or with solitary flowers.

*Flowers*    Bisexual, actinomorphic. Calyx of 5 sepals, free or united into a tube. Corolla of 5 petals, sometimes small or none, often notched at tip. Androecium of 5 or 10 stamens. Gynoecium a compound pistil of 2 to 5 united carpels, with 1 locule, ovules numerous, placentation free central, ovary superior, styles 2 to 5. See Figure 14-2.

*Fruit*    A dry capsule usually opening by valves or teeth.

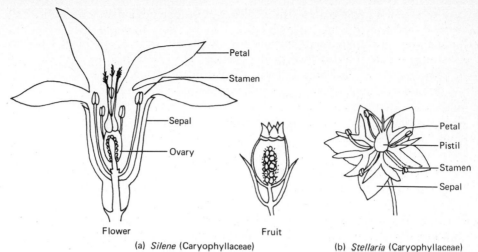

(a) *Silene* (Caryophyllaceae)          (b) *Stellaria* (Caryophyllaceae)

**Figure 14-2** Drawings of floral structures of *(a) Silene* and *(b) Stellaria* (Caryophyllaceae). Longitudinal section of *Silene* flower and fruit; note free-central placentation in fruit. Flower of *Stellaria;* note deeply lobed (pinked) petals.

*Seed*   Embryo usually curved around the perisperm.

**Systematics**   The family is classified into two subfamilies: Alsinoideae, with distinct sepals; and Caryophylloideae, with united sepals.

**Size, Distribution, and General Information**   A family of about 70 genera and 1750 species, mainly of north temperate distribution but cosmopolitan.

Some of the genera are *Stellaria* (120) (chickweed); *Cerastium* (60) (mouse-ear chickweed); *Arenaria* (250) (sandwort); *Silene* (500) (catchfly, campion); *Lychnis* (12) (campion); *Gypsophila* (125) (baby's breath); *Dianthus* (300) (carnation, pink, sweet william); and *Paronychia* (50) (whitlow wort).

**Economic Importance**   The family includes several ornamentals: pinks, sweet william, carnation (*Dianthus*), and baby's breath (*Gypsophila*). Several troublesome weeds occur in the family, including chickweed (*Stellaria media*) and mouse-ear chickweed (*Cerastium*).

## SUBCLASS III, CARYOPHYLLIDAE; ORDER 2, POLYGONALES

### Polygonaceae: Buckwheat, Smartweed, or Knotweed Family

**Field Recognition**   Mostly herbs with swollen nodes; leaves with nodal ocrea, or in their absence, flowers in involucrate heads; calyx petaloid often in two series; $CA^{3+3} CO^0 A^{3+3} \underline{G}^{\text{③}}$ or $CA^5 CO^0 A^{5-8} \underline{G}^{\text{③}}$.

**Description**  *Habit*  Herbs, shrubs, vines, or rarely trees, usually with swollen nodes.

*Leaves*  Alternate, simple, usually with an ocrea (modified stipule) sheathing the stem at the base of the petiole.

*Inflorescences*  Racemose, paniculate, spicate, or capitate.

*Flowers*  Usually bisexual, small, actinomorphic. Calyx of 3 to 6 sepals, imbricate, often enlarged and becoming membranous in fruit. Corolla lacking. Androecium of 3 to 8 stamens, free or united at the base. Gynoecium a compound pistil of 3 united carpels, with 1 locule, ovule solitary, placentation basal, ovary superior, stigmas 2 to 4. See Figure 14-3.

*Fruit*  A 2-sided or triangular nut.

*Seed*  With a curved or straight embryo surrounded by perisperm.

**Size, Distribution, and General Information**  A family of about 40 genera and 800 species chiefly of north temperate distribution, with a few located in the tropics, the arctic, and the Southern Hemisphere.

Some of the larger genera are *Rumex* (200) (dock); *Rheum* (50) (rhubarb); *Polygonum* (300) (smartweed, knotweed); *Fagopyrum* (15) (buckwheat); *Coccoloba* (150) (sea grape); and *Eriogonum* (200) (false buckwheat).

Buckwheat produces grain-like seed and was once grown extensively in the northeastern United States. The bees visiting the flowers of buckwheat produce a dark, uniquely flavored honey. The petioles of rhubarb are used in salads, stewed as a dessert, or used to fill pies. The leaf blades of rhubarb contain oxalic acid and are poisonous, often causing loss of livestock. The sea grape (*Coccoloba uvifera*) grows on sandy beaches in subtropical Florida, the West Indies, and South and Central America. It is often cultivated as an ornamental, and the fruits are used to make jelly.

**Economic Importance**  The family is of little economic importance except for buckwheat (*Fagopyrum esculentum*) and rhubarb (*Rheum rhaponticum*).

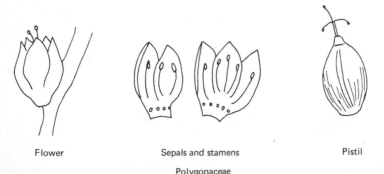

Flower                    Sepals and stamens                    Pistil

Polygonaceae

**Figure 14-3**  Floral structures of family Polygonaceae: flower, sepals with stamens, and pistil.

Some species of *Polygonum* are weeds. The Eurasian *P. orientale,* which is known as *kiss-me-over-the-garden-gate,* is one of the few ornamentals from this family.

## SUBCLASS IV, DILLENIIDAE; ORDER 2, THEALES

### Guttiferae or Clusiaceae: Hypericum or Garcinia Family

**Field Recognition** Herbs, shrubs, or trees; leaves opposite, often glandular dotted; flowers usually yellow; stamens numerous; $CA^{4-5} CO^{4-5} A^{\infty} \underline{G} \underline{(3\,or\,5)}$.

**Description** *Habit* Trees, shrubs, herbs.
*Leaves* Opposite, simple, entire, often glandular-dotted; without stipules.
*Inflorescences* Cymose or subumbellate.
*Flowers* Bisexual or unisexual, actinomorphic. Calyx of 4 to 5 free sepals. Corolla of 4 to 5 free petals. Androecium of numerous stamens, fascicled. Gynoecium a compound pistil of 3 to 5 united carpels, with (1) 3 to 5 locules, ovules numerous, placentation axile, ovary superior.
*Fruit* A capsule or berry, rarely a drupe.

**Systematics** Classified into two subfamilies: Clusioideae, with mostly unisexual flowers; and Hypericoideae, with mostly bisexual flowers.

**Size, Distribution, and General Information** A family of about 40 genera and 1000 species distributed mainly in temperate and tropical regions.
Some of the genera are *Hypericum* (400) (St. John's wort); *Garcinia* (400) (mangosteen); and *Clusia* (145) (autograph tree).

**Economic Importance** The family is of no significant importance, although the fruits of many species of *Garcinia* are edible, especially that of *G. mangostana* (mangosteen). Several species of *Hypericum* are grown as ornamentals. In the western United States, *Hypericum perforatum* (Klamath weed) once caused poisoning of livestock. But Klamath weed was brought under biological control by the introduction of two species of beetles which feed upon it.

## SUBCLASS IV, DILLENIIDAE; ORDER 3, MALVALES

### Malvaceae: Mallow Family

**Field Recognition** Herbs and shrubs; with stellate pubescence; leaves palmately veined; stamens united by their filaments; pistil of many carpels: $CA^{3-5}$ $CO^5 A^\infty G\underline{\textcircled{\infty}}$.

**Description** *Habit* Herbs, shrubs, or rarely small trees.
*Leaves* Alternate, simple, palmately veined; with stipules.
*Inflorescences* Flowers solitary or cymose.
*Flowers* Bisexual, actinomorphic. Calyx of 3 to 5 sepals, more or less united, often subtended by an involucre of bracteoles. Corolla of 5 petals, free, often fused at base to staminal column. Androecium of numerous stamens, filaments united into a tube or column. Gynoecium a compound pistil of 1 to many (often 5) carpels, with 2 to 5 or more locules, ovules numerous, placentation axile, ovary superior, style often branched above. See Figure 14-4.
*Fruit* A capsule, schizocarp or rarely a berry.
*Seed* Embryo curved, usually without endosperm.

**Size, Distribution, and General Information** A family of about 75 genera and 1000 species of tropical and temperate distribution.

Some of the genera are *Althaea* (12) (hollyhock, marshmallow); *Malva* (40) (mallow); *Sida* (200); *Abutilon* (100) (velvet leaf, flowering maple, Indian mallow); *Hibiscus* (300) (okra, rose mallow, flower-of-an-hour); *Gossypium* (20) (cotton).

*Hibiscus rosa-sinensis* (rose of China or hibiscus) has hundreds of cultivars and is enjoyed for its brilliant large flowers. The shrubby rose of Sharon (*H. syriacus*) is hardier than rose of China and is used in colder climates as an ornamental. The roots of *Althaea officinalis* were once an ingredient in marsh-

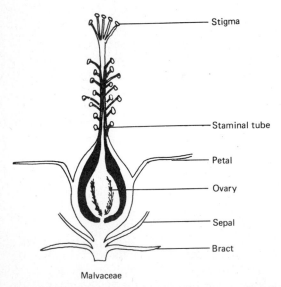

Malvaceae

**Figure 14-4** Diagrammatic longitudinal view of a flower of the family Malvaceae.

mallow confections. *Althaea rosea* is the perennial garden favorite, hollyhock. Cotton, the symbol of the Old South, is grown in favorable climates in many different countries and in numerous varieties. Its cultivated history extends back as far as 3000 B.C. in India, Egypt, and South America.

**Economic Importance** The family is of considerable importance for cotton fibers, cottonseed meal, and cottonseed oil. The cultivated forms of cotton arose from *Gossypium barbadense* and *G. hirsutum,* both of the New World, and *G. arboreum* and *G. herbaceum* from the Old World. People in the southeastern United States enjoy okra (*Hibiscus esculentus*), a vegetable largely unappreciated by Yankees. Several species of *Sida* are weeds. Many Malvaceae are valued as ornamentals. *Abutilon avicennae,* cultivated in China, is the source of a fiber called *China jute.*

## SUBCLASS IV, DILLENIIDAE; ORDER 6, VIOLALES

### Violaceae: Violet family

**Field Recognition** Herbs (ours); flowers zygomorphic, 5-merous; petals 5, the anterior ones spurred; stamens 5, frequently 1 is spurred at the base; $CA^5$ $COZ^5 A^5 \underline{G} ③$.

**Description** *Habit* Annual or usually perennial herbs, or shrubs.
*Leaves* Alternate or rarely opposite, simple; stipules often leafy or small.
*Inflorescences* Flowers solitary, to paniculate; some species have cleistogamous flowers.
*Flowers* Bisexual, zygomorphic or actinomorphic. Calyx of 5 sepals, persistent. Corolla of 5, mostly unequal petals, anterior petal larger and spurred. Androecium of 5 stamens, abaxial stamen often spurred at base. Gynoecium a compound pistil of 3 united carpels, with 1 locule, ovules numerous, placentation parietal, ovary superior, style simple.
*Fruit* A capsule.
*Seed* With endosperm.

**Size, Distribution, and General Information** A family of about 22 genera and 900 species of tropical and temperate distribution.

The two larger genera are *Viola* (500) (violet) and *Hybanthus* (150) (green violet).

During the summer months, several species of violets produce an inconspicuous set of flowers without petals. In some species, these cleistogamous flowers are hidden underground. They never open, but pollinate themselves. In many places in eastern North America, annual violets, called *Johnny-jump-ups,* herald the arrival of spring with a sea of blue flowers.

**Economic Importance**   The family is of little economic importance except for the garden favorites: violets, violas, and pansies.


## SUBCLASS IV, DILLENIIDAE; ORDER 6, VIOLALES

### Cucurbitaceae: Cucurbit or Gourd Family

**Field Recognition**   Coarse, tendril-bearing vines; flowers usually yellow, unisexual; ovary inferior; fruit a berry or pepo; (CA⑤CO⑤A⁵ G⁰) (CA⑤CO⑤ A⁰ $\overline{G}$③).

**Description**   *Habit*   Climbing with spirally coiled tendrils, or prostrate annual herbs.
*Leaves*   Alternate, entire or lobed.
*Inflorescences*   Of various types, often axillary, plants monoecious or dioecious.
*Flowers*   Unisexual, rarely bisexual, actinomorphic. Calyx of 5 sepals, tubular, fused to ovary wall. Corolla of 5 petals. Androecium of 1 to 5, stamens (often 5 stamens, with 2 groups of 2 fused together, giving 2 compound stamens, plus 1 simple stamen, the total appearing as 3 stamens). Gynoecium a compound pistil of 1 to 10 (usually 3) fused carpels, ovules numerous, placentation parietal or free central, ovary inferior, style simple or 3-parted.
*Fruit*   A fleshy, berry-like pepo.
*Seed*   Often flattened.

**Size, Distribution, and General Information**   A family of about 110 genera and 640 species mainly of tropical and subtropical distribution.
Some genera are *Citrullus* (3) (watermelon, citron melon); *Lagenaria* (6) (bottle gourd); *Luffa* (6) (vegetable sponge, loofah); *Sicyos* (15) (nimble kate, bur cucumber); *Cucurbita* (15) (pumpkin, squash, gourd, vegetable marrow); *Cayaponia* (45); *Melothria* (10) (creeping cucumbers); *Cucumis* (25) (cucumber, canteloupe, muskmelon, gherkin); and *Echinocystis* (15) (wild cucumber).
The merits of the various Cucurbitaceae were early recognized by primitive agriculturists. Muskmelons were mentioned in Egyptian records from 2400 B.C. Watermelons have been cultivated for so long in Africa that they are unknown in the wild. Hard-shelled gourds serve as scoops, ladles, and buoyant floats for fish nets. In North America, gourds are sometimes used as birdhouses. American Indians cultivated pumpkins and squashes for food. The fibers from the fruit of the vegetable sponge or loofah are used for fuel filters, air filters, and packing material.

**Economic Importance**   The Cucurbitaceae is important as a source of food: pumpkin, squash, cucumber, gherkin, muskmelon, and watermelon. A few species are grown as ornamentals. Some are weedy.

## SUBCLASS IV, DILLENIIDAE; ORDER 7, SALICALES

### Salicaceae: Willow Family

**Field Recognition** Deciduous trees and shrubs; plants dioecious; catkin-bearing; flowers seemingly subtended by bracts with a cup-like disc or a gland; seeds with tufts of hairs; $(CA^0 \ CO^0 \ A^{2-X} \ G^0) \ (CA^0 \ CO^0 \ A^0 \ \underline{G}^{②})$.

**Description** *Habit* Trees or shrubs.

*Leaves* Deciduous, simple, alternate; stipules small or sometimes folia-ceous and persistent.

*Inflorescences* Plants dioecious; flowers dense in erect or pendulous catkins.

*Flowers* Unisexual, naked. Staminate flowers with 1 to 2 nectariferous glands and 2 to 30 stamens. Pistillate flowers with 1 to 2 nectariferous glands. Gynoecium a compound pistil of 2 carpels, with 1 locule, ovules numerous, placentation parietal or basal, ovary superior, styles 2 to 4.

*Fruit* A capsule with numerous hairy seeds (the hair is restricted to a coma and not scattered over the seed).

**Size, Distribution, and General Information** A family of 3 genera and about 530 species, chiefly of north temperate climates but of wide distribution.

The larger genera are *Salix* (500) and *Populus* (35).

Stems of pussy willow (*Salix discolor*) are used in floral arrangements. They are gathered in late winter and brought inside where the yellow, fluffy, staminate catkins soon appear. Weeping willow is *S. babylonica*. Quaking aspen (*Populus tremuloides*) produces a striking appearance when its leaves flutter in the breeze. *Populus deltoides* (cottonwood) of eastern and central North America and *P. trichocarpa* (black cottonwood) of western North America are large native trees.

**Economic Importance** The Salicaceae is important for several species grown as ornamentals and for use in basket making or for pulpwood. It is inter-esting to note that salycylic acid, the root compound for aspirin, is named for *Salix*. In ancient times, the soft inner bark of *Salix* was widely esteemed as a headache cure. Its chemical derivative (aspirin) still is. Convention has it that willow wood was the traditional material from which polo balls were made.

## SUBCLASS IV, DILLENIIDAE; ORDER 8, CAPPARALES

### Cruciferae or Brassicaceae: Mustard Family

**Field Recognition** Herbs with an odorous, watery juice; flowers of 4 sepals, 4 petals, and 6 stamens; fruit a silique or silicle; $CA^4 \ CO^4 \ A^{4+2} \ \underline{G}^{②}$.

**Description**   *Habit*   Annual, biennial, or perennial herbs with a pungent, watery juice.

*Leaves*   Alternate, simple, often dissected, pubescence of simple to stellate hairs; stipules absent.

*Inflorescences*   Usually racemose, rarely bracteate.

*Flowers*   Bisexual, actinomorphic. Calyx of 4 sepals. Corolla of 4 petals (rarely none) arranged in a cross—hence cruciform; petals often long-clawed. Androecium of 6 stamens, with 4 long and 2 short. Gynoecium a compound pistil of 2 carpels, with 2 locules, ovules many, placentation parietal, ovary superior, stigmas 2. See Figure 14-5.

*Fruit*   A silique (long fruit) or silicle (short fruit), variable, important in classification.

**Size, Distribution, and General Information**   A cosmopolitan family of about 375 genera and 3200 species, occurring mainly in the north temperate zone, particularly in the Mediterranean region.

Some of the genera are *Brassica* (50) (Brussel sprouts, broccoli, mustard, cabbage, cauliflower, and so on); *Barbarea* (20) (winter cress); *Arabis* (120) (rock cress); *Cardamine* (160) (bitter cress); *Alyssum* (150) (madwort); *Draba* (300) (whitlow grass); *Sisymbrium* (90) (tumble mustard); *Capsella* (5) (shep-

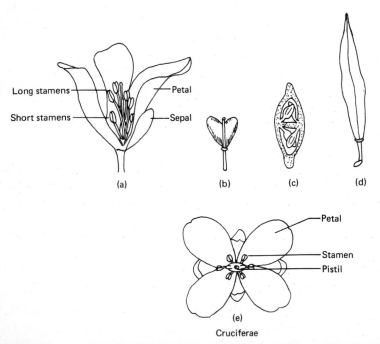

**Figure 14-5**   Flower and fruits of Cruciferae family. *(a)* Longitudinal section of flower; note the two short and the four long stamens. *(b)* Silicle. *(c)* Cross section of a silicle. *(d)* Silique. *(e)* Flower (top view).

herd's purse); *Lepidium* (150) (peppergrass); *Cakile* (15) (sea rocket); *Raphanus* (8) (radish); *Dentaria* (20) (toothwort); and *Iberis* (30) (candytuft).

**Economic Importance**   The family is of considerable economic importance for food crops, ornamentals, and weeds. *Brassica nigra* is black mustard, and *B. oleracea* gave rise to cabbage, cauliflower, broccoli, Brussels sprouts, kohlrabi, and kale. Turnip is *B. campestris,* and *B. napus* is rape. Horseradish is *Cochlearia armoracia,* and radishes are *Raphanus sativus.* Cruciferae produce an abundance of vitamin C. The sulfur compounds of the family help to give them their pungent odor. Black mustard seeds are ground to make the condiment called *mustard* to which horseradish may be added to make a hotter blend.

A blue dyestuff called *woad* was once obtained from *Isatis tinctoria,* whose dried leaves were fermented—a malodorous process requiring several weeks. Woad was replaced by the blue dye indigo. There is a semantic theory that the word *weed* is a permutation of the Anglo-Saxon root word *woad.*

Among the ornamentals are honesty or moneywort (*Lunaria*), stocks (*Matthiola*), candytuft (*Iberis*), and sweet alyssum (*Lobularia*). Troublesome weeds include the weedy mustards (*Brassica, Barbarea,* and *Sisymbrium*), shepherd's purse (*Capsella bursa-pastoris*), and peppergrass (*Lepidium virginicum*).

## SUBCLASS IV, DILLENIIDAE; ORDER 10, ERICALES

### Ericaceae-Pyrolaceae-Monotropaceae

Many manuals traditionally treat this group of three families as one family, the Ericaceae, but most phylogenists now separate the group into three families as treated here.

### Ericaceae: Heath Family

**Field Recognition**   Woody, often shrubby; leaves alternate, evergreen or deciduous; flowers urceolate or campanulate; stamens distinct, often twice as many as the petals, anthers opening by terminal pores; $CA^{(4-5)} CO^{(4-5)} A^{8-10} \underline{G}^{(4-5)}.$

**Description**   *Habit*   Shrubs or rarely small trees.
*Leaves*   Mostly alternate, simple, often evergreen; stipules absent.
*Inflorescences*   Solitary or racemose.
*Flowers*   Bisexual, actinomorphic or slightly zygomorphic. Calyx of 4 to 5 united sepals. Corolla of 4 to 5 united petals, often urceolate or campanulate. Androecium of stamens mostly double the number of corolla lobes, free or

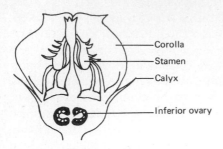

Ericaceae (*Vaccinium*)

**Figure 14-6** Longitudinal section of a *Vaccinium* flower, family Ericaceae.

fused at base; anthers opening by apical pores. Gynoecium a compound pistil of (2) 4 to 5 (10) united carpels, with 4 to 5 locules, ovules numerous, placentation axile, ovary superior or inferior. See Figure 14-6.

**Fruit**   A capsule, berry, or drupe.

**Seed**   Embryo cylindrical, in endosperm.

**Systematics**   The family is sometimes classified into four subfamilies: (1) Rhododendroideae (*Rhododendron*); (2) Ericoideae (*Erica*); (3) Vaccinioideae (*Vaccinium*); and (4) Epigaeoideae (*Epigaea*).

**Size, Distribution, and General Information**   A family of about 50 genera and 1350 species of cosmopolitan distribution except in deserts; confined to high elevations in the tropics. Often grows in acid soils on peat, moors, swamps, and woodlands; associated with endotrophic mycorrhiza.

Some of the chief genera are *Erica* (500) (heath); *Kalmia* (8) (mountain laurel, laurel, ivy); *Rhododendron* (500 to 600) (azalea); *Arctostaphylos* (70) (manzanita); *Vaccinium* (300 to 400) (blueberry, huckleberry, cranberry); *Gaylussacia* (49) (huckleberry); and *Bejaria* (30) (Andes rose).

Horticulturalists have produced many handsome cultivars by crossing the various species of *Rhododendron*. Some rhododendrons are deciduous, others are evergreen.

The anthers of *Kalmia* are held in pockets of the corolla, and the filaments are bent like bows when the flower is open. An insect visiting the flower releases the anthers, which strike the insect, loading it with pollen. When grazing is in short supply, livestock are sometimes forced to eat the leaves of *Kalmia*, which may cause poisoning.

Sourwood (*Oxydendrum arboreum*) is a small tree which is a favorite of honeybees in the southern Appalachians. The Scottish Highlands have a conspicuous covering of heather, *Calluna vulgaris*.

*Menziesia* and *Gaultheria* are strikingly abundant along the Pacific Coast. The former is called *Madrono* or *Madrone*. *Gaultheria* is known as *sallal*, except for a few miles on the Oregon Coast where it is known as *shallal;* both are Indian words.

*Epigaea repens* is the trailing arbutus, a well-loved wild flower.

**Economic Importance** Economically, the family is important for its ornamentals and for blueberries and cranberries. The commercial cranberry is *Vaccinium macrocarpon*, and blueberries are obtained from several species and cultivars. Briar pipes are made from the burls of *Erica scoparia* of southern France. Ornamental shrubs are obtained from *Kalmia, Rhododendron, Erica, Pieris, Leucothoe,* and others.

## Pyrolaceae: Pyrola Family

**Field Recognition** Herbaceous; leaves evergreen; flowers 4- or 5-merous; petals separate; stamens 10, anthers opening by terminal pores; $CA^{4-5}$ $CO^{4-5}$ $A^{8-10}$ $\underline{G}^{\textcircled{4-5}}$.

**Description** *Habit* Perennial herbs.
*Leaves* Usually alternate, sometimes somewhat whorled, evergreen, entire or toothed; stipules absent.
*Inflorescences* Flowers in racemes, umbels, or sometimes corymbs, or scapose and solitary.
*Flowers* Bisexual, actinomorphic, nodding. Calyx of 4 to 5 sepals, persistent. Corolla of 4 or 5 petals, free or united only at base. Androecium of 10 (rarely 8) stamens, free, anthers opening by terminal pores. Gynoecium a compound pistil of 4 or 5 carpels, with 4 to 5 locules, ovules numerous, placentation axile, ovary superior, stigma 5-lobed.
*Fruit* A more or less globose capsule.

**Size, Distribution, and General Information** A family of about 3 genera and 30 species of north temperate and arctic distribution extending in the New World into Mexico and the West Indies.

The genera are *Chimaphila* (8) (wintergreen, pipsissewa); *Moneses* (1); and *Pyrola* (20) (shinleaf).

These are handsome little plants of the forest floor. In the summer, they display attractive white flowers.

**Economic Importance** The family has no economic importance.

## Monotropaceae: Indian Pipe Family

**Field Recognition** White, yellow, brown, or red fleshy herbs; lacking chlorophyll; leaves scaly; anthers opening by longitudinal slits; $CA^{2-6}$ $CO^{\textcircled{3-6}}$ $A^{6-12}$ $\underline{G}^{\textcircled{4-6}}$.

**Description** *Habit* Nongreen saprophytic or parasitic herbs.

*Leaves*   Reduced to scales, alternate, upper ones becoming bract-like.

*Inflorescences*   Flowers solitary to capitate.

*Flowers*   Bisexual, actinomorphic. Calyx of 2 to 6 sepals. Corolla of 3 to 6 petals, free or united into a lobed corolla. Androecium of 6 to 12 stamens, free or united at base, anthers opening lengthwise by longitudinal slits. Gynoecium a compound pistil of 4 to 6 carpels, with 1 to 6 locules, ovules numerous, placentation parietal, ovary superior, stigma capitate.

*Fruit*   A loculicidal capsule opening by valves.

**Size, Distribution, and General Information**   A family of about 12 genera and 21 species of north temperate and tropical mountain distribution.

Some of the genera are *Monotropa* (5) (Indian pipe); *Sarcodes* (1) (snow plant); *Pterospora* (1) (pinedrops); and *Allotropa* (1) (sugar-stick).

These nongreen plants obtain their food through mycorrhizae from decaying organic matter or from the roots of living trees. Snow plant (*Sarcodes sanguinea*) thrusts up its blood-red spike through humus in coniferous woodlands of the middle and upper elevations of the Sierra Nevada. Indian pipe (*Monotropa uniflora*) grows in eastern forests and is waxy white or tinged with pink.

**Economic Importance**   The family has no economic importance.

### SUBCLASS IV, DILLENIIDAE; ORDER 13, PRIMULALES

#### Primulaceae: Primrose Family

**Field Recognition**   Herbs; leaves opposite, whorled, or basal; flowers 5-merous; petals united; stamens opposite the petals; seeds numerous; CA⑤ CO⑤A⁵ G⑤.

**Description**   *Habit*   Perennial herbs, rarely annuals or subshrubs.

*Leaves*   Opposite, whorled, or alternate; sometimes all basal; simple, sometimes lobed.

*Inflorescences*   Terminal, axillary, scapose; variable from solitary to paniculate or umbellate.

*Flowers*   Bisexual, actinomorphic, often heterostyled. Calyx of 5 united sepals, persistent. Corolla of 5 united petals, tubular, sometimes split nearly to base, lobes usually 5. Androecium of 5 stamens, opposite the petals (i.e., each stamen is directly in front of a petal lobe). Gynoecium a compound pistil of 5 carpels, with 1 locule, ovules numerous, placentation free central, ovary usually superior, style simple. See Figure 14-7.

*Fruit*   A capsule, variously dehiscent.

*Seed*   Embryo small, in fleshy or hard endosperm.

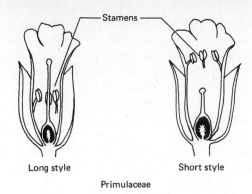

Long style

Short style

Primulaceae

**Figure 14-7** Flowers of *Primula* (Primulaceae); note the heterostyly, or the two style lengths.

**Size, Distribution, and General Information** A family of about 20 genera and 1000 species of chiefly north temperate distribution.

Some of the genera are *Primula* (500) (primrose); *Dodecatheon* (52) (shooting star); *Cyclamen* (15); and *Lysimachia* (200) (loosestrife).

Cyclamens are native plants of the Mediterranean region, the foothills of the Alps, and Asia Minor. Some are grown for the florist trade; others are garden favorites. After pollination, the flower stalk coils in a spiral, pulling the ripe fruit close to the soil.

The largest genus is *Primula,* found largely in Europe and Asia. The flowers show heterostyly—that is, the flowers have two style lengths and two stamen positions. The long style forms have stamens deep in the tube, and the short style form has stamens high on the corolla. Many primroses are grown either in gardens or for the florist trade.

Several species of the yellow-flowered *Lysimachia* are attractive and easily grown perennials for garden culture.

**Economic Importance** The family is important for several ornamentals.


## SUBCLASS V, ROSIDAE; ORDER 1, ROSALES

### Hydrangeaceae: Hydrangea Family

**Field Recognition** Shrubs; leaves opposite or alternate, without stipules; calyx tube more or less fused to ovary; stamens numerous and in several series; ovary wholly or partially inferior; $CA \overline{\textbf{4--10}} CO^{4-10} A^{4-\infty} \overline{G\textbf{2--5}}$.

**Description** *Habit* Herbs, soft-wooded undershrubs, or rarely climbers.
*Leaves* Opposite or alternate, simple; without stipules.
*Inflorescences* Cymose or corymbose.
*Flowers* Bisexual, or outer flowers sterile with large petal-like sepals.

Calyx 4- to 10-lobed or toothed, calyx tube more or less fused to the ovary. Corolla of 4 to 10 petals. Androecium of 4 to numerous stamens (in several series). Gynoecium a compound pistil of 2 to 5 united carpels, with 3 to 6 locules (or incompletely distinct), ovules numerous, placentation axile or parietal, ovary half-inferior to inferior, styles as many as the locules.

*Fruit*   A loculicidal capsule.

*Seed*   Small, sometimes winged and reticulate.

**Size, Distribution, and General Information**   A family of about 17 genera and 250 species mostly in the Northern Hemisphere (eastern Asia and North America but with a few in western South America).

Some of genera are *Hydrangea* (80); *Decumaria* (2) (climbing hydrangea); *Philadelphus* (75) (mock orange); and *Deutzia* (50).

**Economic Importance**   The family is the source of many ornamental shrubs, including *Hydrangea paniculatata,* the florist's hydrangea.

## SUBCLASS V, ROSIDAE; ORDER 1, ROSALES

### Grossulariaceae: Gooseberry Family

**Field Recognition**   Shrubs, often spiny; leaves lobed; flowers 5-merous; calyx petaloid; ovary inferior with 2 parietal placentas; fruit a pulpy berry; $CA^{\circleddash} CO^5 A^5 \overline{G}^{②}$.

**Description**   *Habit*   Shrubs, often spiny.

*Leaves*   Alternate or clustered, simple; stipules absent or fused to petiole.

*Inflorescences*   Racemose or subsolitary.

*Flowers*   Bisexual or unisexual by abortion, actinomorphic. Calyx of 4 to 5 connate sepals, fused to ovary. Corolla of 5 petals mostly small or scale-like. Androecium of 5 stamens, alternate with the petals. Gynoecium a compound pistil of 2 united carpels, with 1 locule, ovules free to numerous, placentation by 2 parietal placentas, ovary inferior, styles 2, free or fused.

*Fruit*   A pulpy berry crowned by persistent calyx.

*Seeds*   Numerous, with endosperm.

**Size, Distribution, and General Information**   A family of about 2 genera and 150 species of the temperate Northern Hemisphere and the Andes of South America.

The genera are *Ribes* (150) (gooseberries, currants) and *Grossularia* (50), which is sometimes included in *Ribes*.

Most edible currants and gooseberries originated in Eurasia and have been intentionally cultivated since the 1600s. Currants are nonspiny bushes whose fruits are used in jellies. Gooseberries are spiny shrubs, and the fruits are used to make pies. Certain species of *Ribes* are the alternate hosts of the white pine blister rust.

**Economic Importance**   The family is important for gooseberries, currants, and some ornamental shrubs.

## SUBCLASS V, ROSIDAE; ORDER 1, ROSALES

### Crassulaceae: Stonecrop Family

**Field Recognition**   Succulent herbs, sometimes undershrubs; leaves without stipules; flowers 4- to 5-merous; carpels the same number as the petals, scale-like gland at the base of each ovary; $CA^{4 \ 5} \ CO^{4-5} \ A^{8-10} \ \underline{G}^{4-5}$.

**Description**   *Habit*   Usually succulent herbs or undershrubs, annuals or perennials.
*Leaves*   Opposite or alternate; without stipules.
*Inflorescences*   Usually cymose.
*Flowers*   Bisexual, actinomorphic. Calyx of usually 4 to 5 sepals, free or united into a tube. Corolla of 4 to 5 petals, the same number as the sepals, free or variously united. Androecium of 4 to 30 stamens, as many or twice as many as the petals, or if fewer, alternate with the petals, slightly perigynous. Gynoecium of simple pistils each with 1 carpel, pistils free or united at base, each with 1 locule, ovules many or rarely few, placentation parietal, ovary superior, style short or elongate, each pistil subtended by a scale-like gland. See Figure 14-8.
*Fruit*   A membranous or leathery follicle.
*Seed*   Small; endosperm none or minute.

**Size, Distribution, and General Information**   A family of about 35 genera

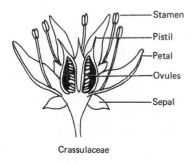

Crassulaceae

Stamen
Pistil
Petal
Ovules
Sepal

**Figure 14-8**  Diagrammatic view of a flower of the family Crassulaceae; note the longitudinal section of two of the carpels showing the ovules.

and 1500 species of cosmopolitan distribution with the largest number in southern Africa. Most are xerophytic perennials.

Some of the genera are *Crassula* (300) (stonecrop, jade plant); *Sedum* (600) (orpine, stonecrop); *Sempervivum* (25) (live-forever, hen and chickens); *Cotyledon* (41); and *Kalanchoë* (200) (bryophyllum, life plant).

Members of this family are familiar rock garden plants. Most have succulent leaves and a waxy cuticle which enable the plants to withstand droughts. *Sedum telephium* was once grown in Europe as a source of a home remedy for skin wounds. *Sedum tectorum* (houseleek or thunderwort) was planted on cottage roofs to help hold the slates in place. It was also planted on thatched roofs in the belief it magically warded off lightning. Several of the bryophyllums produce plantlets from adventitious buds along the margins of their leaves. Other species of *Kalanchoë* are used in the florist trade.

**Economic Importance**   The Crassulaceae is most important as a source of many rock garden plants and houseplants which are often called *succulents*.

## SUBCLASS V, ROSIDAE; ORDER 1, ROSALES

### Saxifragaceae: Saxifrage Family

**Field Recognition**   Perennial herbs, leaves alternate, without stipules; flowers 5-merous, perigynous; $CA^5 \ CO^{5(0)} \ A^{5 \ or \ 10} \ \underline{G^{2-3}}$.

**Description**   *Habit*   Perennial herbs.
*Leaves*   Alternate, without stipules.
*Inflorescences*   Racemose or cymose, rarely solitary.
*Flowers*   Bisexual, actinomorphic, perigynous or rarely epigynous. Calyx of 5 sepals. Corolla of 5 petals, alternate with the sepals or absent, often clawed. Androecium of 5 or 10 stamens, inserted with the petals, filaments free. Gynoecium a compound pistil of 2 carpels, with 1 to 3 locules, ovules numerous, placentation axile, ovary superior, free or united to the tubular receptacle, styles free.
*Fruit*   A capsule.
*Seed*   Small embryo surrounded by endosperm.
**Size, Distribution, and General Information**   A family of about 30 genera and 580 species distributed mainly in cold and temperate regions.

Some of the larger genera are *Saxifraga* (370) (saxifrage); *Heuchera* (50) (alumroot); *Astilbe* (25); *Mitella* (15) (miterwort, bishop's-cap); *Tiarella* (5) (foamflower); and *Chrysosplenium* (55) (golden saxifrage).

The name *saxifrage,* "breaker of stones," is derived from the old doctrine of signatures of the Middle Ages. The clues provided by the habit and morphology of plants suggested their medicinal uses to medieval herbalists. The pale bulbets formed in the leaf axile of *Saxifraga cernua* resembled small pebbles and hence were used to treat kidney stones.

**Economic Importance**   The Saxifragaceae is of little importance except for rock garden or perennial ornamentals.

## SUBCLASS V, ROSIDAE; ORDER 1, ROSALES

### Rosaceae: Rose Family

**Field Recognition**   Herbs, shrubs, and trees; leaves with stipules; flowers actinomorphic; sepals 5; petals 5; stamens numerous; hypanthium often present; $CA^{\circledS} CO^{\circledS} A^{\infty} \underline{G}^1$ or $\underline{G}^{\infty}$ or $\overline{G}^{\circledS}$.

**Description**   *Habit*   Trees, shrubs, and herbs.

*Leaves*   Alternate, simple or compound, sometimes with glandular teeth; with paired stipules which are sometimes adnate to the petiole.

*Inflorescences*   Variable, solitary flowers to racemose and cymose clusters.

*Flowers*   Bisexual, actinomorphic, often perigynous to some degree. Calyx of 5 sepals, united at base. Corolla of 5 petals, attached to the rim of the hypanthium or rarely absent. Androecium of numerous stamens, sometimes only 5 or 10. Gynoecium a simple pistil of 1 to numerous separate carpels or of 2 to 5 carpels united into a compound pistil, often adnate to the calyx tube, ovules usually 2 or more per carpel, ovary superior or inferior. See Figures 14-9 and 14-10.

*Fruit*   An achene, follicle, pome, or drupe, sometimes on a fleshy enlarged receptacle.

*Seed*   Usually without endosperm.

**Systematics**   This family is usually divided into four subfamilies: (1) Spiraeoideae (fruit a follicle or capsule, *Spiraea*); (2) Pyroideae (Pomoideae) (fruit a fleshy pome, ovary inferior, *Pyrus, Malus, Sorbus*); (3) Rosoideae (gynoecium of 10 or more pistils, fruit often dry, *Rubus, Fragaria, Potentilla, Geum, Rosa*); and (4) Prunoideae (gynoecium of a single pistil, rarely 2 to 5; fruit a drupe, *Prunus*).

**Size, Distribution, and General Information**   A family of about 100 genera and 2000 species. Distribution cosmopolitan, but mainly north temperate.

Some important genera are *Spiraea* (100); *Rosa* (250) (rose); *Rubus* (250) (blackberry, dewberry, raspberry); *Potentilla* (500) (five-finger, cinquefoil);

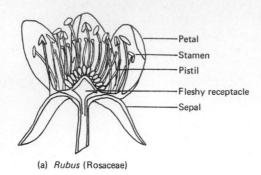

(a)  *Rubus* (Rosaceae)

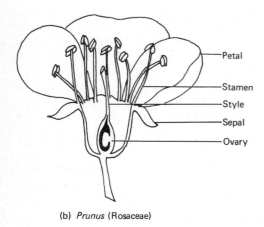

(b)  *Prunus* (Rosaceae)

**Figure 14-9** Longitudinal section of generalized flowers of *(a) Rubus* and *(b) Prunus* of family Rosaceae.

*Fragaria* (15) (strawberry); *Pyrus* (30) (pear); *Malus* (35) (apple); *Sorbus* (100) (mountain ash, chokeberry); *Amelanchier* (25) (serviceberry, shadbush); *Pyracantha* (10) (fire thorn); *Crataegus* (200) (hawthorn); *Prunus* (430) (plum, peach, cherry, apricot, almond); *Cotoneaster* (50); *Cydonia* (1) (quince); and *Chaenomeles* (3) (flowering quince).

The Rosaceae shows great diversity in types of fruit obtained from the same general floral plan. In *Spiraea,* the separate pistils each mature into a follicle with a single seed. In *Prunus* (cherry or plum), the single pistil develops into a *drupe,* a fleshy fruit with a single stony pit containing a single seed. In *Malus* (apple) and *Pyrus* (pear), the calyx and most of the ovary wall becomes fleshy and develops into a pome. In *Fragaria* (strawberry), the receptacle develops into the familiar red, juicy, edible structure with tiny achenes embedded on the surface. *Rubus* (blackberry or raspberry) has a fleshy elongated receptacle upon which druplets are attached. In *Rosa,* the separate pistils are enclosed in a cup formed largely from the calyx. This cup, called a *hip*, enlarges and turns bright red at maturity.

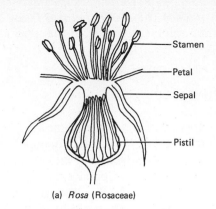

(a) *Rosa* (Rosaceae)

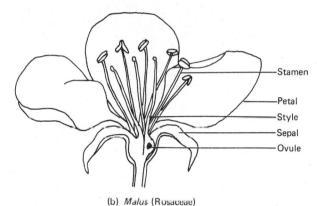

(b) *Malus* (Rosaceae)

**Figure 14-10**    Longitudinal section of generalized flowers of *(a) Rosa* and *(b) Malus* of family Rosaceae.

**Economic Importance**    The Rosaceae are important in temperate regions for fruits and ornamentals. They perhaps rank third in commercial importance in the temperate zone among the families of flowering plants. The ornamentals include *Spiraea prunifolia* (bridal wreath), *S. japonica* (pink spiraea), *Kerria japonica, Pyracantha coccinea* (fire thorn), *Crataegus* (hawthorn), *Chaenomeles lagenaria* (flowering quince), *Rosa damascena* (damask rose), *R. multiflora* (multiflora rose), *R. fragrans* (tea rose), *Prunus caroliniana* (Carolina laurel cherry), *P. triloba* (flowering almond), *P. persica* (flowering peach), *P. serrulata* (Japanese flowering cherry), *Malus* (flowering crabs), *Cotoneaster* and *Sorbus* (which both have several ornamental species), and many more.

Fruit-producing members include *Cydonia oblonga* (quince), *Pyrus communis* (pear), *Malus* (apple), *Rubus* (blackberry and raspberry), *Fragaria* (strawberry), *Prunus spinosa* (sloe plum, which is the source of the flavoring in sloe gin), *P. amygdalus* (almond), *P. persica* (peach), *P. cerasus* (sour cherry), *P. avium* (sweet cherry), *P. domestica* (plum or prune), *Eriobotrya japonica* (loquat), and several others.

## SUBCLASS V, ROSIDAE; ORDER 2, FABALES

The legumes have been traditionally treated as one large, somewhat heterogeneous family, the Leguminosae. Hutchinson (1973) and Cronquist (personal communication) agree that the group is perhaps best split into three separate families of the order Fabales: Mimosaceae, Caesalpiniaceae, and Fabaceae (Papilionaceae). This is the treatment used in this book, although many manuals follow the more traditional approach. In the latter case, these groups are often recognized as subfamilies.

### Description of Order Fabales

Trees, shrubs, and herbs. *Leaves* simple to bipinnate, stipules present or absent. *Flowers* actinomorphic to zygomorphic. *Petals* free or some partially united. *Stamens* numerous to few, various. *Pistil* of a solitary, superior carpel. *Fruit* often but not always a legume. The order has 590 genera and 13,200 species and is cosmopolitan in distribution.

### Key to Families of Order Fabales

1   Flowers actinomorphic                                 Mimosaceae
1   Flowers more or less to distinctly zygomorphic.
    2   The upper petal (standard) positioned within the
        adjacent lateral (wing) petals                 Caesalpiniaceae
    2   The upper petal (standard) positioned outside the
        lateral (wing) petals                        Fabaceae

### Mimosaceae: Mimosa Family

**Field Recognition**   Trees, shrubs, or herbs; leaves often bipinnately compound; flowers actinomorphic, 5-merous; stamens 10 to many, usually extended beyond corolla, filaments often brightly colored; $CA^5 \ CO^5 \ A^{5-\infty} \ \underline{G}^1$.

**Description**   *Habit*   Trees, shrubs, or herbs.
*Leaves*   Mostly bipinnate, rarely pinnately compound.
*Inflorescences*   Flowers often in tight clusters, or capitate, spicate, or racemose.
*Flowers*   Bisexual, actinomorphic. Calyx of 5 sepals, valvate, fused into a 5-lobed tube. Corolla of 5 petals, valvate, free or fused into a tube. Androecium of 5 to numerous stamens. Gynoecium a simple pistil of 1 carpel, with 1 locule, ovary superior. See Figure 14-11.
*Fruit*   A legume, sometimes indehiscent.

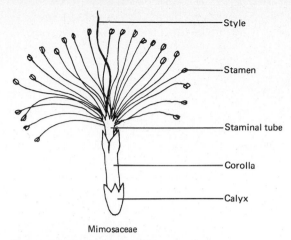

Mimosaceae

**Figure 14-11** Flower of Mimosaceae.

**Size, Distribution, and General Information** A family of about 40 genera and 2000 species. Mostly of tropical or subtropical distribution, with many in dry regions.

Some of the genera are *Prosopis* (40) (mesquite); *Mimosa* (450 to 500) (sensitive plant); *Acacia* (750 to 800); *Albizia* (100 to 150) (mimosa, silk tree); *Schrankia* (30) (sensitive brier); *Desmanthus* (40) (prickle weed); and *Neptunia* (11).

In Mimosaceae, which is represented mostly by woody tropical and subtropical plants, the flowers are small but grouped together into dense clusters. The stamens extend beyond the short petals and give a fluffy soft appearance to the flower cluster. Although the leaves are typically bipinnate, some species have seemingly simple leaves, which are actually phyllodes or leaf-like petioles having no blade.

The mesquite (*Prosopis*) is an important small weedy tree in the arid southwestern United States. The leaves of *Mimosa pudica* (sensitive plant) close with the slightest touch. Acacias are generally native to warm semiarid regions and are often grown as ornamentals or for cut flowers. *Albizia julibrissin* (mimosa tree) is widely cultivated and has become naturalized in the southeastern United States.

**Economic Importance** The family is of little direct economic importance except for some timber species, ornamentals, and gum arabic or gum acacia from *Acacia senegal* or *A. arabica*.

### Caesalpiniaceae: Caesalpinia Family

**Field Recognition** Trees, shrubs, or rarely herbs; flowers more or less zygomorphic, with the large upper petal (standard) inside the two lateral petals (wing); petals 5, stamens 10; $CA^5 \ COZ^5 \ A^{10(\infty)} \ \underline{G}^1$.

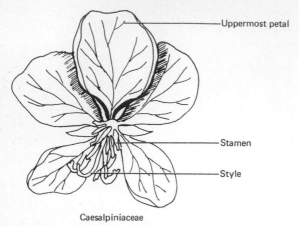

Caesalpiniaceae

**Figure 14-12**   Flower of Caesalpinaceae; note that the uppermost petal is located within the adjacent two lateral petals.

**Description**   *Habit*   Trees, shrubs, or rarely herbs.

*Leaves*   Pinnate or bipinnately compound, rarely simple or 1-foliolate; stipules mostly absent.

*Inflorescences*   Racemes, spikes, or rarely cymose; showy.

*Flowers*   Bisexual, zygomorphic, rarely subactinomorphic. Calyx of 5 sepals, the upper 2 sometimes fused. Corolla of 5 petals or fewer or absent, the upper petal inside the two lateral petals. Androecium of 10 (rarely numerous) stamens, free or fused. Gynoecium a simple pistil of 1 carpel, with 1 locule, ovary superior. See Figure 14-12.

*Fruit*   A legume, sometimes indehiscent; often winged.

**Size, Distribution, and General Information**   A family of about 150 genera and 2200 species mostly of tropical or subtropical distribution.

Some of the genera are *Gleditsia* (11) (honey locust); *Cassia* (500–600) (senna); *Cercis* (7) (redbud, Judas tree); *Gymnocladus* (3) (Kentucky coffee tree); and *Parkinsonia* (2) (palo verde).

**Economic Importance**   The family is important as a source of drugs, dyes, timber, and ornamentals. *Cercis canadensis* (redbud) is a favorite flowering tree in eastern North America. Judas is said to have hanged himself from a tree that we may call *Cercis siliquastrum* (Judas tree). Many species of *Cassia* are cultivated for the leaves which yield the drug senna, which is the base for a laxative. *Bauhina purpurea* (orchid tree) is used as an ornamental in the warmer parts of North America. The heartwood of *Haematoxylom campechianum* (logwood) yields the dye hematoxylin.

**Fabaceae: Bean or Pea Family**

**Field Recognition**   Trees, shrubs, or herbs; leaves pinnately or palmately compound or simple; flowers papilionaceous and distinctly irregular; corolla of 5 petals forming a standard, 2 wings, and a keel (the standard is located to the outside of the 2 wing petals); stamens 10; $CA^{\veebar}COZ^5 A^{10\,or\,9+1} \underline{G}^1$.

**Description** *Habit* Trees, shrubs, or herbs.

*Leaves* Compound or 1-foliate or rarely simple.

*Flowers* Bisexual, zygomorphic. Calyx of 5 sepals, more or less united in a tube. Corolla 5-merous, with a standard petal, 2 lateral wing petals, and 2 keel petals which are more or less fused by their lower margins into a keel. Androecium of 10 stamens, monadelphous or diadelphous. Gynoecium a simple pistil of 1 carpel, with 1 locule, ovary superior. See Figure 14-13.

*Fruit* Usually a legume, sometimes indehiscent; rarely a loment breaking into 1-seeded segments.

*Seed* Usually with food reserves in cotyledons.

**Size, Distribution, and General Information** A family of about 400 genera and 9000 species, the Fabaceae occurs all over the world but particularly in the warm temperate regions of both the Northern and Southern Hemispheres.

Some of the genera are *Crotalaria* (500) (rattlebox); *Lupinus* (200) (lupine); *Cytisus* (25 to 30) (Scotch broom); *Medicago* (100) (bur clover, alfalfa); *Melilotus* (25) (sweet clover); *Trifolium* (300) (clover); *Lotus* (100) (bird's-foot trefoil); *Psoralea* (130) (snakeroot, prairie turnip); *Amorpha* (20) (leadplant); *Indigofera* (700) (indigo); *Wisteria* (10); *Robinia* (20) (black locust); *Sesbania* (50); *Astragalus* (2000) (milk vetch); *Stylosanthes* (50) (pencil flower); *Arachis* (15) (peanut, groundnut); *Desmodium* (450) (beggar-lice); *Lespedeza* (100); *Vicia* (150) (vetch, broad bean); *Lathyrus* (130) (sweet pea); *Abrus* (12) (rosary pea); *Centrosema* (45) (blue pea); *Clitoria* (40) (butterfly pea); *Sophora* (50) (Japanese pagoda tree); *Baptisia* (35) (false indigo); *Glycine* (10) (soybean); *Cicer* (20) (chickpea); *Lens* (10) (lentil); *Pisum* (6)

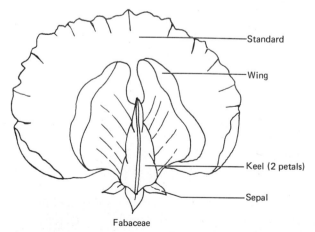

Standard

Wing

Keel (2 petals)

Sepal

Fabaceae

**Figure 14-13** Papilionaceous flower of the family Fabaceae; note that the standard is located outside the wing petals.

(English or garden pea); *Vigna* (80 to 100) (southern pea); *Erythrina* (100) (coral tree, coral bean); *Phaseolus* (200 to 240) (cultivated beans); and *Pueraria* (35) (kudzu).

**Economic Importance**   The family is of considerable importance as a source of high-protein food, oil, and forage as well as ornamentals and other uses. *Glycine max* (soybean or soya) is becoming one of the world's most important crops as a source of oil and high-protein meal. *Phaseolus vulgaris* (kidney bean), *P. lunatus* (lima beans), *P. coccineus* (scarlet runner beans), and many other cultivars of beans have been developed such as pole beans, bush beans, and half-runners. Snap or string beans are immature fruits in which the seeds have not yet ripened. *Vigna unguiculata* (cowpea) is enjoyed by people in the southern states. *Vicia faba* (the broad bean), *Pisum sativum* (English or garden pea), and *Lens esculenta* (lentil) were used by Eurasian peoples before the New World beans were discovered and introduced into the Old World. *Cicer arietinum* (chick-pea) is grown in southern Europe, India, and elsewhere.

Alfalfa (*Medicago sativa*) is one of the world's best forage and hay crops and the various kinds of clover, *Trifolium, Melilotus,* and *Vicia,* are likewise raised to feed cattle. These same plants are good nectar producers and are sought by honeybees. Peanuts (*Arachis*), also known as groundnuts or goobers, produce a valuable crop of edible seeds, oil, meal, and peanut butter. The yellow flower of the peanut develops aboveground, but growth of the flower stalk pushes the ripening ovary into the soil where the fruits mature underground.

Indigo dye is obtained by fermenting and oxidizing the juice from the leaves of the indigo plant (*Indigofera tinctoria*). The hard, red, black-tipped seeds of *Abrus precatorius* are strung into necklaces and rosaries. The seeds are highly poisonous, and the ingestion of one seed can be fatal. The pastel-colored sweet peas (*Lathyrus odoratus*) are ornamental garden favorites. The genus *Lupinus* has contributed the cultivated lupines of the cooler areas of North America, as well as the beloved Texas bluebonnet. *Wisteria* is grown as a vine or pruned into the shape of a tree. The black locust (*Robinia pseudoacacia*) produces a cloud of pendant clusters of white flowers in the spring. Black locust wood is prized for fence posts because it is relatively resistant to decay.

## SUBCLASS V, ROSIDAE; ORDER 5, MYRTALES

### Onagraceae: Evening Primrose Family

**Field Recognition**   Herbs; flowers 4-merous, rarely 2-merous; hypanthium present, and on its rim are inserted sepals, petals, and stamens; ovary inferior; CA ♦ CO ♦ A $\overset{\text{or }8}{4}$ G $\overline{④}$.

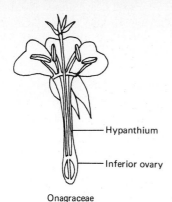

Hypanthium

Inferior ovary

Onagraceae

**Figure 14-14** Longitudinal section of a flower of the family Onagraceae.

**Description**  *Habit*  Herbs or rarely shrubs, sometimes aquatic.

*Leaves*  Alternate or opposite, simple; stipules mostly absent or deciduous.

*Inflorescences*  Spicate, racemose, or paniculate, or the flowers solitary.

*Flowers*  Bisexual, actinomorphic. Hypanthium present. Calyx of (2)4(5) sepals, persistent or deciduous, fused to the ovary. Corolla of (2)4(5) petals, sometimes none. Androecium of 4 stamens, or twice as many as calyx lobes. Gynoecium of 1 pistil with 4 carpels, with 2 to 6 locules, ovules 1 to many, placentation axile, ovary inferior, style simple, but stigma sometimes lobed or parted. See Figure 14-14.

*Fruit*  A capsule, nut, or berry.

*Seed*  With little or no endosperm.

**Size, Distribution, and General Information**  A family of about 21 genera and 640 species of temperate and subtropical distribution.

Chief genera are *Epilobium* (215) (fireweed, willowweed); *Ludwigia* (75) (seedbox, primrose willow); *Clarkia* (36); *Oenothera* (80) (evening primrose); *Fuchsia* (100); *Gaura* (18); *Lopezia* (17); and *Circaea* (12) (enchanter's nightshade).

Named in honor of the sixteenth century herbalist Fuchs, *Fuchsia* is a popular greenhouse or garden plant in temperate climates, and an almost endless array of cultivars have been developed. *Epilobium* quickly colonizes on burned areas in certain regions, giving it the name *fireweed*. The flowers of *Oenothera* are bright yellow or white, and sometimes turning pink or reddish. They are called *evening primroses* because their flowers characteristically open in the late afternoon and close again early the next morning.

**Economic Importance**  The family is of little importance. Species of several genera are cultivated as ornamentals.

## SUBCLASS V, ROSIDAE; ORDER 8, CORNALES

### Cornaceae: Dogwood Family

**Field Recognition**  Trees and shrubs; leaves opposite or alternate, veins curved; flowers 4- to 5-merous; stamens alternate with the petals; ovary inferior; fruit a drupe or berry; $CA^{4-5} CO^{4-5} A^{4-5} \overline{G}^{②}$.

**Description**  *Habit*  Trees and shrubs, or rarely perennial herbs.
*Leaves*  Opposite or alternate, simple, veins curved, without stipules.
*Inflorescences*  Cymose, corymbose, or capitate, sometimes subtended by showy, petal-like bracts; plants rarely dioecious.
*Flowers*  Bisexual or unisexual, actinomorphic, small. Calyx of 4 to 5 sepals, fused to ovary. Corolla of 4 to 5 petals, free, rarely absent. Androecium of 4 to 5 stamens, alternate with petals. Gynoecium a compound pistil of 2 carpels, with 1 to 4 locules, ovule solitary in each locule, placentation axile, ovary inferior, style simple.
*Fruit*  A fleshy drupe or berry.
*Seed*  With endosperm.

**Size, Distribution, and General Information**  A family of about 12 genera and 100 species of north and south temperate and tropical mountain distribution.
Some of the genera are *Cornus* (45) (dogwood) and *Aucuba* (4) (gold-dust plant).
Dogwoods should be called "dagwoods"; in medieval times they were a source of a hardwood used for making wooden daggers. The wood is as hard as horn, hence the generic name *Cornus*, which comes from the Latin *cornu*, "horn." Several of the dogwoods have showy white bracts subtending each inflorescence.

**Economic Importance**  The Cornaceae is important for ornamental shrubs or small trees. *Aucuba* is grown in the warm temperate states. *Cornus florida* is widely grown in eastern North America. Western flowering dogwood (*C. nuttallii*) occurs naturally from California to British Columbia and Idaho and is used as an ornamental. Fruits of the cornelian cherry (*C. mas*) are edible.

## SUBCLASS V, ROSIDAE; ORDER 12, EUPHORBIALES

### Euphorbiaceae: Spurge Family

**Field Recognition**  Herbs, shrubs, or trees; often with milky sap; leaves mostly alternate; flowers unisexual; ovary superior and usually trilocular, ovules often with a caruncle; "Euphorbia-type" flowers: $(CA^0 \, CO^0 \, A^1 \, G^0)$ $(CA^0 \, CO^0 \, A^0 \, \underline{G}^{③})$; "non-Euphorbia"-type flowers: $(CA^{0 \, or \, 5} \, CO^{0 \, or \, 5} \, A^{1-\infty} \, G^0)$ $(CA^{0 \, or \, 5} \, CO^{0 \, or \, 5} \, A^0 \, \underline{G}^{③})$.

**Description**  *Habit*  Herbs, shrubs, or trees; some xerophytic and cactus-like, often with a milky sap.

*Leaves*  Mostly alternate, simple or compound, often reduced or deciduous in xerophytic species; with stipules.

*Inflorescences*  Various, often condensed, hence giving the appearance of a single flower, a cyathium; plants monoecious or dioecious.

*Flowers*  Unisexual, actinomorphic. Calyx of 5 sepals or none. Corolla of 5 petals or usually none. Androecium of 1 to many stamens, free or united, rudimentary ovary often present in the male flowers. Gynoecium a compound pistil of 3 united carpels, with 3 locules, ovules solitary or paired, placentation axile, ovary superior, styles free or united at base. See Figure 14-15.

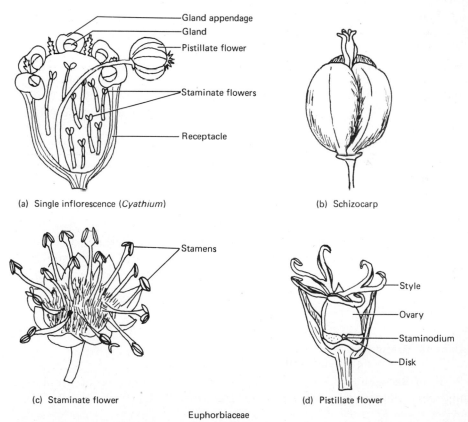

(a) Single inflorescence (*Cyathium*)

(b) Schizocarp

(c) Staminate flower

(d) Pistillate flower

Euphorbiaceae

**Figure 14-15**  Diagrammatic drawings of floral structures typical of family Euphorbiaceae. *(a)* "*Euphorbia*-type" inflorescence. *(b)* Three-lobed schizocarp. *(c)* "Non-*Euphorbia*"-type staminate flower. *(d)* "Non-*Euphorbia*"-type pistillate flower.

*Fruit*   A schizocarp or capsule.
*Seed*   Often with a conspicuous caruncle.

**Size, Distribution, and General Information**   A family of about 290 genera and 7500 species of wide distribution in tropical and temperate regions.

Some of the chief genera are *Euphorbia* (2000) (spurge, poinsettia); *Phyllanthus* (600); *Hevea* (12) (pará or Brazilian rubber tree); *Aleurites* (2) (tung oil tree); *Croton* (750) (croton oil); *Manihot* (170) (cassava, manioc, yuca, tapioca); *Acalypha* (450) (chenille plant, red hot cattail); *Ricinus* (1) (castor bean); *Hippomane* (5) (manchineel tree); *Pedilanthus* (14); *Tragia* (100) (nose-burn); *Sapium* (120) (Chinese tallow tree); and *Jatropha* (175).

Euphorbiaceae is one of the most diverse in habit, habitat, and morphology of the flowering plant families. They range from the tall tree *Hevea* of the Amazonian rain forest to small cactus-like succulents of Africa. The familiar Christmas poinsettia from Mexico was introduced into cultivation in 1828 by Mr. J. R. Poinsett, the United States minister to Mexico. The red or greenish-white leaves of poinsettia are bracts subtending the small flower clusters.

An extract from the seeds of castor bean is used as a purgative. The seeds also yield a fine machine oil. Castor beans contain a highly toxic substance called *ricin* that can cause death when small amounts are ingested. The manchineel tree (*Hippomane mancinella*) of the West Indies has a poisonous sap, and even the rainwater dripping from the tree is enough to cause dermatitis. The roots of *Manihot esculenta* (cassava) yield a starch widely used for human food in tropical countries. Many cultivars of cassava are rich in HCN (hydrocyanic acid), which must be neutralized during preparation of the roots before cooking. Croton oil, once used as a purgative, is obtained from the seeds of *Croton tiglium*.

**Economic Importance**   The family is of considerable importance for the following products: Brazilian rubber (*Hevea*), castor oil (*Ricinus communis*), cassava and tapioca (*Manihot*), florists' poinsettia (*Euphorbia pulcherrima*), tung oil (*Aleurites fordii*), and numerous ornamentals. Many are poisonous, causing sickness or death if ingested or dermatitis if the juice contacts the skin.

## SUBCLASS V, ROSIDAE; ORDER 13, RHAMNALES

### Vitaceae: Grape Family

**Field Recognition**   Woody, climbing by tendrils; inflorescences opposite the leaves; stamens opposite the petals; fruit a berry; CA ⊕⊖CO$^{4-5}$ A$^{4-5}$ G ②.

**Description**   *Habit*   Shrubs climbing by tendrils, or small trees, often with watery juice.

*Leaves* Alternate, simple or compound; stipules petiolar or absent.

*Inflorescences* Racemose, paniculate, cymose, borne opposite the leaves.

*Flowers* Bisexual or unisexual, actinomorphic. Calyx of 4 to 5 small sepals, united. Corolla of 4 to 5 petals, free or united at the tips. Androecium of 4 to 5 stamens, opposite the petals and arising from a disc. Gynoecium a compound pistil of 2 united carpels, with 2 to 6 locules, each with 1 or 2 ovules, ovary superior, style short, capitate.

*Fruit* Berry-like, often with watery juice.

*Seed* With a straight embryo and endosperm.

**Size, Distribution, and General Information** A family of about 12 genera and 700 species of tropic and warm temperate distribution.

Chief genera are *Vitis* (60 to 70) (grape); *Cissus* (350); *Ampelocissus* (95); *Parthenocissus* (15) (Virginia creeper, Boston ivy); and *Ampelopsis* (2) (pepper vine).

The wine grape (*Vitis vinifera*) is probably native to the area around the Caspian Sea. Hieroglyphics from ancient Egypt show details of wine production. Thousands of cultivars have been developed by hybridization and selection. These cultivars are used for various purposes: wines, table grapes, or raisins.

**Economic Importance** Vitaceae is important for the wine grapes, table grapes, grape juice, raisins, and for several ornamentals.

## SUBCLASS V, ROSIDAE; ORDER 14, LINALES

### Linaceae: Flax Family

**Field Recognition** Flowers 5-merous; petals distinct and often clawed; stamens with filaments united at base; fruit a capsule; $CA^5 CO^5 A^{5 \text{ or } 10} \underline{G(\overline{2-5})}$.

**Description** *Habit* Herbs and shrubs, sometimes arborescent.

*Leaves* Alternate, simple, entire; stipules present or absent, sometimes gland-like.

*Inflorescences* Cymose.

*Flowers* Bisexual, actinomorphic. Calyx of (4)5 sepals, free or partially united. Corolla of (4)5 petals, free, often clawed. Androecium of 5, 10, or more stamens; if 5, alternate with the petals or sometimes alternating with small staminodes, filaments united at base. Gynoecium a compound pistil of 2 to 5 carpels, with 2 locules, ovules 2 in each locule, placentation axile, ovary superior, styles 3 to 5, free or united.

*Fruit* A capsule or rarely a drupe.

*Seed* With fleshy endosperm and a straight embryo.

**Size, Distribution, and General Information** A family of about 12 genera and 290 species.

The major genus is *Linum* (230) (flax).

Flax is the stem fiber of *Linum usitatissimum* obtained by retting the softer tissues in water. Linseed oil, used in paints and varnishes, is obtained from the seeds of flax. Oil cake, which is the cake remaining after the oil is extracted, is used as cattle feed. Flax cultivation extends back at least 5000 years; linen cloth has been found in Egyptian tombs. By the 1700s, the use of flax fibers to make cloth for bed sheets became common. We continue to call them linens.

**Economic Importance** The family is of little importance except for flax, linseed oil, and oil cake.

## SUBCLASS V, ROSIDAE; ORDER 15, POLYGALALES

### Polygalaceae: Milkwort Family

**Field Recognition** Perianth greatly modified, often appearing papilionaceous; sepals often petaloid; stamens unusual, united in a split sheath beyond the middle; anthers opening by an apical pore; ovary biloculate; seeds often pilose; $CA^5 \ COZ^{3-5} \ A^{3-8} \ \underline{G}^{②}$.

**Description** *Habit* Herbs, shrubs, climbers, or rarely small trees.

*Leaves* Alternate, rarely opposite, simple; without stipules.

*Inflorescences* Solitary, spicate, or racemose.

*Flowers* Bisexual, zygomorphic. Calyx of 5 sepals, free, the 2 inner often petaloid, wing-like. Corolla of 3 to 5 petals, outer 2 free or united with the lower most, upper 2 free, or minute or scale-like. Androecium of 3 to 8 stamens, united beyond the middle, with a split sheath, sometimes fused to the petals, anthers opening by an apical pore. Gynoecium a compound pistil of 2(5) carpels, with 2(3 to 5) locules, ovules solitary in each locule, placentation axile, ovary superior, style simple.

*Fruit* A capsule or drupe-like seed.

*Seed* Often pilose.

**Size, Distribution, and General Information** A family of about 12 genera and 800 species of temperate and tropic distribution.

The chief genera are *Polygala* (500 to 600) (milkwort, candy flower) and *Monnina* (150).

**Economic Importance** The family is of little economic importance. *Polygala senega* (snakeroot) is medicinal. Several species of *Polygala* are sometimes used as ornamentals.

## SUBCLASS V, ROSIDAE; ORDER 16, SAPINDALES

### Sapindaceae: Soapberry Family

**Field Recognition** Trees or shrubs; leaves pinnately compound; flowers small; petals glandular; stamens with hairy filaments; $CA^5 CO^5 A^{10} \underline{G}^{\circledR}$.

**Description** *Habit* Trees, shrubs, or in some cases tendril-bearing vines.

*Leaves* Alternate or very rarely opposite, simple or pinnately compound; stipules usually absent.

*Inflorescences* Racemose, paniculate, or cymose; the plants commonly polygamodioecious.

*Flowers* Bisexual or unisexual, actinomorphic or zygomorphic. Calyx of 5 sepals, free or united. Corolla of 5 petals, sometimes 3 or absent, equal or unequal. Androecium of (4)10($\infty$) stamens, inserted within a prominent receptacular disc, filaments free and often hairy. Gynoecium a compound pistil of 3 united carpels, with 1 to 4 (often 3) locules, ovules 1 to 2 or many in each locule, placentation parietal, ovary superior, style terminal.

*Fruit* A capsule, nut, berry, drupe, schizocarp, or samara.

*Seed* Often arillate, with a curved embryo and no endosperm.

**Size, Distribution, and General Information** A family of about 150 genera and 2000 species of mainly tropical and subtropical distribution.

Some of the genera are *Sapindus* (13) (soapberry); *Litchi* (2); *Koelreuteria* (8) (golden rain tree); *Blighia* (7) (akee, vegetable marrow); *Cardiospermum* (12) (balloon vine); and *Paullinia* (180) (guarana).

The family takes its name from the soapberry tree (*Sapindus saponaria*), which contains saponin. When saponin is moistened, it forms a lather and may be used as a soap. *Paullinia cupana* (guarana) is cultivated in Brazil. Its seeds are used like cacao to make a drink high in caffeine content. The akee tree (*Blighia sapida*), originally from tropical West Africa and grown in the West Indies, produces handsome orange or red fruits, each about 8 centimeters long. The aril is cooked and eaten, but it must be gathered at the proper time or it is poisonous. *Litchi chinensis* (lychee or litchi nut) is cultivated for its edible fruit, which is a one-seeded nut. It is a favorite of the Cantonese.

**Economic Importance** The family is of some importance as the source of guarana, lychee nuts, and akee, plus some ornamentals.

## SUBCLASS V, ROSIDAE; ORDER 16, SAPINDALES

### Hippocastanaceae: Horse Chestnut or Buckeye Family

**Field Recognition**  Trees and shrubs; leaves opposite, palmately compound; flowers zygomorphic; $CA^{\circleddash}\,COZ^{4-5}\,A^{5-8}\,\underline{G}^{\circledthree}$.

**Description**  *Habit*  Trees and shrubs, with large winter buds.
*Leaves*  Opposite, palmately compound, leaflets serrate or entire, pinnately veined; stipules absent.
*Inflorescences*  In terminal panicles or racemes.
*Flowers*  Bisexual, zygomorphic. Calyx of 5 sepals, usually united at base. Corolla of 4 to 5 petals, free, unequal. Androecium of 5 to 8 stamens, free. Gynoecium a compound pistil of 3 united carpels, with 3 locules or by abortion 1 or 2 locules, ovules 2 in each locule, placentation axile, ovary superior, style elongated with a simple stigma.
*Fruit*  A leathery capsule with 1 or more large seeds, 3-lobed or 1- or 2-lobed by abortion.
*Seed*  Large.

**Size, Distribution, and General Information**  A family of 2 genera and 15 species of north temperate and South American distribution.
The two genera are *Aesculus* (13) (buckeye, horse chestnut) and *Billia* (2).
The horse chestnut (*Aesculus hippocastanum*) is widely planted in Europe and the cooler parts of North America. It is probably the most familiar species in the family. The Ohio buckeye is *A. glabra*.

**Economic Importance**  The family is of importance for several ornamental trees and shrubs. The seeds and apparently the foliage are poisonous.

## SUBCLASS V, ROSIDAE; ORDER 16, SAPINDALES

### Aceraceae: Maple Family

**Field Recognition**  Trees or shrubs; leaves opposite, usually simple with palmate venation; flowers actinomorphic; fruit a samara; $CA^{4-5}\ CO^{4-5}\ A^{4\ or\ 8\ or\ 10}\ \underline{G}^{\circledtwo}$.

**Description**  *Habit*  Trees and shrubs.
*Leaves*  Opposite, simple, usually palmately lobed and veined, or sometimes pinnately compound.
*Inflorescences*  Fasciculate corymbose, racemose, or paniculate; the plants monoecious, polygamous, or dioecious.
*Flowers*  Bisexual or unisexual, actinomorphic. Calyx of 4 to 5 sepals. Co-

rolla of 4 to 5 petals or none. Androecium of 4, 8, or 10 stamens, free, with a prominent receptacular disc; rudimentary ovary often present in male flowers. Gynoecium a compound pistil of 2 united carpels, with 2 locules, ovules 2 in each locule, placentation axile, ovary superior, styles 2.

*Fruit* A 2-winged samara, separating when ripe.

**Size, Distribution, and General Information** A family of 2 genera and 202 species of north temperate and tropical mountain distribution.

The two genera are *Acer* (200) (maple, box elder) and *Dipteronia* (2).

Few trees of eastern North America contribute so much to the autumn coloration as do the various species of maple. The maple wood or "hard maple" used in furniture is either that of sugar maple (*A. saccharum*) or black maple (*A. nigrum*). These two species also yield a watery sap which is boiled to produce maple syrup.

**Economic Importance** The family is important for maple wood used in furniture and charcoal, for maple syrup, and for many ornamental trees.

## SUBCLASS V, ROSIDAE; ORDER 16, SAPINDALES

### Anacardiaceae: Cashew Family

**Field Recognition** Trees or shrubs; flowers 5-merous; stamens inserted beneath a disc surrounding the ovary; ovary unilocular; CA⁵ CO⁵ A¹⁰ G③.

**Description** *Habit* Trees and shrubs, often with a resinous bark and a milky sap.

*Leaves* Alternate, simple or compound; without stipules.

*Inflorescences* Paniculate.

*Flowers* Bisexual or unisexual, actinomorphic. Calyx of 5 sepals, variously divided. Corolla of (3)5(7) petals or absent, free or rarely united. Androecium usually of (5)10 stamens, arising from a disc, filaments free. Gynoecium a compound pistil of (1)3(5) united carpels, with 1(2 to 5) locules, ovule solitary, placentation axile, ovary superior, styles 1 to 3.

*Fruit* Usually a drupe.

*Seed* Embryo curved, endosperm absent.

**Systematics** The family is divided into five tribes: (1) Anacardieae (*Mangifera, Anacardium*); (2) Spondiadeae (*Spondias*); (3) Rooeae (*Rhus*); (4) Semecarpeae (*Semecarpus*); and (5) Dobineeae (*Dobinea*).

**Size, Distribution, and General Information** A family of about 60 genera and 600 species, mostly of tropical distribution but found in temperate areas of the Mediterranean, eastern Asia, and America.

The chief genera are *Rhus* (including *Toxicodendron*) (250) (lacquer tree, sumac, poison ivy); *Pistacia* (10) (Chian turpentine, pistachio nuts, mastic); *Mangifera* (40) (mango); *Anacardium* (15) (cashew nut); *Schinus* (30) (pepper tree, American mastic); and *Cotinus* (3) (smoke tree).

The lacquer tree (*Rhus vernicifera*) of China and Japan has been milked for its poisonous latex for centuries. The viscous sap is applied by brush to various wood carvings; it turns black with oxidation, forming the common lacquer ware of the Orient. Poison ivy, poison oak, and poison sumac are well known in North America. Dermatitis in humans can be caused by touching the plants, having contact with smoke particles from burning parts of the plants, or by having contact with animals that have brushed against the plants. The leather-tanning industry uses Sicilian sumac (*Rhus coriaria*) and the red quebracho (*Schinopsis lorentzii*) of Argentina and Paraguay as sources of tannic acids.

*Mangifera indica* (mango) is a delightful tropical fruit. The cashew tree (*Anacardium occidentale*) of northeastern South America and the southern West Indies bears a strange kidney-shaped fruit on a flower stalk that enlarges to form a fig-like or pear-like structure. The fruit and seed coat of the cashew contain a poison; however, the seeds are edible. *Pistacia vera* of the Mediterranean is the source of pistachio nuts. The first turpentine used by artists came from the terebinth tree (*P. terebinthus*).

**Economic Importance** The family is important for the cashew, pistachio, mango, resins, oils, lacquers, tannic acids, and several ornamentals. Also, it is important as a cause of contact dermatitis in humans.

## SUBCLASS V, ROSIDAE; ORDER 16, SAPINDALES

### Rutaceae: Citrus or Rue Family

**Field Recognition** Shrubs and trees with aromatic oil glands; leaves glandular punctate; outer stamens usually opposite the petals; ovary deeply lobed; $CA^{4-5} CO^{4-5} A^{8-10} \underline{G}^{④-⑤}$.

**Description** *Habit* Shrubs, trees, and very rarely herbs; often xerophytic and aromatic.

*Leaves* Alternate or opposite, simple or compound, glandular dotted; without stipules.

*Inflorescences* Cymose.

*Flowers* Bisexual or rarely unisexual, actinomorphic or rarely zygomorphic. Calyx of 4 to 5 sepals, free or united. Corolla of 4 to 5 petals, sometimes none, mostly free. Androecium of 8 to 10 stamens (rarely many), free or rarely united, attached to a disc. Gynoecium a compound pistil of 4 to 5 united carpels, with 4 to 5 locules, ovules 1 to 2 in each locule, placentation axile, ovary superior.

*Fruit*   A capsule, hesperidium, drupe, or samara, rarely a schizocarp.

*Seed*   Without endosperm.

**Size, Distribution, and General Information**   A family of about 150 genera and 900 species of temperate and tropical distribution, numerous in southern Africa and Australia.

Some of the chief genera are *Citrus* (12); *Zanthoxylum* (20–30) (prickly ash, toothache tree); *Ruta* (7) (rue); *Ptelea* (3) (hop wafer tree); *Murraya* (12) (curry bush, orange jessamine); and *Fortunella* (6) (kumquat).

*Citrus* produces some of the world's most important fruits, the tasty and juicy citrus fruits which are so high in vitamin C: *Citrus medica* (citron), *C. limon* (lemon), *C. aurantiifolia* (lime), *C. sinensis* (orange), *C. paradisi* (grapefruit), and *C. reticulata* (tangerine).

In India the leaves of *Murraya koenigii* are mixed with a little turmeric to make curry powder. Bark of *Cusparia febrifuga* from the rain forests of South America is the main flavoring in Angostura bitters used in beverages. Members of Rutaceae often have high concentration of alkaloids. Several potential anticancer drugs have recently been found in members of the family.

**Economic Importance**   The family is of considerable importance as a source of citrus fruits, curry, bitters, and several ornamentals.

## SUBCLASS V, ROSIDAE; ORDER 16, SAPINDALES

### Zygophyllaceae: Caltrop Family

**Field Recognition**   Leaves usually pinnate or 2-foliolate with paired persistent stipules; flowers with a disc; stamens with basal scales; ovary 4 to 5 locular with a terminal style; $CA^{4-5}$ $CO^{4-5}$ $A^{5 \text{ or } 10 \text{ or } 15}$ $\underline{G}^{\circledS}$.

**Description**   *Habit*   Perennial herbs, shrubs, rarely trees; branches often jointed at the nodes.

*Leaves*   Opposite or rarely alternate, mostly pinnately compound or 2-foliolate; with paired persistent stipules, which are often spiny.

*Inflorescences*   Cymose or flowers solitary.

*Flowers*   Bisexual, actinomorphic or rarely zygomorphic. Calyx of 4 to 5 sepals, free or united at the base. Corolla of (0)4 to 5 petals, free. Androecium of 5, 10, or 15 stamens, free, with a scale at base of each filament. Gynoecium a compound pistil of (2)5(12) carpels, with 4 to 5 locules, ovules 2 or more in each locule, placentation axile, ovary superior, style simple.

*Fruit*   A capsule or rarely a berry.

*Seed*   With or without endosperm.

**Size, Distribution, and General Information**   A family of about 25 genera and 240 species of mainly tropical and subtropical, often arid, regions.

Some of the larger genera are *Zygophyllum* (100) (bean capers); *Guaiacum* (6) (lignum vitae wood); *Larrea* (2) (creosote bush); and *Tribulus* (20) (caltrops, puncture vine).

The Zygophyllaceae is called the *caltrop* family because the spines on the fruit of *Tribulus terrestris* are similar to the sharp iron caltrops once used on battlefields to stab the feet of men or horses.

The lignum vitae tree (*Guaiacum officinale*) produces the hardest and heaviest of commercial woods and yields the medicinal resin guaiacum, once used to treat syphilis. The wood is used for bowling balls, gears, mallets, furniture, rollers, and bearings. The creosote bush (*Larrea*) is familiar in the southwestern United States. Some species of *Zygophyllum* are used as spices: *Z. fabago* buds are used in sauces, and *Z. coccineum* is a substitute for black pepper.

**Economic Importance**   The family is important for lignum vitae wood, spices, and a few ornamentals.


### SUBCLASS V, ROSIDAE; ORDER 17, GERANIALES

#### Oxalidaceae: Oxalis or Sheep Sorrel Family

**Field Recognition**   Leaves palmately compound, with a sour taste; flowers 5-merous; stamens united; styles 5; $CA^5 CO^5 A^{10} G^{\circledS}$.

**Description**   *Habit*   Perennial herbs or shrubs, often with fleshy rhizomes or tubers.

*Leaves*   Alternate, palmately or pinnately compound, without stipules.

*Inflorescences*   Solitary or subumbellate, rarely cymose, or racemose.

*Flowers*   Bisexual, actinomorphic. Calyx of 5 sepals. Corolla of 5 petals, short-clawed, free or slightly united at base. Androecium of 10 stamens, fused at base, sometimes 5 of the stamens are without anthers. Gynoecium a compound pistil of 5 united carpels, with 5 locules, ovules 1 or more in each locule, placentation axile, ovary superior, styles 5, free and persistent.

*Fruit*   A capsule.

*Seed*   With a straight embryo and fleshy endosperm.

**Size, Distribution, and General Information**   A family of 3 genera and 875 species of mostly tropical and subtropical distribution.

The three genera are *Oxalis* (800) (oxalis, sour grass, sorrel); *Biophytum* (70); and *Eichleria* (2).

Indians use the tuberous roots of *Oxalis tuberosa* as food in the high Andes of South America. The roots, called *oca,* are mellowed in the sun to eliminate the calcium oxalate. Children often chew the leaves of sour grass (*Oxalis*) for the pleasant sour taste. Several species of *Oxalis* are troublesome weeds, especially in greenhouses. Some are grown as ornamentals.

**Economic Importance**   The family is of little importance.


## SUBCLASS V, ROSIDAE; ORDER 17, GERANIALES


### Geraniaceae: Geranium Family

**Field Recognition**   Flowers 5-merous; stamens with filaments united at base; fruit with elastic dehiscent schizocarps which curl on the beak; $CA^5 CO^5 A^{5-15} \underline{G}^{\textcircled{5}}$.

**Description**   *Habit*   Mostly herbs, sometimes undershrubs; sometimes aromatic.

*Leaves*   Alternate or opposite, compound or simple, lobed or divided; with paired stipules.

*Inflorescences*   Cymose or umbellate, often with attractive flowers.

*Flowers*   Bisexual, actinomorphic or slightly zygomorphic. Calyx of 5 free sepals, persistent. Corolla of 5 petals, rarely 4, or absent. Androecium of 5 to 15 stamens, filaments united at the base. Gynoecium a compound pistil of 3 to 5 united carpels, with 3 to 5 locules, ovules 1 to 2 in each locule, placentation axile, ovary superior, styles slender and beak-like.

*Fruit*   Capsular, dehiscing into 3 to 5 schizocarps (usually 1 to 2 seeded), the styles usually adhering to the ovarian beak and the basal portion recurving spirally.

*Seed*   With a straight or folded embryo, endosperm.

**Size, Distribution, and General Information**   A family of about 5 genera and 750 species of wide temperate and subtropical distribution.

The larger genera are *Geranium* (400) (cranesbill); *Erodium* (90) (storksbill); and *Pelargonium* (250) (florist geranium).

The well-known geraniums grown in pots or as bedding plants are cultivars of several species of *Pelargonium* from South Africa. Geranium oil is distilled in Algeria from *P. odoratissimum. Pelargonium* has a calyx spur adnate to the pedicel; this structure is absent in *Geranium.* Spotted geranium (*G. maculatum*) is a favorite spring wild flower in deciduous woods of eastern North

America. Several species of *Erodium* and *Geranium* are weedy. The persistent dry style on the fruit of *Erodium* twists into a "corkscrew" and is very hygroscopic; it is sometimes used as a toy to indicate changes in humidity.

**Economic Importance**   The Geraniaceae is important for the florist geranium and geranium oil.


## SUBCLASS V, ROSIDAE; ORDER 18, APIALES

### Umbelliferae or Apiaceae: Carrot or Parsley Family

**Field Recognition**   Aromatic herbs with hollow stems; leaves compound with sheathing bases; inflorescences umbellate; flowers 5-merous, often yellow or white; stamens 5; ovary 2-carpellate, bilocular, inferior; fruit a schizocarp; $CA^{\underline{5}} CO^{\underline{5}} A^5 \overline{G}^{\underline{2}}$.

**Description**   *Habit*   Mostly biennial or perennial herbs, rarely slightly woody, stem often stout with hollow internodes or wide, soft pith.
   *Leaves*   Alternate, much divided, sheathing at the base.
   *Inflorescences*   A simple or compound umbel or rarely capitate.
   *Flowers*   Bisexual, actinomorphic. Calyx of 5 sepals, epigynous. Corolla of 5 petals, epigynous, some deciduous, mostly white or yellow. Androecium of 5 stamens, alternate with the petals. Gynoecium a compound pistil of 2 united carpels, with 2 locules, ovules solitary in each locule, ovary inferior, style swollen at base. See Figure 14-16.
   *Fruit*   A dry schizocarp splitting into 2 mericarps, often ribbed.
   *Seed*   With a small embryo in oily endosperm.

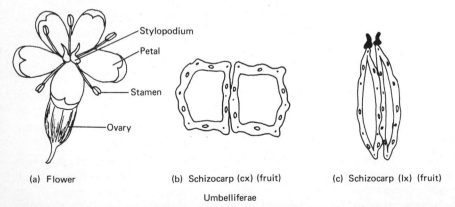

(a) Flower          (b) Schizocarp (cx) (fruit)          (c) Schizocarp (lx) (fruit)

Umbelliferae

**Figure 14-16**   Family Umbelliferae: *(a)* diagrammatic flower, *(b)* cross section of a fruit (schizocarp), and *(c)* longitudinal section of a fruit. The stylopodium is the enlarged base of the style.

**Size, Distribution, and General Information**  A family of about 275 genera and 2850 species chiefly of north temperate distribution but cosmopolitan, found at high elevations in the tropics.

Some of the genera are *Hydrocotyle* (100) (pennywort); *Eryngium* (230) (snakeroot); *Sanicula* (37) (snakeroot); *Apium* (1) (celery); *Chaerophyllum* (40) (chervil); *Angelica* (80); *Daucus* (60) (carrot, Queen Anne's lace); *Conium* (4) (poison hemlock); *Cicuta* (10) (water hemlock); *Petroselinum* (5) (parsley); *Carum* (30) (caraway); *Pimpinella* (150) (anise); *Anethum* (1) (dill); *Ferula* (133) (asafoetida); *Peucedanum* (120) (parsnip); *Coriandrum* (2) (coriander); *Foeniculum* (5) (fennel); and *Pastinaca* (15) (parsnip).

*Conium maculatum* (poison hemlock) is the famous plant used to put to death the Greek philosoper Socrates. A native of Europe and Asia, it has spread into the New World. The North American water hemlock (*Cicuta*) is just as poisonous but not so weedy.

Known since before the time of Greece and Rome, parsnips (*Pastinaca sativa*), carrots (*Daucus carota*), parsley (*Pteroselinum hortense*), and celery (*Apium graveolens*) have been used for food, garnish, flavoring, and medicine. Seeds of Umbelliferae are of importance because of their essential oils. Of these, caraway seeds (*Carum carvi*) used in bread, rolls, and cheeses are the most important. Dill (*Anethum graveolens*) is used in flavoring dill pickles. Coriander (*Coriandrum sativum*) is added to many dishes in China and India and certain Mediterranean countries. Many species have been used for medical purposes.

**Economic Importance**  The family Umbelliferae is important for food, flavoring, and ornamentals, and for some poisonous species.

## SUBCLASS VI, ASTERIDAE; ORDER 1, GENTIANALES

### Gentianaceae: Gentian Family

**Field Recognition**  Herbs; leaves opposite, often connate at the base, without stipules; flowers 4- or 5-merous; stamens the same number as the petals and alternate with them; ovary superior, with 2 carpels, mostly 1 locular, with 2 parietal placentas; $CA^{\textcircled{4-5}}CO^{\textcircled{4-5}}A^{4-5}\ \underline{G}^{\textcircled{2}}$

**Description**  *Habit*  Annual or perennial herbs, rarely woody.
*Leaves*  Opposite, simple, entire, often connate at the base or connected by a line; without stipules.
*Inflorescences*  Cymose.
*Flowers*  Bisexual, actinomorphic, showy. Calyx of 4 to 5(12) sepals, tubular or separate. Corolla of 4 to 5(12) united petals. Androecium of the same number of stamens as the petals, epipetalous, alternate with the corolla lobes.

Gynoecium a compound pistil of 2 carpels, with 1(2) locules, ovules numerous, placentation parietal, ovary superior, style simple.

*Fruit*   A capsule.

*Seed*   Small in size, with endosperm.

**Size, Distribution, and General Information**   A family of about 80 genera and 900 species of cosmopolitan distribution and found in a variety of habitats: arctic, alpine, saline, aquatic, tropical, and temperate.

The chief genera are *Gentiana* (400) (gentian); *Sabatia* (20) (marsh pink, rose pink); *Swertia* (100) (columbo); and *Nymphoides* (20) (water snowflake).

Many species of *Gentiana* grow at high elevations in the temperate zone. The majority produce a distinctive tubular flower of an intense blue color. Since they are so showy, many are cultivated in gardens in the cold temperate zone. An extract of the roots of *G. lutea* is used for a tonic.

*Nymphoides* is an aquatic plant with the habit of water lilies. The flower cluster appears to develop from the top of the petiole, but actually the floating leaf grows from the inflorescence axis.

**Economic Importance**   The family is of importance only for showy ornamentals often grown in rock gardens.

### SUBCLASS VI, ASTERIDAE; ORDER 1, GENTIANALES

#### Apocynaceae: Dogbane Family

**Field Recognition**   Sap milky; leaves opposite or whorled, stipules absent; flowers 5-merous; calyx often glandular inside; fruit usually a follicle; seed usually with a tuft of hairs; $CA^{\circledS}CO^{\circledS}A^5\ \underline{G^{\circled2}}$. (Apocynaceae is likely to be confused with Asclepiadaceae. Apocynaceae lacks the corona, pollinia, and translator-corpusculum of the Ascelepiadaceae. See Figure 14-17.)

**Description**   *Habit*   Trees, shrubs, perennial herbs, lianas; with milky sap.

*Leaves*   Opposite or whorled, rarely alternate, simple, entire; without stipules.

*Inflorescences*   Racemose, cymose, or solitary.

*Flowers*   Bisexual, actinomorphic. Calyx of (4)5 sepals, united, glandular inside. Corolla of 5 petals united into a tube. Androecium of (4)5 stamens, inserted in the tube, filaments free or rarely united. Gynoecium a compound pistil of 2 united carpels, with 1 or 2 locules, placentation parietal, ovary superior or half-inferior, style 1.

*Fruit*   A follicle, sometimes a capsule, drupe, or berry.

*Seed*   Flat often with a crown of hairs, embryo straight.

**Size, Distribution, and General Information**   A family of about 180 genera and 1500 species of mostly tropical and subtropical distribution but with a few species in the temperate zone.

Some of the genera are *Vinca* (5) (periwinkle, myrtle); *Amsonia* (25) (blue star); *Rauvolfia* (100) (Indian snakeroot); *Nerium* (3) (oleander, rosebay); *Apocynum* (7) (dogbane); *Ervatamia* (80) (crape jasmine, clavel de la India); *Allamanda* (15) (cup of gold); and *Catharanthus* (5) (periwinkle).

Many species in the Apocynaceae are notorious for the poisons they contain. In Africa, poisoned arrows are made with an extract of *Acokanthera abyssinica* bark. If ingested, one leaf of *Nerium oleander* can be fatal to an adult human. The yellow oleander (*Thevetia nereifolia*) poses a similar hazard. Both are grown in subtropical areas as ornamentals. Arrow poisons have been prepared from the seeds of the African species of *Strophanthus*. Long used by native healers in India, *Rauvolfia* yields the alkaloid reserpine, which can lower blood pressure and tranquilize mental patients suffering from schizophrenia.

**Economic Importance**   The family is of importance for drugs and for its several ornamentals: *Amsonia, Nerium, Vinca, Allamanda, Ervatamia,* and others. Poisoning of humans, especially children, sometimes occurs. When pasture is poor, livestock may eat the plants.

## SUBCLASS VI, ASTERIDAE; ORDER 1, GENTIANALES

### Asclepiadaceae: Milkweed Family

**Field Recognition**   Sap milky; leaves opposite or whorled, stipules absent; flowers 5-merous with distinctive corona, pollinia, translators, and corpuscula; fruit a follicle; seed with a tuft of silky hairs; $CA^{\circledcirc} CO^{\circledcirc} \widehat{A^5} \underline{G^{\circledcirc}}$.

**Description**   *Habit*   Perennial herb, vines, shrubs, rarely trees, with milky sap; sometimes cactus-like.

*Leaves*   Opposite, simple, entire, pinnately veined; without stipules.

*Inflorescences*   Mostly cymose, but also racemose or umbellate.

*Flowers*   Bisexual, actinomorphic, with an elaborate corona. Calyx of 5 sepals, with a short tube. Corolla of 5 united petals. Androecium of 5 stamens, pollen in waxy pollinia. Gynoecium a compound pistil of 2 weakly united carpels, ovules numerous, ovary superior, styles 2 and united at style apices, stigma 1 with 5 lobes. See Figure 14-17.

The stamens and carpels are united into a complex structure, the gynostegium; the carpels are free below, but united into the single 5-lobed stigma. The pollinia are connected in pairs by the translators and the corpusculum. A corona may be present which consists of 5 hoods, which are sometimes mistaken for petals. A beak may be associated with each hood.

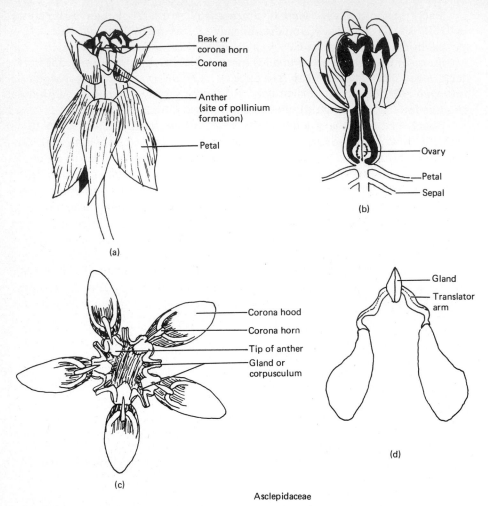

**Figure 14-17**  Diagrammatic views of Asclepidaceae floral structures. *(a)* Flower. *(b)* Longitudinal section showing inferior ovary. *(c)* Top view of the center of a flower. *(d)* Pollinia pair.

> *Fruit*  A pair of follicles.
> *Seed*  Numerous, with tufts of silky hairs called a *coma*.

**Size, Distribution, and General Information**  A family of about 130 genera and 2000 species mainly of tropical and subtropical distribution, but with numerous species in the temperate zone.

Some of the genera are *Asclepias* (120) (milkweed); *Hoya* (200) (wax plant); *Ceropegia* (160); *Stapelia* (75) (carrion flower); *Dischidia* (80); and *Matelea* (130) (milkweed vine).

During World War II, the United States was unable to import kapok, and hair from the seeds of *Asclepias* was used as a substitute in life jackets.

Monarch butterfly larvae feed on the leaves of several species of milkweed from which the larvae obtain compounds toxic to predators.

The flower tube of *Ceropegia,* lined with small hairs which point downward, forms a trap for small flies. The flies which were attracted by the odor cannot escape until the hairs wither. The pollinia of *Ceropegia* are attached to the flies' bodies when they escape. *Dischidia rafflesiana* has curious pitcher-like leaves. Each pitcher is about 10 centimeters deep and contains debris carried by nesting ants. If the plant is disturbed, the ants rush about defending the *Dischidia* as well as their nest.

*Stapelia* (carrion flower) inhabits semiarid regions of Africa. The leaves are reduced to scales, and the stems are thick and fleshy. The flowers are large with dull red or maroon color and have a carrion-like odor which attracts flies. The Asiatic genus *Hoya,* which has large clusters of highly scented waxy flowers, provides several greenhouse plants.

**Economic Importance**   The family is of little importance, except for some ornamentals. Some species cause livestock poisoning. In India, madar fiber is obtained from *Calotropis gigantea* and rajmahal hemp from *Marsdenia tenacissima.*

## SUBCLASS VI, ASTERIDAE; ORDER 2, SOLANALES

### Solanaceae: Nightshade or Potato Family

**Field Recognition**   Leaves alternate, stipules absent; flowers actinomorphic, 5-merous; stamens 5; ovary 2-locular, sometimes falsely divided again; ovules numerous; fruit a berry or capsule; $CA^{\circledS} CO^{\circledS} A^5 \underline{G}^{\circledsmall{2}}$.

**Description**   *Habit*   Herbs, shrubs, or trees, or sometimes vines.
*Leaves*   Alternate, simple; without stipules.
*Inflorescences*   Cymose.
*Flowers*   Bisexual, usually actinomorphic or weakly zygomorphic. Calyx of 5 united sepals, persistent. Corolla of 5 united petals, rotate to tubular. Androecium of 5 stamens, inserted on the corolla tube and alternate with its lobes, anthers opening either lengthwise or by terminal pores, sometimes divided by a false septum. Gynoecium a compound pistil of 2 carpels, with 2 locules, ovules numerous, placentation axile, ovary superior, style terminal. See Figure 14-18.
*Fruit*   A berry or capsule.
*Seed*   With a curved or straight embryo in endosperm.

**Size, Distribution, and General Information**   A family of about 90 genera and 2000 species of tropical and temperate distribution. The greatest number of

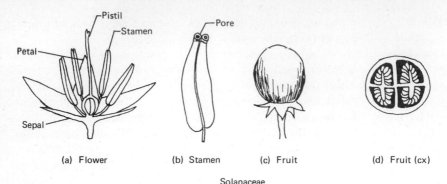

**Figure 14-18**  *(a)* Flower (longitudinal section), *(b)* stamen, *(c)* fruits, and *(d)* fruit in cross section of family Solanaceae. Note the pores at the apex of the stamen.

genera (40) are found in Central and South America; only two or three genera are native to Europe and Asia.

Some of the chief genera are *Solanum* (1700) (potato, nightshade, and so on); *Datura* (10) (jimsonweed, Jamestown weed, angel trumpet); *Nicotiana* (21) (tobacco); *Lycium* (80–90) (kaffir thorn, matrimony vine); *Atropa* (4) (belladonna); *Hyoscyamus* (20) (henbane); *Capsicum* (50) (pepper, chili); *Lycopersicon* (7) (tomato); *Petunia* (40); *Schizanthus* (15) (butterfly flower); *Physalis* (100) (ground-cherry); and *Salpiglossis* (18).

Members of the nightshade family provide drugs and food; some are weedy, some are poisonous, and others are handsome ornamentals. Most important is the white or Irish potato introduced into Europe from the Andean region in the late 1500s. The potato did well in the cooler parts of Europe and became a staple in Ireland. In 1845 and 1846 late blight struck the potato crop in Ireland. So completely dependent on the potato were the Irish that over a million people died of famine, and over a million migrated to North America.

*Solanum melongena* (eggplant) is from northern India. The Jerusalem cherry (*S. pseudocapsicum*) is sold at Christmas in the florist trade. The fruit of Jerusalem cherry is poisonous and is sometimes eaten by children.

Tomato (*Lycopersicon lycopersicum*), the favorite home garden vegetable in North America, was once believed to be poisonous. Tomatoes originated in the Andes and were widely cultivated throughout the Inca empire and northward. Those first seen in Europe were yellow-fruited and were called *golden apples*. Later they were known as *love apples* and supposedly had value as an aphrodisiac. By the late 1700s, tomatoes were widely grown and were introduced into northern North America by Italians and other European immigrants.

*Capsicum annuum* and *C. frutescens* have provided many cultivars of pepper or chili. Peppers are rich in vitamins C and A. The various cultivars provide chili powder, tabasco sauce, bell peppers, sweet and hot peppers, paprika, and pimento. (Note: Black pepper is derived from a member of the Piperaceae.)

*Physalis* (ground-cherry) produces an edible fruit enclosed in a bladder-like persistent calyx called the husk, hence the name *husk tomatoes*. *Physalis alkekengi* (Chinese lantern plant) is grown for its ornamental fruits.

Many Solanaceae contain powerful alkaloids: *Atropa belladonna* (deadly nightshade or belladonna), *Hyoscyamus niger* (henbane), *Nicotiana tabacum* (tobacco), *Datura* (jimsonweed or Jamestown weed), and *Solanum nigrum* (black nightshade). The drug atropine is obtained from belladonna. It is used as an antidote for poisoning from organic phosphate pesticides and nerve gas. Women once put belladonna juice in their eyes to enlarge the pupils to make their eyes attractive. Recently, a fad of chewing the seeds of *Datura* has often had fatal results. *Datura stramonium* is sometimes called *Jamestown weed* because some British soldiers were poisoned in Jamestown in 1676 after eating cooked foliage of *D. stramonium*.

**Economic Importance**   The Solanaceae is of considerable importance for food, drugs, weeds, and poisonous plants.

## SUBCLASS VI, ASTERIDAE; ORDER 2, SOLANALES

### Convolvulaceae: Morning Glory Family

**Field Recognition**   Sap often somewhat milky; leaves alternate; flowers 5-merous; corolla tubular and plicated; stamens 5, epipetalous; gynoecium of 2 carpels; ovules solitary or paired and erect; $CA^5 CO^{\circledS} A^5 \underline{G}^{\circledtwo}$.

**Description**   *Habit*   Herbs, shrubs, or trees, often a climber with milky sap.

*Leaves*   Alternate, simple, variously lobed or entire; without stipules.

*Inflorescences*   An axillary dichasium, or racemose, or paniculate, or flowers solitary.

*Flowers*   Bisexual, actinomorphic. Bracts often large and showy, sometimes forming an involucre. Calyx of 5 sepals, usually free, persistent. Corolla of 5 united petals, plicate. Androecium of 5 stamens, epipetalous, inserted toward base of tube and alternate with the lobes. Gynoecium a compound pistil of 2 united carpels, with 1 to 4 locules, ovules solitary or paired, erect, placentation axile, ovary superior, style terminal.

*Fruit*   Usually a capsule.

*Seed*   With endosperm.

**Size, Distribution, and General Information**   A family of about 55 genera and 1650 species of tropical and temperate distribution.

Some of the genera are *Convolvulus* (250) (bindweed, wild morning glory);

*Dichondra* (4 to 5); *Jacquemontia* (120); *Calystegia* (25); *Ipomoea* (500) (morning glory); and *Evolvulus* (100).

Some species of *Convolvulus* and *Ipomoea* are weedy. Sweet potato (*I. batatas*) is an important food crop in the southern United States and in tropical countries. Some species are grown as ornamentals: *I. purpurea* (morning glory) and *Calonyction aculeatum* (moonvine). The seeds of *I. violacea* contain *d*-lysergic acid amide and are hallucinogenic.

**Economic Importance**   The morning glory family is important as a source of food, for several ornamentals, for drug plants, and as weeds.

## SUBCLASS VI, ASTERIDAE; ORDER 2, SOLANALES

### Polemoniaceae: Phlox Family

**Field Recognition**   Flowers 5-merous; calyx of 5 united sepals; corolla of 5 united petals; stamens 5, attached at various levels within the corolla; ovules numerous; $CA^{\circledS}CO^{\circledS}A^5 \underline{G}^{\circledES}$.

**Description**   *Habit*   Mostly annual to perennial herbs, rarely shrubs.
*Leaves*   Alternate or opposite, simple or compound; without stipules.
*Inflorescences*   Terminal, cymose, corymbose, or capitate, or the flowers solitary.
*Flowers*   Bisexual, actinomorphic, rarely slightly zygomorphic. Calyx of 5 united sepals. Corolla of 5 united petals. Androecium of 5 stamens, epipetalous, alternate with the lobes, free from one another, included or exserted. Gynoecium a compound pistil of 3 united carpels, with 3 locules, ovules numerous, placentation axile, ovary superior, style 1, stigmas 3 or rarely 2.
*Fruit*   Usually a capsule.
*Seed*   With a straight embryo and endosperm.

**Size, Distribution, and General Information**   A family of about 15 genera and 300 species mostly in North America and the Andes of South America, a few in Northern Europe and Asia.

Some of the genera are *Polemonium* (50) (Jacob's ladder); *Phlox* (77); *Gilia* (120); *Cantua* (11); *Collomia* (15); and *Ipomopsis* (24).

Pollinators, such as long-tongued flies, bees, moths, butterflies, and hummingbirds, are attracted to the colorful corolla of the Polemoniaceae. Many of the species of this family have been adopted into flower gardens. *Phlox* is probably the most popular genus. The tall garden phloxes are largely cultivars obtained by crossing *P. carolina*, *P. glaberrima*, *P. maculata*, and *P. paniculata*. The annual *P. drummondii* from Texas is a favorite in both Europe and North America.

**Economic Importance**   The Polemoniaceae is important in flower gardens.


## SUBCLASS VI, ASTERIDAE; ORDER 2, SOLANALES

### Hydrophyllaceae: Waterleaf Family

**Field Recognition**   Inflorescences often helicoid cymes; flowers 5-merous; sepals united, often with appendages between; petals united; gynoecium with 1 locule; ovules numerous on 2 parietal placentas; $CA^{\circledS} CO^{\circledS} A^5 \underline{G}^{\circledtwo}$.

**Description**   *Habit*   Annual or perennial herbs, rarely shrubs, often hairy or sometimes spiny.
*Leaves*   Alternate or rarely opposite, sometimes basal, simple to pinnately or palmately compound.
*Inflorescences*   Often tending toward helicoid cymes or flowers solitary.
*Flowers*   Bisexual, actinomorphic. Calyx of 5 united sepals, often with appendages between the lobes. Corolla of 5 united petals, rotate. Androecium of 5 stamens, epipetalous, alternate with the lobes and often inserted toward the base of the tube. Gynoecium a compound pistil of 2 united carpels, with 1 locule, ovules often numerous, placentation of 2 parietal placentas, or rarely 2-locular with placentas on septum, ovary superior, styles 1 or 2.
*Fruit*   A capsule.
*Seed*   With a small embryo, endosperm.

**Size, Distribution, and General Information**   A family of about 18 genera and 250 species widely distributed, occurring on all continents except Australia, but mainly in North America.
   The chief genera are *Hydrophyllum* (10) (waterleaf); *Nemophila* (13) (baby blue-eyes); *Phacelia* (200); *Nama* (40 to 50); *Hydrolea* (20); and *Ellisia* (1) (water pod).

**Economic Importance**   The family is of little or no importance except for a few garden annuals and a few minor weeds.


## SUBCLASS VI, ASTERIDAE; ORDER 3, LAMIALES

### Boraginaceae: Borage Family

**Field Recognition**   Bristly herbs, stems round; leaves alternate; inflorescences of helicoid cymes; flowers 5-merous and usually actinomorphic; ovary 4-lobed, style arising from among the lobes; fruit of 4 nutlets; $CA^5 CO^{\circledS} A^5 \underline{G}^{\circledtwo}$.

(The Boraginaceae may be confused with Hydrophyllaceae, Labiatae, and Verbenaceae.)

**Description**   *Habit*   Herbs, shrubs, trees, or lianas.

*Leaves*   Usually alternate, simple, often covered by rough hairs; without stipules.

*Inflorescences*   Often a scorpioid (helicoid) cyme.

*Flowers*   Bisexual, usually actinomorphic. Calyx of 5 united sepals. Corolla of 5 united petals, funnel-form or tubular. Androecium of 5 stamens, epipetalous, alternate with the corolla lobes. Gynoecium a compound pistil of 2 united carpels, with 2 (or falsely 4) locules, ovules in pairs, placentation axile sometimes appearing basal, ovary superior, style arising from the middle of the lobes of the ovary. See Figure 14-19.

*Fruit*   Often deeply 4-parted into 1-seeded nutlets.

*Seed*   Embryo erect.

**Size, Distribution, and General Information**   A family of about 100 genera and 2000 species of tropical and temperate distribution but especially abundant in the Mediterranean region.

Some genera are *Tournefortia* (150); *Heliotropium* (250) (heliotrope, turn-sole); *Borago* (3) (borage); *Myosotis* (50) (forget-me-not); *Lithospermum* (60) (gromwell, puccoon); *Cynoglossum* (50–60) (hound's-tongue); *Mertensia* (50) (bluebells); *Amsinckia* (50) (fiddle-neck); *Plagiobothrys* (100); and *Symphytum* (25).

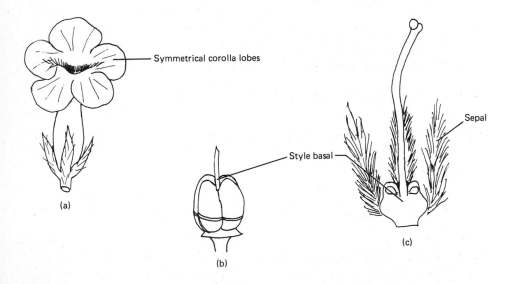

Boraginaceae

**Figure 14-19**   Diagrammatic drawing of the floral structures of Boraginaceae. *(a)* Symmetrical corolla lobes. *(b)* Deeply four-lobed ovary. *(c)* Gynobasic style.

A few species of *Mertensia, Myosotis, Borago, Cynoglossum,* and *Heliotropium* are cultivated in flower gardens. The geiger tree (*Cordia sebestena*) is widely planted for its handsome orange-red flowers and its edible plum-like fruits. Viper's bugloss (*Echium vulgare*) is a noxious herbaceous weed.

It is not known how the forget-me-nots (*Myosotis*) obtained their name, but one legend suggests these were the last words of a gallant gentleman who drowned while crossing a stream to collect a bouquet of these sky-blue flowers for his girl friend.

**Economic Importance** The family is of no significant economic importance except for a few ornamentals. *Symphytum* is a minor forage plant.

## SUBCLASS VI, ASTERIDAE; ORDER 3, LAMIALES

### Verbenaceae: Vervain or Verbena Family

**Field Recognition** Stems often 4-angled; leaves opposite; flowers 5-merous, zygomorphic; style single and terminal; $CA\textcircled{5}COZ\textcircled{5}A^{2+2}\underline{G}\textcircled{2}$. (The Verbenaceae is most easily confused with the Boraginaceae, which have alternate leaves and a 4-lobed ovary, and with the Labiatae, which have opposite leaves and a style arising from the base of the ovary.)

**Description** *Habit* Herbs, shrubs, trees, and a few vines; often with square stems.

*Leaves* Usually opposite, rarely alternate or whorled, simple or compound; without stipules.

*Inflorescences* Racemose or cymose, sometimes with an involucre of colored bracts.

*Flowers* Bisexual, zygomorphic. Calyx of (4)5(8) sepals, lobed or toothed, persistent. Corolla of 5 united petals often 2-lipped, 4 to 5 lobed. Androecium of (2)4(5) stamens, didynamous, epipetalous. Gynoecium a compound pistil usually of 2 carpels, with 2(4 or 5) locules, ovules solitary (or paired), placentation axile, ovary superior, style terminal. See Figure 14-20.

*Fruit* A drupe or berry.

**Size, Distribution, and General Information** A family of about 73 genera and 3000 species, mostly of tropical or subtropical distribution.

Some of the genera are *Verbena* (250) (vervain); *Lantana* (150); *Clerodendrum* (400) (glory-bower); *Vitex* (3) (chaste tree); *Lippia* (220) (frogfruit); *Callicarpa* (140) (beauty-berry); *Tectona* (3) (teak); and *Avicennia* (14) (mangrove).

Verbenaceae are mostly herbaceous in the temperate zone, but in the tropics many are shrubs and trees. Teak (*Tectona grandis*) is a native tree of In-

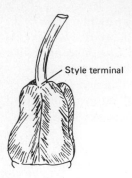

Style terminal

Verbenaceae ovary

**Figure 14-20** Terminal style of family Verbenaceae.

domalaysia. The heartwood is relatively immune to termites and rots. The wood is remarkable also for its resistance to changes in temperature and humidity. It is especially suitable for decking and fine trim on boats, and for flooring, paneling, and furniture.

The European verbain (*Verbena officinalis*) was once regarded as a cure-all. The *Verbena* most often grown in gardens is *V. hybrida* from southern Brazil. Lantanas from tropical America are often grown as greenhouse or bedding plants. In many subtropical areas, *Lantana* has escaped and has become a weedy pest which can cause livestock poisoning. *Lippia citriodora* (lemon-scented verbena) yields oil of verbena. In temperate areas, several species of *Clerodendrum* are grown as greenhouse ornamentals including *C. paniculatum* (pagoda flower) and *C. thomsoniae* (bleeding heart). *Callicarpa americana* (beauty-berry), a woody shrub, is grown for its ornamental fruits. Another woody shrub known for its blue or white flowers is the lilac chaste tree (*Vitex agnus-castus*).

**Economic Importance**  The family is important for teak lumber and for many ornamental herbs and shrubs.

## SUBCLASS VI, ASTERIDAE; ORDER 3, LAMIALES

### Labiatae or Lamiaceae: Mint Family

**Field Recognition**  Herbs and shrubs with square stems; aromatic; leaves opposite; inflorescences axillary or whorled; flowers 5-merous, zygomorphic; stamens 2 or 4; ovary deeply 4-lobed, style basally attached between the 4 lobes (gynobasic); fruit of 4 nutlets; $CA^{\circledS}COZ^{\circledS}A^{2or4}\underline{G}^{\circledtwo}$. (This family may be confused with the Boraginaceae, the Verbenaceae, or the Scrophulariaceae.)

**Description**  *Habit*  Usually herbs, sometimes shrubs or trees; stems square; with aromatic oils.

*Leaves*   Opposite or whorled, simple; without stipules.

*Inflorescences*   Axillary or whorled.

*Flowers*   Usually bisexual, zygomorphic. Calyx of 5 united sepals, often ribbed, persistent, often 2-lipped. Corolla of 5 united petals, 2-lipped. Androecium of 2 or 4 stamens, often 2 pairs of different lengths (didynamous), epipetalous. Gynoecium a compound pistil of 2 carpels, falsely 4-locular, placentation basal, ovary superior, style attached to base of lobes (gynobasic), stigma mostly bifid. See Figure 14-21.

*Fruit*   Usually a group of 4 nutlets, each with 1 seed.

*Seed*   With little or no endosperm.

**Systematics**   The Labiatae are closely allied morphologically with the Verbenaceae.

**Size, Distribution, and General Information**   A family of about 180 genera and 3500 species of cosmopolitan distribution but centered in the Mediterranean region.

The chief genera are *Lamium* (40 to 50) (henbit, dead nettle); *Ajuga* (40) (bugleweed); *Teucrium* (300) (germander); *Scutellaria* (300) (skullcap); *Marrubium* (40) (horehound); *Nepeta* (250) (catnip); *Prunella* (7) (self-heal); *Stachys* (300) (betony, Chinese artichoke); *Salvia* (700) (sage); *Monarda* (12) (horsemint, bee balm); *Origanum* (15 to 20) (marjoram); *Thymus* (300 to 400) (thyme); *Mentha* (25) (mint, peppermint, spearmint); *Rosmarinus* (3) (rosemary); *Coleus* (150) (Jacob's coat); *Hyptis* (400); *Ocimum* (150) (basil); *Lavandula* (28) (lavender); *Molucella* (4) (bells of Ireland); and *Glechoma* (10 to 12) (ground ivy).

Many plants from the mint family are used in cooking as herbs and spices: sage (*Salvia officinalis*), thyme (*Thymus vulgaris*), pot marjoram or oregano (*Origanum vulgare*), sweet marjoram (*Origanum majorana*), spearmint

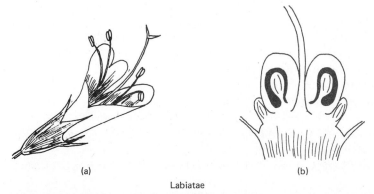

(a)                                          (b)

Labiatae

**Figure 14-21**   Diagrammatic drawings of the floral structures of Labitae. *(a)* Bilabiate corolla. *(b)* Longitudinal section showing gynobasic style (the ovary is deeply four-lobed).

(*Mentha spicata*), peppermint (*Mentha piperita*), basil (*Clinopodium vulgaris*), and sweet basil or tulsi (*Ocimum basilicum*).

Lavender (*Lavandula angustifolia*) and rosemary (*Rosmarinus officinalis*) have long been popular in perfumes. The essential oils are obtained by distillation of the inflorescences of these species.

The 150 species of *Coleus* have contributed countless cultivars for greenhouse or garden plantings. *Coleus* is sometimes called *Jacob's coat* because of the multicolored, variegated color patterns of the leaves.

Cats are fond of catnip (*Nepeta cataria*) and often will nibble or roll over the plants, sometimes killing the plants. In the 1960s many people tried smoking catnip; however, it does not have any intoxicating effects.

The family has many attractive ornamentals: *Salvia, Ajuga, Monarda* (horsemint), *Scutellaria* (skullcap), and *Dracocephalum* (dragonhead).

**Economic Importance**   The family is a source of aromatic essential oils, many ornamentals, and important culinary herbs.

## SUBCLASS VI, ASTERIDAE; ORDER 5, PLANTAGINALES

### Plantaginaceae: Plantago Family

**Field Recognition**   Herbs; leaves in a basal rosette, prominent veins are parallel; inflorescences spicate or capitate on stout or wiry scapes; flowers 4-merous; corollas membranous; stamens often exserted; $CA\textcircled{4} CO\textcircled{4} A^4 \underline{G}\textcircled{2}$.

**Description**   *Habit*   Herbs.
*Leaves*   Basal, alternate or opposite, simple, sometimes reduced, often sheathing at base.
*Inflorescences*   Scapose, capitate, or spicate on stout or wiry scapes.
*Flowers*   Usually bisexual, actinomorphic, inconspicuous. Calyx of 4 united sepals, herbaceous. Corolla of 4 united petals, scarious, 3- to 4-lobed. Androecium of 4 stamens, inserted on corolla tube. Gynoecium a compound pistil of 2 united carpels with 1 to 4 locules, ovules 1 or more in each locule, placentation axile or basal; ovary superior, style simple.
*Fruit*   A capsule or bony nutlet.
*Seed*   With a straight embryo in fleshy endosperm.

**Size, Distribution, and General Information**   A family of 3 genera and 269 species of cosmopolitan distribution.

The genera are *Plantago* (265) (plantain); *Littorella* (3); and *Bougueria* (1).

**Economic Importance**   The family is of no significant importance, although seeds of *Plantago psyllium* are used as a laxative. A number of species of *Plantago* are troublesome weeds, especially on lawns.

## SUBCLASS VI, ASTERIDAE; ORDER 6, SCROPHULARIALES

### Oleaceae: Olive Family

**Field Recognition**   Leaves opposite; flowers 4-merous; stamens 2; ovary 2-locular; seeds usually 2 per locule; $CA^{④}CO^{④}A^2 \underline{G}^{②}$.

**Description**   *Habit*   Shrubs and trees, sometimes climbing.
*Leaves*   Opposite or rarely alternate, simple or pinnately compound; stipules absent.
*Inflorescences*   Axillary or terminal, racemose, or paniculate.
*Flowers*   Usually bisexual, actinomorphic. Calyx of 4 united sepals, 4-lobed, rarely absent. Corolla of 4 united petals, 4-lobed, sometimes absent. Androecium of 2 stamens. Gynoecium a compound pistil of 2 carpels, with 2 locules, placentation axile, ovary superior, style simple, with a capitate or bifid stigma.
*Fruit*   Berry-like, a capsule, or a drupe.
*Seed*   Embryo straight, endosperm may or may not be present.

**Size, Distribution, and General Information**   A family of about 29 genera and 600 species of cosmopolitan distribution but with the greatest diversity in temperate and tropical Asia.

The chief genera are *Olea* (20) (olive); *Syringa* (30) (lilac); *Ligustrum* (40 to 50) (privet hedge); *Chionanthus* (2) (granny graybeard, fringe tree); *Fraxinus* (70) (ash); *Forsythia* (7) (golden bells); *Jasminum* (300) (jasmine); and *Osmanthus* (15) (tea olive, sweet olive, devilwood).

The cultivated olive is *Olea europaea,* a valuable tree of the Mediterranean region. Olives are also grown in California, Argentina, Chile, South Africa, and Australia. When freshly picked, the fruit has a bitter glycoside that renders it inedible until the glycoside is neutralized. Green olives and black olives differ in their method of commercial preparation. Ripe olives contain 20 to 30 percent oil, which is removed by pressing.

Golden bells (*Forsythia*) is a handsome yellow flowering shrub. Cultivars of *Syringa* (lilac) are fragrant flowering shrubs and are available in a range of colors from purple to white and red. Several species of *Ligustrum* are used for hedges or for foundation plantings around homes. *Osmanthus* has numerous handsome evergreen shrubs. *Osmanthus fragrans* provides the extra pleasure of a distinctive perfume in the fall. Its leaves are used to perfume tea. The wood of ash (*Fraxinus*) is valuable for being both firm and elastic.

The flowers of *Jasminum grandiflorum* are used in fine perfumes. It is cultivated along the French Riviera and the jasmine oil is obtained by enfleurage rather than distillation, since heating lowers the quality of the oil. Many of the 300 species of *Jasminum* are cultivated as ornamentals.

**Economic Importance**   The Oleaceae is of considerable importance as a source of olive oil, olives, ash lumber, and many ornamental shrubs.

## SUBCLASS VI, ASTERIDAE; ORDER 6, SCROPHULARIALES

### Scrophulariaceae: Figwort or Snapdragon Family

**Field Recognition**   Flowers 5-merous, zygomorphic; corolla 2-lipped; stamens 4, sometimes with a fifth staminode; ovary 2-locular, style terminal, ovules numerous; $CA⑤COZ⑤A^{2or4}G②$. (The Scrophulariaceae may be confused with the Labiatae, which have a gynobasic style and deeply 4-lobed ovary, or with the Solanaceae, which have actinomorphic flowers containing 5 functional stamens.)

**Description**   *Habit*   Mostly herbs or undershrubs, rarely trees, several climbers, and some parasitic on roots.
*Leaves*   Alternate, opposite or whorled, simple; without stipules.
*Inflorescences*   Variable.
*Flowers*   Bisexual, zygomorphic. Calyx of 5 united sepals. Corolla of 5 united petals, typically 2-lipped, sometimes nearly regular, limb 4- to 5-lobed or rarely 6- to 8-lobed. Androecium of 2 or 4(5) stamens, didynamous, fifth stamen sometimes a staminode inserted on the corolla. Gynoecium a compound pistil of 2 carpels, with 2 locules, ovules numerous, placentation axile, ovary superior, style terminal. See Figure 14-22.
*Fruit*   A capsule, rarely a berry.
*Seed*   Numerous, small, with endosperm.

**Size, Distribution, and General Information**   A family of about 220 genera and 3000 species of general distribution.
Some of the genera are *Scrophularia* (300) (figwort); *Verbascum* (360) (mullein); *Calceolaria* (300 to 400); *Linaria* (150) (toadflax, butter-and-eggs, spurred snapdragon); *Antirrhinum* (42) (snapdragon); *Chelone* (4) (turtlehead); *Penstemon* (252) (beardtongue); *Paulownia* (17) (princess tree, imperial tree, cottonwood, royal paulownia); *Mimulus* (100) (monkey flower); *Gratiola* (20) (hedge hyssop); *Veronica* (300) (speedwell); *Digitalis* (20 to 30) (foxglove); *Agalinis* (60) (false foxglove); *Castilleja* (200) (Indian paintbrush); *Pedicularis* (500) (lousewort); *Torenia* (50) (monkey face); and *Striga* (40) (witchweed).
Most of the species in this large family have flowers somewhat similar to

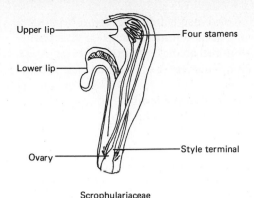

Upper lip — 

Lower lip — 

Ovary — 

Four stamens

Style terminal

Scrophulariaceae

**Figure 14-22** Corolla, stamens, and pistil of family Scrophulariaceae.

the familiar cultivated snapdragons (*Antirrhinum*). Foxglove (*Digitalis purpurea*) is a source of drugs used in the treatment of heart conditions and is also appreciated for the beauty of its flowers.

Many members of this family are parasitic on the roots of other plants and can cause serious crop damage. *Striga lutea* (witchweed) is parasitic on corn roots and has recently become established in the Carolinas. Other parasites include *Agalinis, Castilleja,* and *Pedicularis.*

*Paulownia tomentosa* is unusual in the family since it is a tree obtaining a height of 15 to 20 meters. Producing great clusters of large bluish flowers in the spring, it is widely naturalized in the southeastern United States where it is called *cottonwood.* This name is derived from the capsule which reminds some Southerners of the fruit of cotton.

**Economic Importance**   The family is of little importance except for several ornamentals, root parasites, and *Digitalis.*

## SUBCLASS VI, ASTERIDAE; ORDER 6, SCROPHULARIALES

### Bignoniaceae: Bignonia Family

**Field Recognition**   Usually woody vines or trees; leaves usually opposite, often compound; flowers showy and more or less zygomorphic; fruit often a woody capsule; seeds often conspicuously winged; CA⑤ COZ⑤ A⁴ G②.

**Description**   *Habit*   Mostly woody vines, trees, or rarely herbs.
*Leaves*   Opposite, usually pinnately or palmately compound, sometimes simple, tendrils sometimes replacing terminal leaflet; without stipules.
*Inflorescences*   Cymose.
*Flowers*   Bisexual, more or less zygomorphic, often showy. Calyx of 5

united sepals, campanulate, 5-toothed. Corolla of 5 united petals, the 5 lobes often forming 2 lips, the upper lip of 2 lobes. Androecium of 4 stamens, epipetalous, didynamous, alternate with corolla lobes, staminode often representing the fifth stamen. Gynoecium a compound pistil of 2 united carpels, with 2 locules, each with a placenta; or 1-locular with 2 parietal bifid placentas, ovules numerous, ovary superior, style terminal, 2-lipped.

*Fruit*   A capsule, often woody, or fleshy and indehiscent.

*Seed*   Often winged.

**Size, Distribution, and General Information**   A family of about 120 genera and 650 species, mostly of tropical and subtropical distribution but with a few temperate zone species.

Some of the genera are *Bignonia* (1) (cross vine); *Jacaranda* (50); *Catalpa* (11) (Indian bean, fish bait tree); *Campsis* (2) (trumpet creeper); *Crescentia* (5) (calabash tree); *Kigelia* (1) (sausage tree); and *Spathodea* (20) (African tulip tree).

The Indian bean (*Catalpa*) of eastern North America is one of the few members of *Bignoniaceae* to be found in the temperate zone. Catalpa "worms" (Lepidoptera larvae) feed on the leaves. The trees are sometimes planted around fish ponds, and the "worms" used for bait.

Many members of the family are woody vines or lianas, such as the temperate zone *Campsis radicans* and *Bignonia capreolata*.

*Crescentia cujete* of tropical America is the calabash tree. It produces a woody gourd-like berry which forms a watertight container, or calabash, after the central pulp is removed.

**Economic Importance**   The Bignoniaceae is important for its woody ornamental vines and trees.

**SUBCLASS VI, ASTERIDAE; ORDER 8, RUBIALES**

**Rubiaceae: Coffee or Madder Family**

**Field Recognition**   Herbs, shrubs, and trees; leaves opposite or whorled, usually entire, stipules often inter- or intrapetiolar; flowers 4- or 5-merous; stamens as many as corolla lobes and alternate with them; ovary inferior; CA $\underset{4-5}{\smile}$ CO $\overset{4-5}{\frown}$ A$^{4-5}$ $\overline{G}$②. (Rubiaceae resemble Caprifoliaceae, and members of the two families may be confused. Among other differences, Rubiaceae have stipules, which are lacking in Caprifoliaceae.)

**Description**   *Habit*   Trees, shrubs, or rarely herbs (in North America mostly herbaceous).

*Leaves*   Opposite or whorled, usually entire, simple; with stipules which are often united and as large as the leaves so that the leaves appear whorled.

*Inflorescences*   Cymose, or solitary to almost capitate.

*Flowers*   Bisexual, actinomorphic, rarely slightly zygomorphic. Calyx of 4 to 5 sepals, fused to ovary, with (0)4 to 5 lobes. Corolla usually of 4 to 5(10) united petals, epigynous, more or less tubular, lobes ranging in number from 4 to 10. Androecium of 4 to 5(10) stamens, epipetalous (inserted in the tube or at its mouth) alternate with the corolla lobes. Gynoecium a compound pistil of (1)2 united carpels, with 2 or more locules, ovules 1 to many per locule, placentation usually axile, apical or basal, ovary inferior, style often slender, stigma capitate or variously lobed.

*Fruit*   A capsule, berry, or drupe.

*Seed*   With a small embryo in endosperm.

**Size, Distribution, and General Information**   A family of about 500 genera and 6000 species, largely of tropical and subtropical distribution but some are temperate, and there are even a few arctic species.

Some of the genera are *Rubia* (60) (madder, manjit); *Hedyotis* (including *Houstonia*) (200) (bluets); *Cinchona* (40) (quinine, fever tree, Jesuit's bark); *Gardenia* (250) (cape jasmine); *Cephalanthus* (17) (buttonbush); *Coffea* (40) (coffee); *Galium* (400) (bedstraw, cleavers); *Pinckneya* (1) (fever tree); *Richardia* (10); and *Mitchella* (2) (partridgeberry, twinflower).

Coffee (*Coffea arabica*) is probably the Western world's most popular nonalcoholic beverage. It originated in Ethiopia and was introduced to the rest of the world by Arab traders. Before about 1870, the British drank much coffee; however, when disease wiped out the coffee plantations of the British Empire, they switched largely to tea. Recently, diseases and frost have caused problems in the coffee plantations of South America, and the price of coffee has increased.

Several species of *Cinchona* from the Andes yield quinine used in the prevention and treatment of malaria. *Cinchona* plantations were developed in Java by the Dutch. During World War II when Java was occupied by the Japanese, the supply of quinine to the Allied forces was cut off. Numerous plant taxonomists from the United States were sent to the Andes to search for native *Cinchona*. Toward the end of the war, synthetic drugs were developed which have largely replaced quinine.

*Hedyotis caerulea* (bluets) is a favorite spring wild flower. Twinflowers and two-lobed scarlet fruits of a trailing evergreen plant called *partridge berry* (*Mitchella repens*) are attractive.

In coastal Georgia, Florida, and South Carolina, the numerous bright pink foliaceous sepals make the beautiful fever tree (*Pinckneya pubens*) conspicuous on the edge of swamps. The early settlers used it as a remedy for malaria.

*Rubia tinctorum* (madder) was formerly cultivated for its red dye (alizarin), which is now prepared artificially. Some of the bedstraws (*Galium*) have been used as sources of dyes. Some bedstraws are sweet-scented and were once used to stuff mattresses. Certain bedstraws are called *catchweed* or *cleavers* because they cling so readily to any surface. Gardenias (*Gardenia*) are evergreen shrubs that produce fragrant white flowers and are often used as wedding gifts. *Uragoga ipecacuanha* from Brazil is the source of ipecac used in medicine.

**Economic Importance**   The family is of significant importance for coffee and quinine as well as several ornamentals.

## SUBCLASS VI, ASTERIDAE; ORDER 9, DIPSACALES

### Caprifoliaceae: Honeysuckle Family

**Field Recognition**   Shrubs, sometimes lianas; leaves opposite; flowers 4- or 5-merous, epigynous; calyx fused to ovary; gynoecium multicarpellate, ovary inferior; CA $\widehat{4-5}$ CO $\widehat{4-5}$ A $^{4-5}$ $\overline{G}$ $\widehat{2 \text{ or } 5 \text{ or } 8}$.

**Description**   *Habit*   Shrubs, sometimes lianas, rarely herbs.
*Leaves*   Opposite, usually simple; without stipules or else stipules very small.
*Inflorescences*   Cymose.
*Flowers*   Bisexual, actinomorphic or zygomorphic. Calyx of 4 to 5 united sepals, fused to ovary. Corolla of 4 to 5 united petals, epigynous, sometimes 2-lipped. Androecium of 5 stamens, epipetalous, inserted on corolla tube and alternate with its lobes. Gynoecium a compound pistil of 2, 5, or 8 carpels, with 2 to 5 locules, ovules 1 or more, placentation pendulous or axile, ovary inferior, style terminal, often slender.
*Fruit*   A berry or drupe.
*Seed*   With endosperm.

**Size, Distribution, and General Information**   A family of about 12 genera and 450 species mostly of north temperate and tropical mountain distribution.
Chief genera are *Viburnum* (200) (snowball); *Abelia* (30); *Diervilla* (3) (bush honeysuckle); *Symphoricarpos* (18) (snowberry, coralberry); *Sambucus* (40) (elderberry); *Lonicera* (200) (honeysuckle); *Linnaea* (1 to 3) (twinflower); *Kolkwitzia* (1) (beauty bush); and *Triosteum* (10) (horse gentian).
Children of countless generations have enjoyed the sweet nectar from the flowers of honeysuckle (*Lonicera*). Many species of *Lonicera* are used as ornamental shrubs or vines. The introduced *L. japonica* is a troublesome weed in the eastern United States.
The fruit of *Sambucus* is used to make elderberry wine or elderberry pies or preserves.

Many *Viburnum* species and cultivars are used as showy ornamental shrubs. All have white flowers; some are evergreen, others deciduous. In some cultivars, the fruits are attractive. The outer flowers of the cymose corymb of some species and cultivars of *Viburnum* are sterile; in other cultivars all the flowers are sterile. *Viburnum trilobum* (highbush cranberry) grows all across North America from New England to British Columbia, and the fruits are used in jelly.

**Economic Importance**    The family is important for a number of ornamental shrubs and vines.

## SUBCLASS VI, ASTERIDAE; ORDER 10, ASTERALES

### Compositae or Asteraceae: Sunflower or Aster Family

**Field Recognition**    Inflorescences of involucrate heads (capitula); pappus often present; corolla united and 5-lobed; stamens 5, forming a cylinder around the style; ovary bicarpellate and uniloculate with a single ovule; the basic floral plan for the family is $CA^X CO^{\textcircled{5}} A^{\textcircled{5}} \overline{G}^{\textcircled{2}}$.

**Description**    *Habit*    Herbs, shrubs, infrequently trees and vines. (Largely herbaceous in the temperate zone.)
*Leaves*    Alternate, opposite, or rarely whorled, simple but rarely truly compound, sometimes greatly reduced; without stipules.
*Inflorescences*    An involucrate head or capitulum with 1 to $\infty$ florets (flowers) on a common receptacle surrounded by an involucre of phyllaries. The heads are variously arranged.
*Flowers*    Bisexual or unisexual, actinomorphic or zygomorphic. Calyx epigynous, represented by the variable pappus. Corolla of 5 united petals, 5-lobed and tubular, or ligulate with 3 to 5 teeth or 2-lipped with 3-lobed upper lip and 2-lobed lower lip. Androecium of (4)5 stamens, epipetalous, with the anthers nearly always united, anthers mostly included within the corolla tube, filaments free. Gynoecium a compound pistil of 2 united carpels, with 1 locule, ovule 1, ovary inferior, style with two branches of various shapes. See Figures 14-23 and 14-24.
The florets within the heads or capitula are of five basic plans: (1) tubular (disc) florets, corolla regular, florets perfect; (2) ligulate or ray florets, zygomorphic, pistillate; (3) ligulate or ray florets, zygomorphic, sterile; (4) ligulate or ray florets, zygomorphic, perfect; and (5) tubular florets, corolla regular, florets pistillate. A head or capitulum may contain: (1) ligulate florets only; (2) tubular florets only; or (3) both ligulate and tubular florets.
*Fruit*    An achene (sometimes called a *cypsela*).

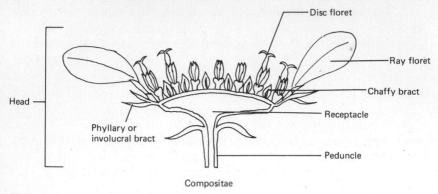

**Figure 14-23**  Longitudinal diagrammatic view of Compositae head.

**Systematics**    Although a large family, it is recognized by a number of common characters which are distinctive. The family is usually divided into two subfamilies and 12 tribes (Willis, 1973).

I   Asteroideae (Tubuliflorae): at least some tubular florets present in each head; sap not usually milky.
   1   Heliantheae (including Helenieae): Style with crown of hairs below the branches; anthers rounded at base; pappus not hairy; phyllaries not membranous at margins; receptacle naked or chaffy. A few are wind-pollinated with separate stamens and unisexual heads.
   2   Astereae: All or only the central florets tubular; anthers blunt at base;

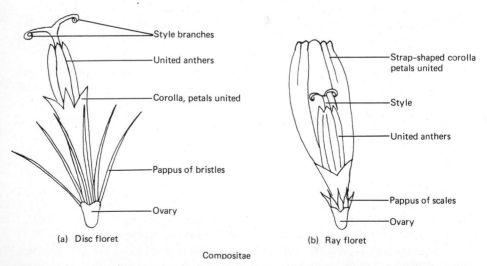

**Figure 14-24**  Diagrammatic drawings of (a) disc floret and (b) ray floret of the family Compositae.

stigmas flattened with marginal row of papillae and a terminal, hairy, segment.

3  Anthemideae: As in tribe Heliantheae, but with phyllaries membranous at tip and edges.

4  Arctotideae: Style thickened or with a ring of hairs below branches; anthers acute at base or with a longer or shorter point and with filaments inserted above the base.

5  Inuleae: As in tribe Astereae, with tubular florets 4- to 5-lobed; anthers tailed at base.

6  Senecioneae: As in Heliantheae, but with a hairy pappus.

7  Calenduleae: Heads with pistillate ray florets and usually staminate tubular florets; style undivided; anthers pointed at base; pappus absent; receptacle not chaffy.

8  Eupatorieae: Florets all tubular, never clearly yellow; anthers blunt at base; stigmas long, but blunt or flattened at tip, with short hairs, stigmatic papillae in marginal rows.

9  Vernonieae: Florets tubular, not yellow (in ours); anthers acute, acuminate or rarely tailed; stigmas semicylindrical, long, pointed, hairy outside; stigmatic papillae inside style branches.

10  Cynareae: Florets tubular, outer florets usually sterile (i.e., disciform); anthers usually tailed; styles thickened or with a circle of hairs below stigma; receptacle usually chaffy.

11  Mutisieae: Florets tubular or sometimes with 2-lipped rays, disc florets actinomorphic or 2-lipped.

II  Lactucoideae (Liguliflorae): Florets all ligulate; sap milky.

   1  Lactuceae (Cichorieae).

**Size, Distribution, and General Information**  A family of about 900 genera and 22,000 species worldwide in distribution. One of the largest families, if not the largest family, of flowering plants.

Chief genera are the following:

1  Heliantheae: *Helianthus* (110) (sunflowers); *Silphium* (15) (rosinweed, compass plant); *Xanthium* (30) (cocklebur); *Ambrosia* (35 to 40) (ragweed); *Zinnia* (20); *Dahlia* (27); *Bidens* (230) (tickweed); *Cosmos* (25); *Helenium* (40) (bitterweed, sneezeweed); *Tagetes* (50) (marigold); *Rudbeckia* (25) (coneflower, black-eyed Susan); and *Gaillardia* (28) (blanketflower).

2  Astereae: *Solidago* (100) (goldenrod); *Bellis* (15) (English daisy); *Aster* (500); *Erigeron* (200) (fleabane); *Baccharis* (400) (groundsel tree); and *Chrysothamnus* (12) (rabbit bush).

3  Anthemideae: *Achillea* (200) (yarrow, milfoil); *Anthemis* (200) (dog fennel); *Chrysanthemum* (200); *Matricaria* (52) (feverfew); *Tanacetum* (50–60) (tansy); and *Artemisia* (400) (wormwood, sagebrush).

4  Arctotideae: *Arctotis* (65); and *Gazania* (40).

5  Inuleae: *Inula* (200); *Antennaria* (100) (pussytoes); and *Gnaphalium* (200) (cudweed, rabbit tobacco).

6  Senecioneae: *Senecio* (2000 to 3000) (groundsel, ragwort).

7   Calenduleae: *Calendula* (20–30) (pot marigold); and *Dimorphotheca* (7) (cape marigold).

8   Eupatorieae: *Eupatorium* (1200) (boneset, joe-pye); *Mikania* (250); and *Liatris* (40) (blazing star, gay flower).

9   Vernonieae: *Vernonia* (1000 +) (ironweed); *Elephantopus* (32) (elephant's foot); and *Stokesia* (1).

10   Cynareae: *Cynara* (14) (artichoke); *Echinops* (100) (globe thistle); *Arctium* (5) (burdock); *Carduus* (100) (thistle); *Cirsium* (150) (thistle); *Centaurea* (600) (cornflower, bachelor's button); and *Carthamus* (safflower).

11   Mutisieae: *Mutisia* (60); and *Gerbera* (70) (gerbera daisy, Transvaal daisy).

12   Lactuceae (Cichorieae): *Cichorium* (9) (chicory, endive); *Crepis* (200) (hawk's-beard); *Hieracium* (1000) (hawkweed); *Taraxacum* (62) (dandelion); *Lactuca* (100) (lettuce); *Tragopogon* (50) (goatsbeard, salsify); and *Sonchus* (50) (sow thistle).

**Economic Importance**   Relative to its numbers, Compositae is not as economically important as several other families are. However, many Compositae are valuable ornamentals—e.g., *Aster, Callistephus, Chrysanthemum, Dahlia, Coreopsis, Gerbera, Helichrysum, Tagetes, Zinnia, Rudbeckia.*

Several are sources of food: *Lactuca, Cynara, Cichorium, Tragopogon, Helianthus,* and *Carthamus.*

Ragweed (*Ambrosia*) pollen is a major cause of hayfever. Some Compositae have sesquiterpene lactones that cause contact dermatitis in humans. Many Compositae are weedy.

*Chrysanthemum coccineum* is the source of the natural insecticide pyrethrum. The red dye safflower, obtained from *Carthamus tinctoria,* is used for coloring butter, liqueurs, candles, and rouge. Eurasian wormwood (*Artemisia absinthium*) was once used to flavor a liqueur; however, absinthe caused addiction, and its use has been declared illegal.

# CLASS LILIOPSIDA: Monocots

## SUBCLASS I, ALISMATIDAE; ORDER 1, ALISMATALES

### Alismataceae: Water Plantain or Arrowhead Family

**Field Recognition**   Aquatic or wetland herbs; flowers bisexual, often whorled, perianth in 2 series; sepals 3, green and sepal-like; petals 3, white; gynoecium apocarpous, usually with many free carpels; $CA^3 \, CO^3 \, A^{6-\infty} \, \underline{G}^{6-\infty}$.

**Description**   *Habit*   Annual or perennial, aquatic or wetland herbs from stout rhizomes; erect or rarely with floating leaves.

*Leaves*   Basal, long-petioled, sheathing, blades linear-lanceolate or ovate, bases sometimes hastate or sagittate, veins parallel with the margins and converging at apex of blade.

*Inflorescences*   Flowers often whorled or in racemes or panicles.

*Flowers*   Bisexual or rarely polygamous, actinomorphic. Calyx of 3 green sepals. Corolla of 3 petals, white, larger than sepals, deciduous or rarely absent. Androecium of $(3)^{6-\infty}$ stamens, free. Gynoecium of 6 or more simple pistils each having 1 carpel, with 1 locule, ovary superior, style persistent, receptacle flat to globose.

*Fruit*   A spiral or ring of achenes.

*Seed*   With a curved embryo.

**Size, Distribution, and General Information**   A family of about 13 genera and 90 species of cosmopolitan distribution.

The chief genera are *Alisma* (10) (water plaintain); *Echinodorus* (30) (burhead); and *Sagittaria* (20) (wapata, arrowhead, swamp potato).

In North America, the Indians once dug the rhizomes of *Sagittaria* for food, hence the name *swamp potato*. A Eurasian species (*S. sagittifolia*) is occasionally cultivated.

**Economic Importance**   The family is of little importance.

### SUBCLASS I, ALISMATIDAE; ORDER 3, NAJADALES

### Potamogetonaceae: Pondweed Family

**Field Recognition**   Aquatic herbs often submerged or with both floating and submerged leaves; flowers with perianth of 4 segments, rudimentary, or none; stamens 1 to 4; gynoecium of 1 to 4 separate carpels, 1-loculed, with 1 seed per ovary; $CA^{0 \text{ or } 4} CO^0 A^4 \underline{G}^{1-4}$.

**Description**   *Habit*   Aquatic, freshwater herbs, with creeping rhizomes, stems mostly submerged.

*Leaves*   Submerged or floating, sheathing, alternate or opposite.

*Inflorescences*   Axillary spikes, peduncle surrounded by a sheath at the base.

*Flowers*   Bisexual. Perianth of 4 free, rudimentary, short-clawed segments, or none. Androecium of 4 stamens, inserted at or on the claws of the perianth segments. Gynoecium of 4 simple pistils, each having 1 carpel and 1 locule, ovary superior, stigmas sessile or on short styles.

*Fruit*   Sessile, free, 1-seeded, indehiscent, drupe-like when fresh.

*Seed*   With no endosperm.

**Size, Distribution, and General Information**   A family of 2 genera and about 100 species of cosmopolitan distribution.

The two genera are *Potamogeton* (100) (pondweed) and *Groenlandia* (1).

The sago pondweed (*Potamogeton pectinatus*) fruits in autumn during the fall migration of waterfowl and provides much food for ducks. Some species are troublesome weeds in ponds and lakes.

**Economic Importance**   The family is of little importance.

## SUBCLASS II, ARECIDAE; ORDER 1, ARECALES

### Palmae or Arecaceae: Palm Family

**Field Recognition**   Woody shrubs or trees; leaves large, fan-shaped or pinnately compound, with sheathing bases; inflorescence paniculate, large, spathes often present; flowers small; $CA^3 CO^3 A^6 \underline{G}^{3 \text{ or } ③}$ or $CA^3 CO^3 A^6 G^0$ and $CA^3 CO^3 A^0 \underline{G}^{3 \text{ or } ③}$.

**Description**   *Habit*   Trees or shrubs, sometimes climbing, stem stout to slender, short to over 30 meters tall, trees usually unbranched.

*Leaves*   Blades palmate, pinnate, or simple, often very large, petiole sheathing at base, leaves usually in a terminal cluster.

*Inflorescences*   Large, paniculate, among or below the leaves, subtended by one or more spathes, plants monoecious or dioecious or flowers bisexual.

*Flowers*   Unisexual or bisexual, actinomorphic. Calyx of 3 sepals, small, separate or united. Corolla of 3 petals, small, separate or united. Androecium of usually 6 stamens, in 2 series. Gynoecium usually a compound pistil having 3 united carpels, with 1 to 3 locules, ovule solitary, ovary superior, rudimentary in staminate flowers.

*Fruit*   Berries or drupes, exocarp fleshy or leathery.

*Seed*   Endosperm relatively large, sometimes milky, embryo small.

**Size, Distribution, and General Information**   A family of about 215 genera and 2500+ species of wide tropical and subtropical distribution, a few in warm temperate regions. The palms form a characteristic feature of tropical vegetation.

Some of the genera are *Areca*(54) (betal nut); *Cocos* (1) (coconut palm); *Lodoicea* (1) (double coconut, coco-de-mer); *Phoenix* (17) (date palm); *Phytelephas* (15) (ivory-nut palm or vegetable ivory); *Elaeis* (2) (oil palm); *Calamus* (375) (rattan, rattan cane); *Sabal* (25) (palmetto, cabbage plam, cabbage tree); *Serenoa* (1) (saw palmetto); *Washingtonia* (2) (fan palm); *Copernicia* (30) (wax palm); *Metroxylon* (15) (sago palm); *Thirnax* (12) (thatch palm); *Coccothrinax* (50) (Biscayne palm); and *Roystonea* (17) (royal palm).

The date palm (*Phoenix dactylifera*) has been cultivated for at least 5000 years. Each pistillate tree can yield as much as 50 kilograms of dates each year.

The date palm is dioecious, and pollen from staminate trees must be hand-transferred to the pistillate trees. Other species of *Phoenix* are tapped for a sweet sap which can be made into palm sugar, or jaggery, or fermented to make palm "wine." Various species of the genus *Metroxylon,* native to India and Malaysia, yield starch or sago made from the pith of the stems.

*Areca catechu* is cultivated in tropical Asia for its seeds called *betel nuts.* The seed is cut into slices and rolled up with a little slaked lime in a leaf of betel pepper (*Piper betle*). When chewed, it turns the saliva bright red; it stimulates the digestive organs and is enjoyed by millions of people.

Oil palm (*Elaeis guineensis*), native to West Africa, yields a valuable oil that is both edible and useful as a lubricant. Increasing numbers of oil palm plantations are being planted in tropical countries, causing some commercial competition for soybean growers.

Coconuts (*Cocos nucifera*) are widely grown in tropical areas, especially near the sea. The young fruit contains a pint or more of a sweetish fluid. The kernels are eaten raw or cooked; oil is extracted by boiling or pressure. The coconut has literally hundreds of uses in areas where it is grown. *Lodoicea maldivica* (the double coconut or coco-de-mer) is one of the largest fruits known and reputedly takes 10 years to ripen. The terminal buds of a number of species of palms are used for food and are called *hearts of palm.* Each time a heart of palm is obtained, a palm tree must be killed. Cabbage palm (*Sabal palmetto*) was once much used in this fashion by Seminole Indians in Florida.

**Economic Importance** The palm family is of considerable importance in tropical and subtropical regions, where it is used for food, shelter, clothing, copra, coconuts, oils, dates, rattan, ivory nuts, and wax. In addition, the palm family is important everywhere because the products of tropical palms enter temperate-zone commerce.

## SUBCLASS II, ARECIDAE; ORDER 4, ARALES

### Araceae: Arum, Philodendron, or Aroid Family

**Field Recognition** North American species are herbaceous; plants often of wet habitats; inflorescence a spadix enveloped or subtended by a single spathe; flowers small or minute; $CA^0 \ CO^0 \ A^{6 \text{ or less}} \ \underline{G}^{2 \text{ or } 3}$.

**Description** *Habit* Variable, herbs with aerial stems, tubers, or rhizomes; climbing shrubs, epiphytes; or marsh plants. Sap often contains raphides (calcium oxalate crystals).

*Leaves* Variable, solitary or few, mostly basal (in ours), or alternate (in tropical species); entire to variously arranged, often hastate or sagittate, with a sheath at the base of the petiole.

*Inflorescences*   A spadix usually subtended by a spathe.

*Flowers*   Bisexual or unisexual, small, often bad-smelling and attracting carrion flies. Perianth absent or small and scale-like of 4 to 6 parts. Androecium of 6 or fewer stamens, opposite the perianth segments, staminodes sometimes present. Gynoecium a compound pistil (1)2 to 3(9) united carpels, with 1 to several locules, placentation parietal, axile, basal or apical, ovary superior or immersed in the spadix, style various, sometimes greatly reduced. See Figure 14-25.

*Fruit*   A berry, often many clustered closely together on the spadix and resembling a multiple fruit. Fruit sometimes leathery and rupturing, 1 to many seeds.

*Seed*   With or without endosperm.

**Size, Distribution, and General Information**   A family of about 115 genera and 2000 species, primarily of tropical and subtropical distribution but with about 8 percent of its species in temperate regions.

Some of the genera are *Arum* (15) (arrowroot); *Anthurium* (550); *Arisaema* (150) (jack-in-the-pulpit, Indian turnip); *Peltandra* (4) (arrow arum); *Symplocarpus* (1) (eastern skunk cabbage); *Orontium* (1) (golden club, never-wet); *Acorus* (2) (sweet flag); *Philodendron* (255); *Pistia* (1) (water lettuce); *Lysichiton* (2) (western skunk cabbage); *Zantedeschia* (8 to 9) (florist's calla lily); *Pothos* (75); *Dieffenbachia* (30) (dumb cane); *Caladium* (15); *Monstera* (50); *Aglaonema* (21) (Chinese evergreen); *Colocasia* (8) (taro); and *Alocasia* (70) (elephant ear).

In rich, moist woods of eastern North America, there are several species of *Arisaema*, often called *jack-in-the-pulpit* or *Indian turnip*. It has an upright spadix as "jack" and a spathe as the "pulpit." The rhizomes of *Arisaema* may be eaten if boiled and dried to eliminate the oxalates. In Southeast Asia and the Pacific Islands, taro (*Colocasia esculenta*) is cultivated for its rhizome which when boiled loses its poisonous nature and forms valuable starchy food. It is the *poi* of the Hawaiian culture.

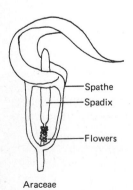

Spathe

Spadix

Flowers

Araceae

**Figure 14-25**   Longitudinal section of an inflorescence from the family Araceae. The flowers are clustered into a grouping called a *spadix* and subtended by a bract called a *spathe*. (Note: The spathe is cut away to show the spadix.)

The western skunk cabbage (*Lysichiton americanum*) and the eastern skunk cabbage (*Symplocarpus foetidus*) are unusual wild flowers of cold, wet places. The fetid odor of the flower attracts flies in large numbers. Never-wet or golden club (*Orontium aquaticum*) is a lovely wild flower of aquatic habitats in the coastal plain of the Eastern states. The golden yellow spadix does not have an obvious spathe.

Dumb cane (*Dieffenbachia*), often grown as a foliage plant, gets its name from the prompt irritation and swelling of the tongue and mouth that develops when a piece of leaf is chewed. *Monstera deliciosa*, with giant slotted leaves, is grown as a popular foliage plant. The ripened fruits are delicious when chilled and served as a dessert. Many species of *Philodendron* are grown as houseplants.

Many cultivars of *Caladium*, often called *fancy-leaved caladiums*, are grown as greenhouse and bedding plants. The white spathe of *Caladium* is hidden in the foliage and usually overlooked because of the attractive, brightly colored leaves. *Anthurium* cultivars have large, often bright red, spathes with a slender spadix. The calla lily of florists is *Zantedeschia aethiopica*.

The aroid family includes the free-floating aquatic water lettuce (*Pistia stratiotes*). It is sometimes used in aquaria, has escaped cultivation, and may be a weed in the warmer parts of the United States.

**Economic Importance**   The aroid family is of considerable importance as a source of ornamentals and a starchy food (taro).


## SUBCLASS III, COMMELINIDAE; ORDER 1, COMMELINALES

### Commelinaceae. Spiderwort Family

**Field Recognition**   Herbs with succulent stems; leaves with a closed basal sheath; sepals 3, green; petals 3, often blue; filaments usually hairy; ovary superior, style simple; $CA^3 \ CO^3 \ A^6 \ \underline{G}^{\circledß}$.

**Description**   *Habit*   Herbaceous annuals or perennials; stems jointed.
*Leaves*   Sheathing at the base, alternate, parallel-veined, entire.
*Inflorescences*   Axillary clusters or terminal cymes or panicles.
*Flowers*   Bisexual, usually actinomorphic (zygomorphic in *Commelina*). Calyx of 3 sepals, green, free. Corolla of 3 petals, equal or unequal, usually ephemeral, rarely united. Androecium of 6 stamens, sometimes 3 with 3 staminodes or reduced to 1 or 2 fertile stamens, filaments often hairy. Gynoecium a compound pistil of 3 united carpels, with 3 locules, ovules few to solitary, placentation axile, ovary superior, style terminal and simple.
*Fruit*   A capsule.
*Seed*   With endosperm.

**Size, Distribution, and General Information**    A family of about 38 genera and 500 species of mostly tropical and subtropical distribution, but also in warm temperate regions.

Some of the chief genera are *Commelina* (230) (dayflower); *Aneilema* (100); *Tradescantia* (60) (spiderwort); *Rhoeo* (1) (Moses-in-the-bulrushes); and *Zebrina* (4 to 5) (wandering Jew).

*Tradescantia* is sometimes called *spiderwort* because the soft, stringy mucilaginous material can be pulled from the broken ends of the stem. The material will harden into a cobweb-like thread after exposure to the air. *Commelina* has two large blue petals and a third quite small petal. The flowers are in tight clusters; each day one flower opens then closes during the day. The next day, a new flower will open—hence the name *day flower.*

*Zebrina pendula* (wandering Jew) is a favorite, easily grown houseplant. *Rhoeo spathacea* produces clusters of flowers located between a pair of large bracts almost hidden in some rather stiff leaves. This suggests the common name *Moses-in-the-bulrushes.*

**Economic Importance**    The family is of little importance except for several ornamentals.

## SUBCLASS III, COMMELINIDAE; ORDER 4, JUNCALES

### Juncaceae: Rush Family

**Field Recognition**    Tufted herbs; leaves reduced; perianth in 2 series with 6 segments; stamens 6; fruit a capsule; $CA^3 CO^3 A^6 \underline{G}^{\circled{3}}$.

**Description**    *Habit*    Perennial or annual tufted herbs, rarely shrub-like, stems usually leafy only at the base.

*Leaves*    Basal, cylindrical to flat, linear or filiform, sheathing basally or reduced to only a sheath.

*Inflorescences*    Flowers in cymes grouped in a panicle, corymb, or head, or flowers solitary.

*Flowers*    Usually bisexual, actinomorphic, small. Perianth segments 6, in 2 series, scale-like, greenish or brownish. Androecium of (3)6 stamens, opposite the perianth segments. Gynoecium a compound pistil of 3 united carpels, 1-locular or with 3 locules, ovules 3 to numerous, placentation parietal, ovary superior, styles 3.

*Fruit*    A capsule.

*Seed*    With endosperm and a straight embryo.

**Size, Distribution, and General Information**    A family of about 9 genera

and 400 species of temperate, arctic, or tropical mountain distribution, in moist, cool places.

The chief genera are *Juncus* (300) (rush); and *Luzula* (80) (wood rush).

The tussocks of rushes often provide stepping-stones for crossing a wet meadow, as well as hiding places for many kinds of birds and mammals. In some countries, rushes are woven into mats, baskets, or chair seats. The pith has been used for candlewicks.

**Economic Importance** The Juncaceae are of little importance.

## SUBCLASS III, COMMELINIDAE; ORDER 5, CYPERALES

### Cyperaceae: Sedge Family

**Field Recognition** Herbs, stems often triangular, solid; leaves 3-ranked, sheaths closed; ligule absent; flowers subtended by a single bract, lodicule absent; perianth represented by bristles, scales or absent; $CA^0 CO^0 A^{1-3} \underline{G}^{\textcircled{2 \text{ or } 3}}$ or $CA^0 CO^0 A^{1-3} G^0$ and $CA^0 CO^0 A^0 \underline{G}^{\textcircled{2 \text{ or } 3}}$. (The Cyperaceae are easily confused with the Gramineae, but are readily distinguished by the combination of characters listed above.)

**Description** *Habit* Perennial or annual herbs, often of damp or marshy habitats, with solid, usually triangular stems.

*Leaves* Three ranked, with a closed sheath, the ligule absent, often in a basal tuft or crowded on the lower part of stem, blades narrow and grass-like.

*Inflorescences* Spicate, racemose, paniculate, or umbellate; a spikelet is often the basic unit (spikelet unlike that of family Gramineae); spikelets usually solitary within a bract; bracts distichously or spirally arranged; plants monoecious or dioecious; inflorescence often subtended by one or more bracts.

*Flowers* Bisexual or unisexual, inconspicuous. Perianth represented by bristles or scales, often absent. Androecium of 1 to 3(6) stamens. Gynoecium a compound pistil of 2 or 3 united carpels, with 1 locule, ovule solitary, ovary superior, with 2 or 3 style branches. See Figures 14-26 and 14-27.

*Fruit* A triangular or lens-shaped achene or nutlet.

**Systematics** Tribes of Cyperaceae and representative genera: Rhynchosporeae (*Rhynchospora*); Scirpeae (*Scirpus*); Cypereae (*Cyperus*); Hypolytreae (*Hypolytrum*); Sclerieae (*Scleria*); Cryptangieae (*Cryptangium*); and Cariceae (*Carex*).

**Size, Distribution, and General Information** A family of about 90 genera

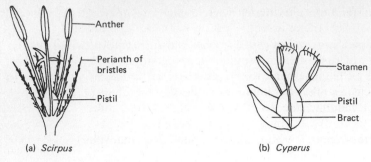

Cyperaceae

**Figure 14-26** Longitudinal views of flowers of *(a) Scirpus* and *(b) Cyperus* (Cyperaceae). The perianth is replaced by bristles or is absent in *Cyperus.*

and 4000 species of worldwide distribution, especially in temperate and cold regions.

Some of the chief genera are *Cyperus* (550) (sedge); *Eriophorum* (20) (cotton grass); *Scirpus* (300) (bulrush, tule); *Eleocharis* (200) (spike rush); *Fimbristylis* (300); *Rhynchospora* (200) (beak rush); *Scleria* (200) (nut rush); *Carex* (1500–2000) (carex, sedge); and *Dichromena* (60) (whitetop sedge).

Perhaps as early as 2400 B.C., Egyptians made a paper (papyrus) from *Cyperus papyrus*, a plant of marshy or riverside habitats. The stems were split into thin strips which were pressed together while still wet. *Cyperus papyrus* is often grown as an ornamental.

Nut grass (*Cyperus rotundus*) is a serious weed that is extremely difficult

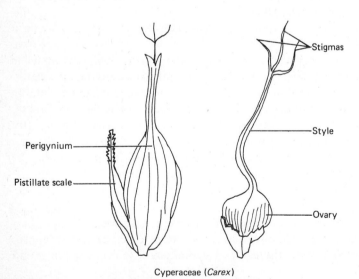

Cyperaceae (*Carex*)

**Figure 14-27** Longitudinal views of pistillate floral structures of *Carex* (Cyperaceae). The drawing to the right has the perigynium partially removed.

to control in lawns and cultivated fields. The closely related *C. esculentus* (chufa or Zulu nuts) produces an edible tuber. Chufas are sometimes planted as a wildlife food plant. *Cyperus alternifolius* (umbrella plant) is often grown as an ornamental.

The name *bulrush* is applied to the plants of the genus *Scirpus* which are found in wetlands or aquatic habitats. The fruits and rhizomes are eaten by waterfowl and muskrats. At Lake Titicaca in the high Andes, Indians build fishing canoes from *S. totora*. Cotton grass (*Eriophorum*) inhabits poorly drained areas in cold zones of both the Northern and Southern Hemispheres. *Carex* is a large genus of much taxonomic complexity.

**Economic Importance**   The Cyperaceae is of no significant importance. A few are grown as ornamentals and some are noxious weeds. Some sedge hay is produced for fodder in cold, wet regions, but it is inferior to grass hay.

## SUBCLASS III, COMMELINIDAE; ORDER 5, CYPERALES

### Gramineae or Poaceae: Grass Family

**Field Recognition**   Stems round, internodes usually hollow; leaves 2-ranked, sheaths open or sometimes closed, ligule usually present; bracts of glumes, lemma and palea; perianth of 0–3 lodicules; stamens usually 3; fruit a grain or caryopsis; $CA^0$ $CO^0$ $A^3$ $\underline{G}^{2 \text{ or } 3}$. (The Gramineae may be confused with the Cyperaceae and Juncaceae. The grasses have round stems, the sedges have triangular stems, and the rushes have many-seeded capsules and round stems.)

**Description**  *Habit*   Annual or perennial herbs, rarely shrubs or trees; stems (culms) erect, ascending, prostrate, or creeping; culms round, usually hollow at the internodes, solid at nodes, a few are solid throughout.

*Leaves*   Two-ranked, alternate, parallel-veined, consisting of sheath, ligule, and blade; sheath encircling the culm.

*Inflorescences*   Composed of units called *spikelets,* variously arranged in spikes, racemes, or panicles; each spikelet has one or more florets, florets arranged on an axis (rachilla), 2 glumes subtend each spikelet.

*Flowers*   Bisexual or sometimes unisexual. Each floret has 2 bracts: the outer (lemma) and the inner (palea). Perianth reduced to 2 or 3 lodicules located just above the palea, or sometimes absent. Androecium of (1)3(6) stamens. Gynoecium a compound pistil of 2 or 3 united carpels, with 1 locule, ovary superior, (1)2(3) styles, stigmas usually feather-like. See Figures 14-28 and 14-29.

*Fruit*   A caryopsis (grain), rarely a nut, berry, or a utricle.

*Seed*   With starchy endosperm.

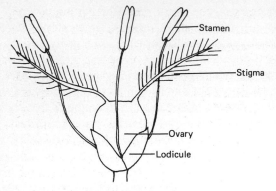

Grass flower (Gramineae)

**Figure 14-28** Longitudinal view of a grass flower (family Gramineae). The lodicules are analogous to the perianth.

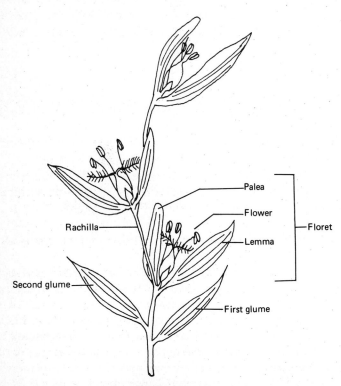

Grass spiklet (Gramineae)

**Figure 14-29** Grass spikelet (family Gramineae). A typical grass spikelet is subtended by the first and second glumes (bracts) and contains one or more florets, each subtended by an outer lemma (bract) and an inner palea (bract).

**Systematics**   The classification of the family Gramineae has been inten-
sively studied in recent years. The treatment presented here (Gould, 1968)
differs substantially from that of Hitchcock (1950). The changes largely reflect
new information from cytogenetics and comparative plant anatomy.

**Size, Distribution, and General Information**   A family of about 620 genera
and 10,000 species forming one of the larger families of flowering plants; widely dispersed in all regions of the world wherever vascular plants can survive,
often dominating the aspect of the vegetation.

Some of the genera are:

Subfamily Festucoideae: Tribe 1. Festuceae: *Bromus* (50) (bromegrass);
*Festuca* (80) (fescue); *Lolium* (12) (ryegrass); *Poa* (300) (bluegrass); and
*Dactylis* (5) (orchard grass). Tribe 2. Aveneae: *Avena* (70) (oats); *Phalaris* (20)
(canary grass); *Ammophila* (2) (beach grass); *Agrostis* (150 to 200) (bent
grass); and *Phleum* (15) (timothy). Tribe 3. Triticeae: *Triticum* (20) (wheat);
*Agropyron* (100 to 150) (wheatgrass); *Secale* (5) (rye); *Elymus* (70) (wild rye);
and *Hordeum* (20) (barley). Tribe 4. Meliceae: *Melica* (70) (melic grass); and
*Glyceria* (40) (manna grass). Tribe 5. Stipeae: *Stipa* (300) (needlegrass); and
*Oryzopsis* (50), (rice grass). Tribe 6. Brachyelytreae: *Brachyelytrum* (4). Tribe
7. Diarrheneae: *Diarrhena* (1). Tribe 8. Nardeae: *Nardus* (1). Tribe 9. Moner-
meae: *Monerma* (1).

Subfamily Panicoideae: Tribe 10. Paniceae: *Panicum* (500) (millet, panic
grass); *Digitaria* (380) (crabgrass); *Paspalum* (250) (Dallis grass, Bahia grass);
*Echinochloa* (30) (barnyard grass, jungle rice); and *Setaria* (140) (foxtail,
Italian millet). Tribe 11. Andropogoneae: *Andropogon* (113) (bluestem, broom
sedge); *Saccharum* (5) (sugarcane); *Schizachyrium* (50) (little bluestem);
*Sorghum* (60) (milo, sorghum, Johnson grass, Sudan grass); and *Zea* (1) (corn,
maize).

Subfamily Eragrostoideae: Tribe 12. Eragrosteae: *Eragrostis* (300) (love
grass); *Muhlenbergia* (100) (muhly grass, bull grass); and *Sporobolus* (150)
(dropseed, smut grass). Tribe 13. Chlorideae: *Eleusine* (9) (goose grass);
*Cynodon* (10) (Bermuda grass); *Bouteloua* (40) (side oats grama, grama grass);
*Buchlöe* (1) (buffalo grass); and *Spartina* (16) (cord grass). Tribe 14. Zoysieae:
*Zoysia* (10). Tribe 15. Aeluropodeae: *Distichlis* (13) (salt grass). Tribe 16.
Unioleae: *Uniola* (2) (sea oats). Tribe 17. Pappophoreae: *Pappophorum* (7)
(pappus grass). Tribe 18. Orcuttieae: *Orcuttia* (4). Tribe 19. Aristideae: *Aris-
tida* (330) (wire grass, three-awn grass).

Subfamily Bambusoideae: Tribe 20. Bambuseae: *Arundinaria* (150) (giant
cane, switch cane). Tribe 21. Phareae: *Pharus* (7 to 8).

Subfamily Oryzoideae: Tribe 22. Oryzeae: *Oryza* (25) (rice); *Zizania* (3)
(wild rice).

Subfamily Arundinoideae: Tribe 23. Arundineae: *Phragmites* (3) (common reed). Tribe 24. Danthonieae: *Danthonia* (10) (poverty oat grass). Tribe
25. Centotheceae: *Chasmanthium* (5).

**Economic Importance**   The grass family is of greater importance than any other family of flowering plants, as indicated by the following: (1) food crops for human consumption—rice, wheat, corn, barley, millet, rye, oats, milo, and cane sugar; (2) forage and grain for domesticated animals; (3) range forage (in North America—big bluestem, little bluestem, Indian grass, switch grass, side oats grama, blue grama, and buffalo grass); (4) industrial uses—ethyl alcohol, starch, and so on; (5) shelter—bamboo, thatch; (6) soil conservation; (7) turf—Bermuda grass, St. Augustine grass, centipede grass, bluegrasses, bent grasses, fescues, ryegrass; (8) ornamentals; and (9) wildlife food.

## SUBCLASS IV, ZINGIBERIDAE; ORDER 1, BROMELIALES

### Bromeliaceae: Pineapple Family

**Field Recognition**   Usually epiphytic with stiff leaves which are often brightly colored towards the base; highly colored 3-merous flowers subtended by brightly colored bracts; $CA^3 CO^3 A^6 \overline{G}^{③}$ or $\underline{G}^{③}$.

**Description**   *Habit*   Mostly short-stemmed epiphytes.

*Leaves*   Mostly basal and rosette forming, usually stiff and often spiny, often brightly colored toward the base.

*Inflorescences*   Flowers in a terminal head, spike, or panicle, often with highly colored bracts.

*Flowers*   Bisexual or rarely unisexual, actinomorphic, hypogynous to epigynous. Perianth in 2 series, outer whorl of 3 calyx-like parts and inner of 3 corolla-like segments. Androecium of 6 stamens, mostly inserted at the base of the perianth, free or partially fused to them. Gynoecium a compound pistil of 3 united carpels, 3-locular, placentation axile, ovary superior or inferior.

*Fruit*   A berry, capsule, or multiple.

*Seed*   With a small embryo in mealy endosperm.

**Size, Distribution, and General Information**   A family of about 44 genera and 1400+ species mostly of tropical America and the West Indies. One species occurs in west tropical Africa.

The chief genera are *Navia* (60); *Pitcairnia* (250); *Puya* (120); *Tillandsia* (500); *Bromelia* (40); and *Ananas* (5).

The pineapple (*Ananas comosus*) was long cultivated by American Indians. It was introduced to the Old World by Spanish and Portuguese travelers. Seedless cultivars were developed by the Europeans and reintroduced into tropical America. The bracts, ovaries, pedicels, and stalk develop and fuse into one sweet, juicy mass.

Most members of the family grow as epiphytes upon trees in the tropical forests of the New World. A few occur in dry areas and some terrestrial species in the Andes have flower stalks 10 to 12 feet tall. *Tillandsia usneoides* (Spanish moss) is a familiar epiphyte hanging in gray masses from trees and telephone wires in the warmer parts of the southeastern states.

**Economic Importance** The family is of importance for the pineapple. Several species are sources of cordage and fiber for fabrics. Many are grown in greenhouses or out-of-doors in tropical areas as ornamentals.

## SUBCLASS V, LILIIDAE; ORDER 1, LILIALES

### Liliaceae (Including Amaryllidaceae): Lily Family

**Field Recognition** Herbaceous perennials; flowers often showy, actinomorphic; sepals 3, petaloid; petals 3; stamens 6, anthers opening by lengthwise slits; ovary superior or inferior; fruit a capsule or berry; $CA^3 CO^3 A^6 \underline{G}^{\textcircled{3}}$ or $\overline{G}^{\textcircled{3}}$.

**Description** *Habit* Mostly perennial herbs, from rhizomes, corms, or bulbs; stems erect or climbing, leafy or scapose.

*Leaves* Alternate, simple, linear, with parallel venation, leaves sometimes few and basal, appearing before or after the flowers.

*Inflorescences* Racemose, axillary, umbellate, or solitary.

*Flowers* Bisexual, usually actinomorphic. Calyx of 3 sepals, nearly always petaloid. Corolla of 3 petals. Androecium of (3)6(12) stamens, filaments free or variously united, anthers opening by a slit lengthwise. Gynoecium a compound pistil of 3 united carpels, with 3 locules, ovules numerous, placentation axile or rarely 1-locular with parietal placenta, ovary superior or inferior. See Figure 14-30.

*Fruit* A capsule or a fleshy berry.

*Seed* With endosperm, embryo straight or curved.

**Systematics** Cronquist (1968) considers separation of the Amaryllidaceae and Liliaceae on the basis of ovary position to be unreasonable. Hutchinson (1973) admits that the character of superior versus inferior ovary has been overemphasized in the monocots, although he treats the families separately. The treatment here follows that of Cronquist with the two families combined. Some genera traditionally included in the Liliaceae are removed to other families: e.g., *Smilax* (Smilacaceae), *Aloe* (Aloeaceae), and so on.

**Size, Distribution, and General Information** A family of about 335 genera and 4800 species of worldwide distribution. Considered separately, the Liliaceae have 250 genera and 3700 species and the Amaryllidaceae 85 genera and 1100 species.

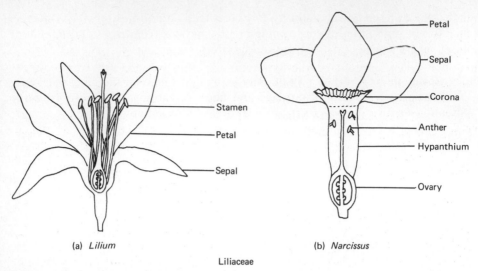

(a) *Lilium*  (b) *Narcissus*

Liliaceae

**Figure 14-30** Longitudinal sections of diagrammatic Liliaceae flowers. *(a) Lilium* with a superior ovary. *(b) Narcissus* with an inferior ovary and a corona. *Narcissus* is placed in Amaryllidaceae by some authors.

Some of the genera are: *Lilium* (80) (lily); *Uvularia* (4) (bellwort); *Hemerocallis* (20) (day lily); *Colchicum* (65) (autumn crocus); *Gloriosa* (5) (gloriosa lily); *Tulipa* (100) (tulip); *Hyacinthus* (30) (hyacinth); *Convallaria* (1) (lily of the valley); *Smilacina* (25) (false Solomon's-seal); *Asparagus* (300); *Amaryllis* (1); *Hippeastrum* (75) (amaryllis); *Crinum* (100 to 110) (swamp lily); *Hymenocallis* (50) (spider lily); *Allium* (450) (onion, chive, garlic, leek, shallot); *Narcissus* (60) (jonquil, daffodil); *Zephyranthes* (35 to 40) (rain lilies); *Lycoris* (10) (surprise lily); *Hypoxis* (100) (yellow star grass); *Erythronium* (25) (fawn lily); *Ornithogalum* (150) (star-of-Bethlehem); *Scilla* (80) (squill); *Trillium* (30) (wake-robin); and *Polygonatum* (50) (Solomon's-seal).

Many species and cultivars of *Lilium* are cultivated: *L. candidum* (Madonna lily); *L. longiflorum* (Easter or white trumpet lily); *L. tigrinum* (tiger lily); *L. auratum* (golden-banded lily); and *L. superbum* (Turk's-cap lily), and so on.

The autumn crocus (*Colchicum autumnale*) produces a poisonous alkaloid, colchicine, which is used in plant breeding to produce polyploids. Colchicine interferes with cell division, resulting in cells with an abnormal double number of chromosomes.

In the large genus *Asparagus*, the leaves are reduced to scales. The young stems of *A. officinalis* provide a delicious vegetable. The asparagus fern (*A. setaceus*) is used in the florist's trade. Sprengeri (*A. densiflorus*) is an ideal plant for pots or hanging baskets.

There are some 450 species of *Allium*, many of which are used either as ornamentals or in cooking. They are famous for their characteristically pungent odor. Best known are *A. cepa* (onion); *A. sativum* (garlic); *A. ascalonicum* (shallot); *A. porrum* (leek); *A. schoenoprasum* (chive); and *A. vineale* (wild onion), which is sometimes a problem to the dairy industry, because cows grazing on *A. vineale* produce garlic-flavored milk. In many species, some or all of the flowers are replaced by bulbils.

Holland is known for its cultivated hyacinths (*Hyacinthus orientalis*), originally from Greece and Asia Minor, and tulips (*Tulipa gesneriana*), introduced from Turkey. *Narcissus* (daffodil, jonquil) are universal spring favorites. Most of the species were natives to southern Europe. Narcissus was a youth in Greek mythology who is said to have changed into a flower by sitting motionless (the root of the word is related to *narcotic*) while pining away for love of his own reflection in a pond.

A number of Liliaceae produce poisonous bulbs: star-of-Bethlehem (*Ornithogalum umbellatum*); lily of the valley (*Convallaria majalis*); squill (*Scilla*); and red squill (*Urginea scilla*). Red squill was once used as an effective rat poison. *Zigadenus,* death camas, is very well known to sheep raisers throughout the West since it is especially poisonous to sheep.

**Economic Importance**    The Liliaceae family is important for the large number of genera used in ornamental horticulture. *Asparagus* and *Allium* are important vegetable crops. Several drugs or poisons are obtained from the family.

## SUBCLASS V, LILIIDAE; ORDER 1, LILIALES

### Iridaceae: Iris Family

**Field Recognition**    Herbs with equitant leaves; sepals 3, petaloid; petals 3; stamens 3; ovary inferior; $CA^3 \ CO^3 \ A^3 \ \overline{G}^\textcircled{3}$.

**Description**    *Habit*    Herbaceous, from rhizomes, bulbs, or corms; stems solitary or several, or plants scapose.

*Leaves*    Mostly basal, often crowded at the base of the stem, mostly linear or broadly linear, flattened at the sides, sheathing at the base, and equitant (folded and overlapping).

*Inflorescences*    Racemose or paniculate, or solitary.

*Flowers*    Bisexual, actinomorphic or zygomorphic, showy. Calyx of 3 sepals, petaloid. Corolla of 3 petals often similar to the sepals. Perianth united below into a tube. Androecium of 3 stamens, opposite the sepals. Gynoecium a compound pistil of 3 united carpels, with 3 locules, ovules usually numerous, placentation axile or 1-locular with 3 parietal placentas, ovary inferior, rarely superior, style 3-lobed, sometimes petaloid. See Figure 14-31.

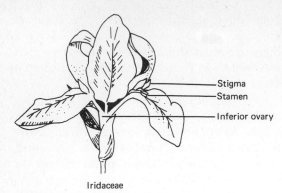

Iridaceae

**Figure 14-31** *Iris* (Iridaceae) flower; note the three-petaloid style branches with the stigma located beneath (the labeled style branch is raised). In turn, the stamens are beneath the style branches.

*Fruit*   A capsule dehiscent by valves.
*Seed*   With a small embryo in hard endosperm.

**Size, Distribution, and General Information**   A family of about 60 genera and 800 species of tropical and temperate distribution with the chief centers of distribution in South Africa and tropical America.

Some of the genera are *Iris* (300) (flag, fleur-de-lis); *Crocus* (75); *Sisyrinchium* (100) (blue-eyed grass); *Tritonia* (55) (montbretia); *Gladiolus* (300); *Freesia* (20); *Belamcanda* (2) (blackberry lily); *Ixia* (1); and *Tigridia* (12) (tiger flower).

The dried styles of *Crocus sativus* are used to make saffron, which was once used as a dye but is now used chiefly in Spanish and Iranian cooking to give color to cooked rice. Around 100,000 styles are required to produce 1 kilogram of saffron. Spring-flowering crocuses (*C. vernus*) herald the arrival of spring with their clumps of brightly colored flowers.

There are countless cultivars of *Iris* and *Gladiolus* available in the horticultural trade. *Iris germanica* yields orrisroot, which is used in perfume and in dentifrices.

**Economic Importance**   Most of the genera in the Iridaceae contain species that are used or could be used as ornamentals.

**SUBCLASS V, LILIIDAE; ORDER 2, ORCHIDALES**

**Orchidaceae: Orchid Family**

**Field Recognition**   Herbaceous perennials; leaves 2-ranked; flowers

3-merous, zygomorphic; petals 2-lateral and one a labellum or lip; pollinia often produced; ovary inferior; $CA^3 \ COZ^{2+1} \ A^{1-2} \ \overline{G}^{③}.$

**Description**   *Habit*   Perennial herbs, terrestrial, epiphytic or saprophytic, with rhizomes, tuberous roots, or rootstock; stem leafy or scapose, often with pseudobulbs and aerial roots.

*Leaves*   Usually opposite, simple, sometimes reduced to scales, often fleshy, sheathing basally.

*Inflorescences*   Spicate, racemose, paniculate, or solitary.

*Flowers*   Usually bisexual, zygomorphic, showy and colorful or small and inconspicuous. Calyx of 3 sepals, green or colored. Corolla of 2 lateral petals and a middle labellum which is often larger. Androecium of 1 or 2 stamens, joined to the style in the column, anther 1, pollen often in waxy pollinia. Gynoecium a compound pistil of 3 carpels, with 1 locule and placentation parietal, or rarely 3-locular, placentation axile, ovary inferior, stigmas 3, the lateral 2 often fertile and the other sterile forming a beak or rostellum. See Figures 14-32 and 14-33.

*Fruit*   A capsule.

*Seed*   Small, numerous, without endosperm.

*Embryo*   Not differentiated.

**Size, Distribution, and General Information**   A family of about 735 genera and 17,000 species; one of the largest families of flowering plants, of worldwide distribution but most abundant in tropical areas and rare in arctic regions.

Some of the genera are *Orchis* (35) (showy orchid); *Cypripedium* (50) (lady's slipper, moccasin flower); *Goodyera* (40) (rattlesnake plantain); *Spiranthes* (25) (ladies' tresses); *Ophrys* (30) (bee orchid); *Habenaria* (600)

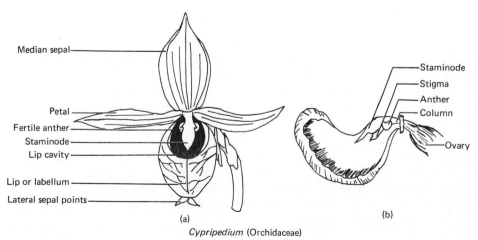

*Cypripedium* (Orchidaceae)

**Figure 14-32**   *Cypripedium* (Orchidaceae). *(a)* Flower and *(b)* longitudinal view of reproductive features.

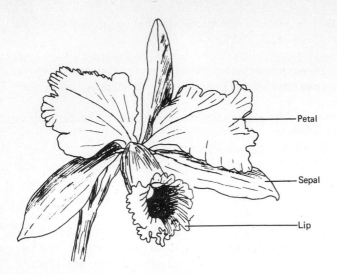

Petal

Sepal

Lip

*Cattleya* (Orchidaceae)

**Figure 14-33**   *Cattleya* (Orchidaceae) flower.

(fringe orchid); *Vanilla* (90); *Epidendrum* (400) (greenfly orchid); *Cattleya* (60) (florist's orchid); *Cymbidium* (40); *Odontoglossum* (200); *Liparis* (250) (twayblades); *Corallorhiza* (15) (coralroot); *Vanda* (60); *Calopogon* (4) (grass pink); *Phalaenopsis* (35); *Dendrobium* (1400); *Malaxis* (300) (adder's mouth); *Tipularia* (4) (cranefly orchid).

Probably no other family contains so great a variety of plants and flowers. The beauty, the strange shapes, and the longevity of many orchid flowers, together with the comparative ease with which they can be cultivated, have made them favorites of horticulturists. Hybrids of *Cypripedium, Cymbidium, Cattleya,* and *Odontoglossum* are generally used, but among the most popular for corsages are cultivars of *Cattleya.*

Many orchids show great specialization in pollination relationships with insects and often depend solely on particular insects for pollination. In the bee orchid (*Ophrys insectifera*) of southern Europe, the flower provides a combination of appearance and odor that attracts male wasps. The male wasps attempt to copulate with the orchid flower, and in doing so, effect pollination.

Some orchids, such as coralroot (*Corallorhiza*), are saprophytic and obtain food from decaying organic matter. Many species of orchids are epiphytic. Lady's slipper (*Cypripedium*) is among the loveliest of all our wild flowers.

The flavoring vanilla is obtained from the pods of certain species in the genus *Vanilla*. Europeans learned of vanilla from the Aztecs of Mexico. The vanilla fruits, known as beans, are fermented and cured in order to produce the familiar flavoring.

**Economic Importance**   The orchid family is of importance primarily for the ornamentals used in the florist trade.

## SELECTED REFERENCES

Bailey, L. H. (compiler): *Hortus Third,* revised by the staff of Liberty Hyde Bailey Hortorium, Macmillan, Riverside, N.J., 1976.

Cronquist, A.: *The Evolution and Classification of Flowering Plants,* Houghton Mifflin, Boston, 1968.

Gould, F. W.: *Grass Systematics,* McGraw-Hill, New York, 1968.

Hitchcock, A. S.: *Manual of Grasses of the United States,* 2d ed., revised by A. Chase, U.S. Department of Agriculture Miscellaneous Publication, no. 200, 1950.

Hutchinson, J.: *The Families of Flowering Plants,* 2d ed., 2 vols., Clarendon, Oxford, 1973.

Lawrence, G. H. M.: *Taxonomy of Vascular Plants,* Macmillan, New York, 1951.

Uphof, J. C. Th.: *Dictionary of Economic Plants,* 2d ed., Stechert-Hafner, Inc., New York, 1968.

Willis, J. C.: *A Dictionary of the Flowering Plants and Ferns,* 8th ed., revised by H. K. A. Shaw, University Press, Cambridge, England, 1973.

# Appendix 1

# Glossary
# of Greek and Latin Words

## SOME COMMON SPECIFIC EPITHETS AND THEIR MEANINGS

Specific epithets may be formed from nouns, adjectives, present participles, etc., and are often combined with prefixes and suffixes. When an adjective is used as the specific epithet, it must agree with the generic name in gender. For more complete listings and for feminine and neuter endings, either Featherly (1954) or Stearn (1973) should be consulted.

*acanthus*   thorn (Gk)*
*acaulis*   stemless (L)*
*acerifolius*   maple-leaved (L/L)*
*acicularis*   needle-like (L)
*aestivalis*   summer (L)
*affinis*   related (L)
*agrestis*   related to fields, in the
      country (L)
*alatus*   winged (L)

*albiflorus*   white-flowered (L/L)
*albus*   white (L)
*allegheniensis*   of the Alleghenies
*alpinus*   alpine (L)
*alternifolius*   alternate-leaved (L)
*altus*   tall (L)
*americanus*   of America
*amoenus*   charming, pleasant (L)
*amphibius*   amphibious (Gk)

*L = Latin; Gk = Greek; Gk → L = Greek root but taken into Medieval Latin and treated as Latin; L/L = both roots for a compound word are Latin; Gk/Gk = both roots for a compound word are Greek.

334

*amplexicaulis*   clasping the stem, embracing the stem (L/L)
*angularis*   angular (L)
*angustifolius*   narrow-leaved (L/L)
*annuus*   annual (L)
*apetalus*   without petals (Gk)
*apiculatus*   pointed at the tip (L)
*aquaticus*   aquatic (L)
*aquifolius*   with pointed leaves (L/L)
*arborescens*   tree-like, becoming tree-like (L)
*arenarius*   of sand (L)
*aristatus*   awned (L)
*aromaticus*   aromatic (Gk)
*articulatus*   with joints (L)
*arvensis*   of cultivated fields (L)
*ascendens*   ascending (L)
*asper*   rough (L)
*aureus*   golden (L)
*auriculatus*   eared (L)
*australis*   southern (L)
*autumnalis*   of autumn (L)
*axillaris*   growing in an axil (L)

*baccatus*   berry-like, pearl-like (L)
*barbatus*   barbed or bearded (L)
*bellus*   beautiful (L)
*bicolor*   two-colored (L/L)
*bidens*   with two teeth (L/L)
*biennis*   biennial (L)
*biflorus*   two-flowered (L/L)
*borealis*   northern (L but derived from Gk mythology)
*bracteosus*   bearing bracts (L)
*brevicaulis*   short-stemmed (L/L)
*brevis*   short (L)
*bulbosus*   bulbous (L)

*caeruleus*   blue (L)
*caespitosus*   tufted (L)
*calcareus*   chalky, limy (L)
*campanulatus*   bell-shaped (L)
*campestris*   growing in fields (L)
*canadensis*   of Canada
*candidus*   white (L)
*canescens*   grayish, becoming gray (L)
*capillaris*   hair-like (L)
*capitatus*   growing in heads (L)
*capreolatus*   twining, tendril-producing (L)

*cardinalis*   cardinal-red (L)
*carneus*   flesh-colored (L)
*carolinianus*   of the Carolinas
*caudatus*   tailed (L)
*caulescens*   becoming a stem (Gk→L)
*cernuus*   drooping, nodding (L)
*chinensis*   of China
*chrysanthus*   golden-flowered (Gk/Gk)
*ciliatus*   ciliate, like an eyelash (L)
*cinnamomeus*   cinnamon-brown (L)
*clavatus*   club-shaped (L)
*coccineus*   scarlet (L)
*communis*   growing in common (L)
*commutatus*   changing (L)
*concolor*   uniform in color (L/L)
*confertus*   crowded (L)
*conglomeratus*   crowded (L)
*contortus*   twisted (L/L)
*convolvulus*   twining (L/L)
*cordatus*   heart-shaped (L)
*coriaceus*   leathery (L)
*corniculatus*   horned, with a little horn (L)
*coronatus*   crowned (L)
*corymbosus*   corymb-like (Gk)
*crenatus*   scalloped (L)
*crispus*   curled (L)
*cristatus*   crested (L)
*cruciformis*   cross-shaped (L/L)
*cuneifolius*   wedge-shaped leaves (L/L)
*cyaneus*   blue (Gk)
*cymosus*   bearing cymes (Gk)

*dactyloides*   finger-shaped (Gk)
*debilis*   weak, disabled (L)
*deciduus*   deciduous (L)
*decumbens*   reclining (L)
*decurrens*   extending down (L)
*deltoides*   triangular (L)
*densus*   thick (L)
*dentatus*   toothed (L)
*depauperatus*   stunted (L)
*didymus*   in pairs (Gk)
*diffusus*   loosely branching (L)
*digitatus*   finger-shaped (L)
*discolor*   of two colors (Gk/L → L)
*dissectus*   deeply cut (Gk/L → L)
*distichus*   two-ranked (Gk/Gk)
*diurnus*   relating to the day (Gk/L → L)

*divaricatus*  spreading (Gk/L → L)
*divergens*  wide-spreading, turning
     (Gk/L → L)
*dulcis*  sweet (L)
*dumosus*  bushy habit (L)

*echinatus*  spiny (Gk)
*edulis*  edible (L)
*effusus*  loose-spreading (L)
*elatus*  tall (L)
*elegans*  elegant (L)
*ellipticus*  elliptical (Gk)
*elongatus*  long (L)
*emarginatus*  with a notch at apex (L)
*erectus*  upright (L)
*ericoides*  heath-like (Gk/L)
*erythrocarpus*  red-fruited (Gk/Gk)
*esculentus*  edible (L)

*falcatus*  sickle-shaped (L)
*farinosus*  mealy (L)
*fasciculatus*  clustered, bundled (L)
*ferrugineus*  rusty-colored (L/L)
*fertilis*  fertile (L)
*fibrosus*  fibrous (L)
*filamentosus*  filamentous (L)
*filiformis*  thread-like (L/L)
*fimbriatus*  fringed (L)
*fistulosus*  hollow, cylindrical (L)
*flabelliformis*  fan-shaped (L/L)
*flaccidus*  soft, flabby (L)
*flavens*  yellowish (L)
*flexuosus*  flexible (L)
*floribundus*  free-flowering (L)
*floridanus*  flowering or of Florida (L)
*foetidus*  having a bad odor (L)
*foliosus*  leafy (L)
*fragilis*  fragile (L)
*fragrans*  fragrant (L)
*frondosus*  full of leaves (L)
*fruticosus*  shrubby (L)
*fulvus*  brownish-yellow (L)
*fusiformis*  spindle-shaped (L/L)

*geminatus*  twin (L)
*gibbosus*  swollen on one side,
     humped (L)
*giganteus*  very large (Gk → L)

*glabellus*  smooth (L)
*glabratus*  smooth (L)
*glandulosus*  glandular (L)
*glaucescens*  waxy (Gk)
*globosus*  shaped like a globe (L)
*glomeratus*  dense clusters (L)
*glutinosus*  gluey or sticky (L)
*gramineus*  grassy (L)
*graminifolius*  with grass-like leaves (L/L)
*grandiflorus*  with large flowers (L/L)
*grandifolius*  with large leaves (L/L)

*hastatus*  spear-shaped (L)
*herbaceus*  not woody (L)
*heterophyllus*  with leaves of several
     shapes (Gk/Gk)
*hirsutus*  hairy (L)
*hispidus*  bristly (L)
*humifusus*  sprawling (L)
*humilis*  dwarf (L)
*hybridus*  hybrid (L)
*hyemalis*  of winter (L)

*ilicifolius*  holly-leaved (L/L)
*imbricatus*  overlapping, shingled (L)
*induratus*  hardened (L)
*inequalis*  unequal (L)
*inflatus*  inflated (L)
*inodorus*  without odor (L)
*integrifolius*  with entire leaves (L/L)

*junceus*  rush-like (L)

*kewensis*  relating to Kew Gardens

*labiatus*  with a lip (L)
*laciniatus*  cut, torn (Gk → L)
*lactatus*  milky (L)
*laevigatus*  smooth (L)
*laevis*  smooth (L)
*lanceolatus*  lance-shaped (L)
*lancifolius*  leaves lance-shaped (L/L)
*lanosus*  wooly (L)
*latiflorus*  broad-flowered (L/L)
*laxiflorus*  loose-flowered (L/L)
*laxus*  loose (L)
*lenticularis*  like a lens, like a lentil (L)
*lepidotus*  with small scales (Gk)

*leucanthus*  white-flowered (Gk/Gk)

*lignosus*  like wood (L)

*ligularis*  strap-shaped (L)

*limosus*  of muddy places (L)

*linearifolius*  with long, slender
    leaves (L/L)

*litoralis*  of the seashore (L)

*lobatus*  lobed (L)

*longiflorus*  long-flowered (L/L)

*longifolius*  long-leaved (L/L)

*lucidus*  shining (L)

*ludovicianus*  of Louisiana

*lunulatus*  crescent-shaped (L)

*luteus*  yellow, mud-colored (L)

*lyratus*  lyre-shaped (Gk → L)


*macrophyllus*  large-leaved (Gk/Gk)

*maculatus*  spotted (L)

*major*  larger (L)

*marginalis*  marginal (L)

*marilandicus*  of Maryland

*maritimus*  growing near the sea (L)

*maximus*  the largest, greatest (L)

*medius*  medium (L)

*microcarpus*  small-fruited (Gk/Gk)

*microphyllus*  small-leaved (Gk/Gk)

*miliaceus*  millet-like (L)

*millefolius*  many-leaved (L/L)

*minimus*  smallest (L)

*minor*  smaller (L)

*mirabilis*  wonderful (L)

*mollis*  soft (L)

*montanus*  of mountains (L)

*monticola*  growing in the mountains (L)

*multicaulis*  with many stems (L/L)

*multiflorus*  many-flowered (L/L)

*muricatus*  roughened by hard points (L)

*mutabilis*  variable (L)


*nanus*  dwarf (Gk)

*natans*  floating (L)

*neglectus*  neglected (L)

*nemoralis*  growing in woods (L)

*nervosus*  nerved (L)

*niger*  black (L)

*nitens*  shining (L)

*niveus*  snow-white (L)

*nobilis*  noble, or well known (L)

*noctiflorus*  night-flowering (L/L)

*nodosus*  with nodes (L)

*notatus*  marked (L)

*novae-angliae*  of New England

*noveboracensis*  of New York

*nudatus*  nude (L)

*nudicaulis*  naked-stemmed (L/L)

*nudiflorus*  naked-flowered (L/L)

*nutans*  nodding (L)

*nyctagineus*  night-flowering (L)


*obtusifolius*  with blunt leaves (L/L)

*occidentalis*  western (L)

*odoratissimus*  very fragrant (L)

*odoratus*  with an odor (L)

*officinalis*  a formally recognized
    medicinal (L)

*oleraceus*  from a vegetable garden (L)

*orientalis*  eastern (L)


*pallens*  pale (L)

*palmatus*  palmate (L)

*paludosus*  marshy (L)

*paniculatus*  in panicles (L)

*papyrifer*  paper-bearing (L)

*parviflorus*  small-flowered (L/L)

*parvifolius*  small-leaved (L/L)

*patens*  spreading (L)

*pauciflorus*  few-flowered (L/L)

*pedatus*  palmately divided (L)

*pedicularis*  with a stalk (L)

*peduncularis*  with peduncles (L)

*peltatus*  shield-shaped (Gk)

*pendulus*  hanging down (L)

*perennis*  perennial (L)

*petiolaris*  relating to the petiole (L)

*plumosus*  feathery (L)

*praealtus*  very tall (L/L)

*praecox*  precocious, premature (L/L)

*pratensis*  growing in meadows (L)

*procumbens*  prostrate (L)

*pubens*  downy (L)

*pubescens*  with soft hairs, becoming
    downy (L)

*pulchellus*  beautiful (L)

*pumilus*  dwarf, small (L)

*punctatus*  marked with dots (L)

*puniceus*  reddish-purple (L)

*purpureus*  purple (L)
*pusillus*  insignificant or very small (L)

*quadrangularis*  four-angled (L/L)
*quadrifolius*  with four leaves (L/L)
*quercifolius*  oak-leaved (L/L)
*quinquefolius*  with five leaves (L/L)

*racemosus*  in racemes (L)
*radiatus*  with rays, spreading from
    center (L)
*radicans*  rooting (L)
*ramosus*  branched (L)
*recurvus*  curved back (L/L)
*regalis*  royal (L)
*repens*  creeping (L)
*reptans*  crawling (L)
*resinosus*  resinous (L)
*reticulatus*  net-like (L)
*revolutus*  rolled backward or inward (L)
*rigidus*  stiff (L)
*riparius*  growing along a river (L)
*roseus*  rose-colored (L)
*rotundifolius*  round-leaved (L/L)
*rubens*  red (L)
*ruderalis*  growing among rubbish (L)
*rugosus*  wrinkled (L)
*rupestris*  growing on rocks (L)

*saccharoides*  sweet (Gk)
*sagittifolius*  arrow-leaved (L/L)
*salicifolius*  willow-leaved (L/L)
*sativus*  cultivated (L)
*saxicola*  dweller among rocks (L/L)
*scaber*  rough (L)
*scandens*  climbing (L)
*scoparius*  broom-like, like a bundle of
    twigs (L)
*secundus*  on the side (L)
*semperflorens*  ever-blooming (L/L)
*sempervirens*  evergreen (L/L)
*sensibilis*  sensible (L)
*septentrionalis*  northern (L/L)
*sericeus*  silky (Gk)
*sessiliflorus*  flowers without stems (L/L)
*sessilifolius*  leaves without petioles (L/L)
*sessilis*  apparently stemless (L)
*setosus*  bristly (L)

*silvaticus*  pertaining to woods (L)
*simplex*  unbranched (L)
*sinensis*  of China
*speciosus*  beautiful (L)
*spectabilis*  spectacular, visible (L)
*spicatus*  with spikes (L)
*spinosus*  with spines (L)
*splendens*  splendid (L)
*squarrosus*  spreading at tip, rough
    because of spreading bracts (L)
*stamineus*  with prominent stamens (L)
*stellatus*  star-like (L)
*stolonifer*  with stolons (L)
*stramineus*  straw-colored (L)
*strictus*  stiff, upright, drawn together (L)
*strigosus*  with stiff bristles, with a few
    stiff bristles (L)
*subulatus*  awl-shaped (L/L)
*suffruticosus*  shrubby (L/L)
*sylvaticus*  of forests (L)
*sylvestris*  growing in woods (L)

*tenellus*  slender (L)
*tenuiflorus*  slender flowers (L/L)
*tenuifolius*  slender-leaved (L/L)
*tenuis*  slender, thin (L)
*teres*  round in cross section (L)
*ternatus*  arranged in threes (L)
*terrestris*  growing in dry ground (L)
*tinctorius*  used for dyeing (L)
*tomentosus*  felty (L)
*tortuosus*  twisted, winding (L)
*tricanthus*  three-spined (L/Gk)
*tridens*  with three teeth (L/L)
*trifoliatus*  three-leaved (L/L)
*tuberosus*  with tubers (L)

*umbellatus*  with umbels (L)
*uniflorus*  one-flowered (L/L)
*urens*  stinging (L)
*utricularis*  bladder-shaped, leather-bag-
    shaped (L)

*vacillans*  swaying, wavering (L)
*variabilis*  variable, mottled (L)
*velutinus*  velvety (L)
*venosus*  with veins (L)
*vernalis*  spring flowering (L)

*versicolor*  variously colored (L/L)
*verticillatus*  whorled (L)
*villosus*  with soft hairs (L)
*violaceus*  violet (L)
*virgatus*  twiggy (L)
*virginianus*  of Virginia
*viridis*  green (L)

*viscosus*  sticky (L)
*vulgaris*  common (L)

*xanthinus*  yellow (Gk)
*xanthorrhizus*  yellow-rooted (Gk/Gk)

*zebrinus*  zebra-striped

## SOME COMMON GREEK AND LATIN PREFIXES

A prefix is added to the beginning of a word in order to modify its meaning. For more complete lists refer to Featherly (1954) or Stearn (1973).

*a-, ab-:*  away from (L)
*a-, an-:*  without (Gk)
*actino-:*  star-like (Gk)
*amphi-:*  of two kinds (Gk)
*andro-:*  male (Gk)
*angusti-:*  narrow (L)
*bi-:*  two (L)
*brevi-:*  short (L)
*chloro-:*  green (Gk)
*chori-:*  separate (Gk)
*chryso-:*  golden (Gk)
*co-, col-, com-:*  together (L)
*cyath-:*  cup-like (Gk)
*di-:*  two, separate, apart (Gk)
*e-, ef-, ex-:*  lacking (L)
*endo-:*  within (Gk)
*epi:*  upon (Gk)
*erythro-:*  reddish (Gk)
*fili-:*  thread-like (L)
*fimbri-:*  fringed (L)
*flavi-:*  yellowish (L)
*grandi-:*  large (L)
*gymno-:*  naked (Gk)
*gyno-:*  female (Gk)
*hirsuti-:*  hairy (L)
*laevi-:*  smooth (L)
*lanci-:*  lance-shaped (L)
*lati-:*  broad (L)
*laxi-:*  loose (L)
*leio-:*  smooth (Gk)
*lepido-:*  scaly (Gk)

*leuco-:*  white (Gk)
*macro-:*  large (Gk)
*mega-:*  very large (Gk)
*melano-:*  black (Gk)
*micro-:*  small (Gk)
*mono-:*  one, single (Gk)
*multi-:*  many (L)
*nudi-:*  naked (L)
*octo-:*  eight (Gk)
*odonto-:*  toothed (Gk)
*pauci-:*  few (L)
*peri-:*  around (Gk)
*phyllo-:*  relating to a leaf (Gk)
*pleio-:*  many (Gk)
*poly-:*  many (Gk)
*pseudo-:*  false (Gk)
*ptero-:*  winged (Gk)
*quadr-:*  four (L)
*rhizo-:*  relating to a root (Gk)
*schiz-:*  split, divided (Gk)
*semper-:*  always (L)
*sessili-:*  sessile (L)
*tenui-:*  slender (L)
*tri-:*  three (L)
*tetra-:*  four (Gk)
*uni-:*  one (L)
*viridi-:*  green (L)
*xantho-:*  yellow (Gk)
*xero-:*  dry (Gk)
*xylo-:*  woody (Gk)
*zygo-:*  fused, a yoke (Gk)

## COMMON BOTANICAL LATIN ABBREVIATIONS

These abbreviations are often used in systematic botany.

**aff.**  *affinis,* "related to"

**ca.**  *circa,* "around, about"

**comb. nov.**  *combinatio nova,* "new combination"

**cv.**  *cultivarietas,* "cultivar"

**det.**  *determinavit,* "identified by"

**exs.**  *exsiccatus,* "dried [herbarium] specimen"

**f.**  *filius,* "son" (after a personal noun)

**f.**  *forma,* "form" (before an epithet)

**hab.**  *habitat,* place where a plant grows

**herb.**  *herbarium,* "herbarium"

**hort.**  *hortorum,* "from gardens"

**s.l.**  *sensu lato,* "in a broad sense"

**s. str.**  *sensu stricto,* "in a narrow sense"

**s.n.**  *sine numero,* "without a [collection] number"

**sp.**  *species,* "species" (singular)

**spp.**  *species,* "species" (plural)

**ssp.**  *subspecies,* "subspecies"

**var.**  *varietas,* "variety"

**sp. nov.**  *species nova,* "new species"

## REFERENCES FOR APPENDIX 1

Borror, D. J.: *Dictionary of Word Roots and Combining Forms,* Mayfield Publishing Company, Palo Alto, Calif., 1960. (A handy, inexpensive paperback book.)

Featherly, H. I.: *Taxonomic Terminology of the Higher Plants,* The Iowa State College Press, Ames, 1954.

Radford, L. S.: "Botanical Names," in A. E. Radford et al., *Vascular Plant Systematics,* Harper and Row, New York, 1974, pp. 57–71.

Stearn, W. T.: *Botanical Latin,* 2d ed., David and Charles, Newton Abbot, England, 1973.

# Selected Literature for the Identification of Vascular Plants in North America North of Mexico

There are three major types of identification aids: floras, manuals, and guides. A *flora* provides an inventory of the plants of an area and is often restricted to the vascular plants. Floras may contain keys and descriptions. A *manual* always has keys for identification and may contain all the vascular plants of a given area or may treat limited groups, such as the grasses of the United States or the aquatic plants of the southeastern states. In practice, the distinctions between a flora and a manual are not sharp, and the two terms are often used interchangeably. A *guide* is a brief treatment of plants in a particular area and consists of keys and a list of taxa. There is no complete flora, manual, or guide of native plants for North America north of Mexico, although many exist for restricted parts of the area.

Selected references for plant identification are listed here. No attempt was made to include all manuals and floras that might be of historical importance. References are provided for special groups such as aquatics, cacti, ferns, grasses, legumes, orchids, poisonous plants, weeds, and woody plants. A section on selected references helpful for identifying cultivated or exotic plants is included. In addition, a section is provided for popular wild flower treatments.

## MANUALS, FLORAS, AND GUIDES

Abrams, L.: *An Illustrated Flora of the Pacific States,* 4 vols., vol. 4 by R. Ferris, Stanford University Press, Stanford, Calif., 1923–1960.

Anderson, J. P. See Welsh, S. L.

Barkley, T. M.: *A Manual of the Flowering Plants of Kansas,* Kansas State University Endowment Association, Manhattan, 1968.

Batson, W. T.: *Genera of Eastern Plants,* 3d ed., Wiley, New York, 1977. (Pocket-size.)

Blomquist, H. L., and H. J. Oosting: *A Guide to the Spring and Early Summer Flora of the Piedmont, North Carolina,* 6th ed., published by the authors, Durham, N.C., 1959.

Clovis, J. F., et al.: *Common Vascular Plants of the Mid-Appalachian Region,* Book Exchange, Morgantown, W.Va., 1972.

Correll, D. S., and M. C. Johnston: *Manual of the Vascular Plants of Texas,* Texas Research Foundation, Renner, Tex., 1970.

Cronquist, A., et al.: *Intermountain Flora; Vascular Plants of the Intermountain West, U.S.A.,* Hafner, New York, 1972– . (Projected six volumes.)

Davis, R. J.: *Flora of Idaho,* Brown, Dubuque, Iowa, 1952.

Deam, C. C.: *Flora of Indiana,* Indiana Department of Conservation, Indianapolis, 1940.

Fassett, N. C.: *Spring Flora of Wisconsin,* 4th ed., University of Wisconsin Press, Madison, 1976.

Fernald, M. L. See Gray, A.

Gilkey, H. M., and L. R. J. Dennis: *Handbook of Northwestern Plants,* Oregon State University Bookstore, Corvallis, 1967.

Gleason, H. A.: *New Britton and Brown Illustrated Flora of the Northeastern United States and Adjacent Canada,* 3 vols., New York Botanical Garden, New York, 1952. (Illustrations; excellent keys.)

——— and A. Cronquist: *Manual of the Vascular Plants of the Northeastern United States and Adjacent Canada,* Van Nostrand, Princeton, N.J., 1963. (Keys are excellent; one of the standard floras of northeastern North America.)

Goodman, G. J.: *Spring Flora of Central Oklahoma,* University of Oklahoma Duplicating Service, Norman, 1958.

Gray, A.: *Manual of Botany,* 8th ed., M. L. Fernald (ed.), American Book, New York, 1950. (One of the standard floras for the northeastern United States and adjacent Canada; the family keys, although workable, are technical and difficult for the beginner.)

Great Plains Flora Association: *Atlas of the Flora of the Great Plains,* Iowa State University Press, Ames, 1977.

Harrington, H. D.: *Manual of the Plants of Colorado,* Sage, Denver, Colo., 1954.

Harvill, A. M.: *Spring Flora of Virginia,* McClain, Parsons, W. Va., 1970.

Hitchcock, C. L., et al.: *Vascular Plants of the Pacific Northwest,* parts 1–5, University of Washington Press, Seattle, 1955–1969. (Excellent family keys.)

——— and A. Cronquist: *Flora of the Pacific Northwest; an Illustrated Manual,* University of Washington Press, Seattle, 1973.

Hultén, E.: *Flora of the Aleutian Islands,* 2d ed., J. Cramer, Weinheim, West Germany, 1960.

———: *Flora of Alaska and Neighboring Territories; A Manual of the Vascular Plants,* Stanford University Press, Stanford, Calif., 1968.

Jepson, W. L.: *A Manual of the Flowering Plants of California,* University of California Press, Berkeley, 1925.

Jones, F. B.: *Flora of the Texas Coastal Bend,* Welder Wildlife Foundation, Sinton, Tex., 1975.

Jones, G. N.: *Flora of Illinois,* 3d ed., University of Notre Dame Press, Notre Dame, Ind., 1963.

Kearney, T. H., and R. H. Peebles, *Arizona Flora,* 2d ed., University of California Press, Berkeley, 1960.

Lakela, O.: *A Flora of Northeastern Minnesota,* University of Minnesota Press, Minneapolis, 1965.

Lloyd, R., and R. S. Mitchell: *A Flora of the White Mountains, California and Nevada,* University of California Press, Berkeley, 1973.

Long, R. W., and O. Lakela: *A Flora of Tropical Florida,* University of Miami Press, Coral Gables, Fla., 1971. (For southern Florida.)

Marie-Victorin, F.: *Flore Laurentienne (Québec),* 2d ed., Presses de l'Université de Montréal, Montréal, 1964. (In French.)

McDougall, W. B.: *Seed Plants of Northern Arizona,* Museum of Northern Arizona, Flagstaff, 1973.

Mohlenbrock, R. H.: *Flowering Plants: Flowering Rush to Rushes,* Southern Illinois University Press, Carbondale, 1970.

————: *Flowering Plants: Lilies to Orchids,* Southern Illinois University Press, Carbondale, 1970.

————: *Guide to the Vascular Flora of Illinois,* Southern Illinois University Press, Carbondale, 1975.

————: *Sedges,* Southern Illinois University Press, Carbondale, 1976.

———— and J. W. Voight: *A Flora of Southern Illinois,* Southern Illinois University Press, Carbondale, 1959.

Morley, T.: *Spring Flora of Minnesota,* University of Minnesota Press, Minneapolis, 1969.

Moss, E. H.: *Flora of Alberta,* University of Toronto Press, Toronto, 1959.

Munz, P. A., in collaboration with D. D. Keck: *A California Flora,* University of California Press, Berkeley, 1959, supplement 1968, combined edition 1973. (Lists major community types; selected illustrations; the standard flora for California.)

————: *A Flora of Southern California,* University of California Press, Berkeley, 1974.

Pease, A. S.: *A Flora of Northern New Hampshire,* New England Botanical Club, Cambridge, Mass., 1964.

Peck, M. E.: *A Manual of the Higher Plants of Oregon,* 2d ed., Binford and Mort, Portland, Ore. 1961.

Polunin, N.: *Circumpolar Arctic Flora,* Clarendon, Oxford, 1959.

Porsild, A. E.: *Illustrated Flora of the Canadian Arctic Archipelago,* 2d ed. rev., National Museum of Canada Bulletin, no. 146, 1964.

Porter, C. L.: *A Flora of Wyoming,* University of Wyoming Agricultural Experiment Station Bulletin nos. 402, 404, 418, and 434, 1962–65.

Radford, A. E., H. E. Ahles, and C. R. Bell: *Manual of the Vascular Flora of the Carolinas,* University of North Carolina Press, Chapel Hill, 1968. (Useful in southeastern United States; family keys difficult; selected illustrations.)

Roland, A. E., and E. C. Smith: "Flora of Nova Scotia," *Proc. N. S. Inst. Sci.,* **26:**3–238, 277–743, 1962–1969.

Rydberg, P. A.: *Flora of the Rocky Mountains and Adjacent Plains,* 2d ed., published by the author, New York, 1922. (Facsimile, Hafner, New York, 1954; many of the names are now obsolete.)

————: *Flora of the Prairies and Plains of Central North America,* New York Botanical Garden, New York, 1932. (Many of the names are now obsolete.)

Scoggan, H. J.: *Flora of Manitoba,* National Museum of Canada Bulletin, no. 140, 1957.

————: *The Flora of Canada,* National Museum of Natural Science Publications in Botany, no. 7, 1978– . (Treats ferns, conifers, and flowering plants, keys. Projected four-volume set.)

Seymour, F. C.: *The Flora of New England,* Tuttle, Rutland, Vt., 1969.

Shinners, L. H.: *Shinners' Spring Flora of the Dallas-Fort Worth Area, Texas,* 2d ed., W. H. Mahler (ed.), Prestige Press, Ft. Worth, 1972.

Shreve, F., and I. L. Wiggins: *Vegetation and Flora of the Sonoran Desert,* 2 vols., Stanford University Press, Stanford, Calif., 1964.

Small, J.K.: *Manual of the Southeastern Flora,* published by the author, 1933. (Facsimile, Hafner, New York, 1972. Many of the names are obsolete; good illustrations of each genus.)

Stevens, O. A.: *Handbook of North Dakota Plants,* North Dakota Agricultural College, Fargo, 1963.

Steyermark, J. A.: *Flora of Missouri,* The Iowa State University Press, Ames, 1963. (Well illustrated and has excellent keys.)

Strausbaugh, P. D., and E. L. Core: *Flora of West Virginia,* introduction and parts I-IV, West Virginia University, Morgantown, 1952–1964. (Parts I and II revised in 1970, 1971.)

Taylor, R. L., and B. MacBryde: *Vascular Plants of British Columbia,* University of British Columbia Press, Vancouver, Canada, 1977. (An inventory printed from computer-stored information.)

Thomas, J. H.: *Flora of the Santa Cruz Mountains of California, A Manual of Vascular Plants,* Stanford University Press, Stanford, Calif., 1961.

Tidestrom, I., and T. R. Kittell: *A Flora of Arizona and New Mexico,* 2 vols., Catholic University of America Press, Washington, D.C., 1941.

Van Bruggen, T.: *The Vascular Plants of South Dakota,* Iowa State University Press, Ames, 1976.

Voss, E. G.: *Michigan Flora,* part 1, *Gymnosperms and Monocots,* Cranbrook Institute of Science Bulletin, no. 55, 1972.

Waterfall, U. T.: *Keys to the Flora of Oklahoma,* 5th ed., published by the author, Oklahoma State University Bookstore, Stillwater, 1972.

Weber, W. A.: *Rocky Mountain Flora; a Field Guide for the Identification of the Ferns, Conifers, and Flowering Plants,* Colorado Associate University Press, Boulder, 1976.

Welsh, S. L.: *Anderson's Flora of Alaska and Adjacent Parts of Canada,* Brigham Young University Press, Provo, Utah, 1974.

Wiggins, I. L., and J. H. Thomas: *A Flora of the Alaskan Arctic Slope,* Arctic Institute of North America Special Publication, no. 4, 1962.

Wooten, E. O., and P. C. Standley: *Flora of New Mexico,* Contributions from the United States National Herbarium, no. 19, 1915. (Some of the nomenclature is now obsolete.)

## AQUATIC PLANTS

Beal, E. O.: *A Manual of Marsh and Aquatic Vascular Plants of North Carolina; with Habitat Data,* North Carolina Agricultural Experiment Station Technical Bulletin, no. 247, 1977.

Carlton, J. M.: *A Guide to Common Florida Salt Marsh and Mangrove Vegetation,* Florida Marine Research Publications, no. 6, 1975.

Cook, C. D. K., et al.: *Water Plants of the World,* Junk, The Hague, 1974.

Correll, D. S., and H. B. Correll: *Aquatic and Wetland Plants of Southwestern United States,* E.P.A., U.S. Government Printing Office, Washington, D.C., 1972. (Facsimile edition, Stanford University Press, Stanford, Calif., 1975; keys, descriptions, excellent illustrations.)

Godfrey, R. K., and J. Wooten: *Aquatic and Wetland Plants of the Southeastern United States,* 2 vols., University of Georgia Press, Athens, in press. (Illustrated.)

Eyles, D. E., and J. L. Robertson: *A Guide and Key to the Aquatic Plants of the Southeastern United States,* U.S. Bureau of Sport Fisheries and Wildlife Circular, no. 158, 1963. (A reprint of U.S. Public Health Bulletin, no. 286, 1944; easily used by the nonprofessional.)

Fassett, N. C.: *A Manual of Aquatic Plants,* University of Wisconsin Press, Madison, 1957. (With revised appendix by E. C. Ogden.)

Hotchkiss, N.: *Common Marsh and Floating-leaved Plants of the United States and Canada,* Dover, New York, 1972. (Reprint of U.S. Bureau of Sport Fisheries and Wildlife Resources Publication, nos. 44 and 93. Illustrated.)

Mason, H. L.: *A Flora of the Marshes of California,* University of California Press, Berkeley, 1957. (Illustrations; excellent illustrated glossary.)

Matsumara, Y., and H. D. Harrington: *The True Aquatic Vascular Plants of Colorado,* Colorado Agricultural Experiment Station Technical Bulletin, no. 57, 1955

Muenscher, W. C.: *Aquatic Plants of the United States,* Comstock, Ithaca, New York, 1944. (An excellent manual covering the entire United States.)

Prescott, G. W.: *How to Know the Aquatic Plants,* Brown, Dubuque, Iowa, 1969. (Keys; line drawings.)

Steward, A. N., L. J. Dennis, and H. M. Gilkey: *Aquatic Plants of the Pacific Northwest with Vegetative Keys,* 2d ed., Oregon State University Press, Corvallis, 1963.

Winterringer, G. S., and A. C. Lopinot: *Aquatic Plants of Illinois,* Illinois State Museum Popular Science Series, no. 6, 1966.


## CACTI

Benson, L.: *The Cacti of Arizona,* 3d ed., University of Arizona Press, Tucson, 1969.
———: *The Native Cacti of California,* Stanford University Press, Stanford, Calif., 1969. (Designed for use by professionals and amateurs.)

Britton, N. L., and J. N. Rose: *The Cactaceae: Descriptions and Illustrations of Plants of the Cactus Family,* 4 vols., Carnegie Institution of Washington Publication, no. 248, 1919–1923. (Keys, descriptions, illustrations.)

Dawson, E. Y.: *How to Know the Cacti,* Brown, Dubuque, Iowa, 1963. (Includes descriptions, keys, line drawings, black-and-white photographs.)
———: *The Cacti of California,* University of California Press, Berkeley, 1966. (A popular illustrated guide to the native cacti of the state.)

Weniger, D.: *Cacti of the Southwest,* University of Texas Press, Austin, 1969.


## CULTIVATED PLANTS AND ORNAMENTALS

Bailey, L. H.: *The Standard Cyclopedia of Horticulture,* 3 vols., Macmillan, New York, 1928. (First published in 1900 as *Cyclopedia of American Horticulture,* it remains an excellent source of information on horticultural plants, including some help with identification.)

————: *Manual of Cultivated Plants,* rev. ed., Macmillan, New York, 1949. (A manual for identification of plants cultivated in the United States; the keys are generally unsatisfactory.)

————, compiler: *Hortus Third,* revised by the staff of Liberty Hyde Bailey Hortorium, Macmillan, Riverside, N. J., 1976. (A concise dictionary of plants cultivated in the United States and Canada; indispensable to those working with cultivated plants.)

Bean, W. J.: *Trees and Shrubs Hardy in the British Isles,* 3 vols., Murray, London, 1914–1933. (Descriptions, illustrations.)

————: *Trees and Shrubs Hardy in the British Isles,* 8th ed., Murray, London, 1970– . (A multivolume set.)

Chittenden, F. J. (ed.): *Dictionary of Gardening,* 4 vols., Clarendon, Oxford, 1951. Supplement, P. M. Synge (ed)., 1969. (Descriptions, line drawings.)

Dirr, M. A.: *Manual of Woody Landscape Plants,* rev. ed., Stipes, Champaign, Ill., 1977.

Graf, A. B.: *Exotic Plant Manual,* 2d ed., Roehrs, Rutherford, N.J., 1970. (Illustrated with 3600 photographs.)

————: *Exotica Series 3; Pictorial Cyclopedia of Exotic Plants,* 9th ed., Roehrs, Rutherford, N.J., 1976. (Illustrated with 12,000 photographs; plants arranged by families; useful for identification of tropical foliage plants, cacti, succulents, and so on.)

Halfacre, R. G., and A. R. Shawcroft: *Carolina Landscape Plants,* 2d ed., Sparks Press, Raleigh, N.C., 1975. (Line drawings, horticultural notes.)

Hillier, H. G.: *Hillier's Manual of Trees and Shrubs,* Barnes, Cranbury, N.J., 1973. (Descriptions, horticultural notes.)

Ouden, P. D., and B. K. Boom: *Manual of Cultivated Conifers; Hardy in the Cold- and Warm-Temperate Zone,* Martinus Nijhoff, The Hague, Netherlands, 1965. (Technical, black-and-white photos, descriptions.)

Perry, F. (ed.): *Simon & Schuster's Complete Guide to Plants and Flowers,* Simon and Schuster, New York, 1974. (Horticultural information, color photos.)

Rehder, A.: *Manual of Cultivated Trees and Shrubs Hardy in North America,* 2d ed., Macmillan, New York, 1940. (Keys to taxa are included.)

————: *Bibliography of Cultivated Trees and Shrubs Hardy in the Cooler Temperate Regions of the Northern Hemisphere,* Arnold Arboretum of Harvard University, Jamaica Plain, Mass., 1949. (Provides references to the sources of scientific names.)

Reisch, K. W., P. C. Kozel, and G. A. Weinstein: *Woody Ornamentals for the Midwest (Deciduous),* Kendall/Hunt, Dubuque, Iowa, 1975. (Descriptions, line drawings.)

Uphof, J. C. Th.: *Dictionary of Economic Plants,* 2d ed., Stechert-Hafner, New York, 1968.

Whitcomb, C. E.: *Know It and Grow It: A Guide to the Identification and Use of Landscape Plants in the Southern States,* published by the author, Tulsa, Okla., 1976.

Wigginton, B. E.: *Trees and Shrubs for the Southeast,* University of Georgia Press, Athens, 1963. (Black-and-white photos.)

Wilson, E. H., and A. Rehder: *A Monograph of Azaleas,* University Press, Cambridge, England, 1921.

Wyman, D.: *Trees for American Gardens,* Macmillan, New York, 1959. (Black-and-white photos, descriptions.)

———: *Shrubs and Vines for American Gardens*, rev. ed., Macmillan, New York, 1969. (Black-and-white photos, descriptions.)

———: *Wyman's Gardening Encyclopedia*, rev. ed., Macmillan, New York, 1977. (An encyclopedia format with horticultural information.)

## PTERIDOPHYTES

Billington, C.: *Ferns of Michigan*, Cranbrook Institute of Science Bulletin, no. 32, 1952.

Blomquist, H. L.: *Ferns of North Carolina*, Duke University Press, Durham, 1934.

Brown, C. A., and D. S. Correll: *Ferns and Fern Allies of Louisiana*, Louisiana State University Press, Baton Rouge, 1942.

Chrysler, M. A., and J. L. Edwards: *The Ferns of New Jersey*, Rutgers University Press, New Brunswick, N.J., 1947.

Cobb, B.: *A Field Guide to the Ferns*, Houghton Mifflin, Boston, 1956. (Peterson Field Guide Series.)

Correll, D. S.: *Ferns and Fern Allies of Texas*, Contributions from the Texas Research Foundation, vol. 2, 1956.

Grillos, S. J.: *Ferns and Fern Allies of California*, University of California Press, Berkeley, 1966.

Harrington, H. D., and L. W. Durrell: *Colorado Ferns and Fern Allies*, Colorado Agricultural Research Foundation, Fort Collins, 1950.

Lakela, O., and R. W. Long: *Ferns of Florida*, Banyan Books, Miami, 1976.

Massey, A. B.: *The Ferns and Fern Allies of Virginia*, 2d ed., Virginia Polytechnic Institute Agriculture Extension Service Bulletin, no. 256, 1958.

McVaugh, R., and J. H. Pyron: *Ferns of Georgia*, University of Georgia Press, Athens, 1951.

Mohlenbrock, R. H.: *Ferns*, Southern Illinois University Press, Carbondale, 1967.

Small, J. K.: *Ferns of the Southeastern States*, Science Press, Lancaster, Pa., 1938. (Many of the names used are now obsolete.)

Taylor, T. M. C.: *Pacific Northwest Ferns and Their Allies*, University of Toronto Press, Toronto, 1970.

Tryon, R. M., et al.: *The Ferns and Fern Allies of Wisconsin*, 2d ed, University of Wisconsin Press, Madison, 1953.

———: *The Ferns and Fern Allies of Minnesota*, University of Minnesota Press, Minneapolis, 1954.

Wherry, E. T.: *The Fern Guide, Northeastern and Midland United States and Adjacent Canada*, Doubleday, Garden City, N.Y., 1961. (Keys, descriptions, line drawings.)

———: *The Southern Fern Guide*, Doubleday, Garden City, N.Y., 1964. (Line drawings, descriptions.)

## GRASSES

Blomquist, H. L.: *The Grasses of North Carolina*, Duke University Press, Durham, 1948. (Excellent illustrations, a vegetative key, photographs.)

Deam, C. C.: *Grasses of Indiana,* Indiana Department of Conservation Publication, no. 82, 1929. (Illustrated.)

Gould, F. W.: *Grasses of Southwestern United States,* University of Arizona, Tucson, 1951.

————: *Grass Systematics,* McGraw-Hill, New York, 1968. (A textbook on all aspects of grasses including a revised system of classification.)

————: *The Grasses of Texas,* Texas A&M University Press, College Station, 1975.

———— and T. D. Box: *Grasses of the Texas Coastal Bend,* Texas A&M University Press, College Station, 1965.

Harrington, H. D.: *How to Identify Grasses and Grasslike Plants,* Swallow, Chicago, Ill., 1977.

Hitchcock, A. S.: *Manual of Grasses of the United States,* 2d ed., revised by A. Chase, U.S. Department of Agriculture Miscellaneous Publication, no. 200, 1950. (A basic illustrated reference to the grasses; classification system is now out of date. Reprinted by Dover Publications, New York.)

Mohlenbrock, R. H.: *Grasses: Bromus to Paspalum,* Southern Illinois University Press, Carbondale, 1972.

————: *Grasses: Panicum to Danthonia,* Southern Illinois University Press, Carbondale, 1973.

Pohl, R. W.: "The Grasses of Iowa," *Iowa State J. Sci.* **40:**341–566, 1966.

————: *How to Know the Grasses,* rev. ed., Brown, Dubuque, Iowa, 1968.

## LEGUMES

Fassett, N. C.: *The Leguminous Plants of Wisconsin,* University of Wisconsin Press, Madison, 1939.

Gambill, W. G.: "The Leguminosae of Illinois," *Ill. Biol. Monogr.,* **22**(4):1–117, 1953.

Isley, D.: "The Leguminosae of the North-Central States," *Iowa State J. Sci.,* **25:**439–482, 1951; **30:**33–118, 1955; **32:**355–393, 1958; **37:**103–162, 1962.

————: "Leguminosae of the United States: I. Subfamily Mimosoideae; II. Subfamily Caesalpinoideae," *Mem. N.Y. Bot. Gard.,* **25**(1):1–152, 1973; **25**(2):1–228, 1975.

Lasseigne, A.: *Louisiana Legumes,* University of Southwestern Louisiana, Lafayette, 1973.

Mahler, W. F.: "Manual of the Legumes of Tennessee," *J. Tenn. Acad. Sci.,* **45:**65–96, 1970.

Turner, B. L.: *The Legumes of Texas,* University of Texas Press, Austin, 1959.

Wilbur, R. L.: *Leguminous Plants of North Carolina,* North Carolina Agricultural Experiment Station Technical Bulletin, no. 151, 1963. (Illustrations, keys, and descriptions.)

## ORCHIDS

Case, F. W.: *Orchids of the Western Great Lakes Region,* Cranbrook Institute of Science Bulletin, no. 48, 1964. (Keys, descriptions, black-and-white and color photographs.)

Correll, D. S.: *Native Orchids of North America North of Mexico,* Chronica Botanica Company, Waltham, Mass., 1950.

Luer, C. A.: *The Native Orchids of Florida,* New York Botanical Garden, Bronx, 1972. (High-quality, excellent reproduction of color photographs.)

————: *The Native Orchids of the United States and Canada Excluding Florida,* New York Botanical Garden, Bronx, 1975. (Excellent, high-quality color reproduction.)


## POISONOUS PLANTS

Hardin, J. W., and J. M. Arena: *Human Poisoning from Native and Cultivated Plants,* 2d rev. ed., Duke University Press, Durham, N.C., 1974.

Kingsbury, J. M.: *Poisonous Plants of the United States and Canada,* Prentice-Hall, Englewood Cliffs, N.J., 1964. (Contains well-documented information.)


## WEEDS*

Buchholtz, K. P., et al. (eds.): *Weeds of North Central States,* North Central Regional Publication, no. 36, 1954. (Illustrated with line drawings.)

Delorit, R. J.: *An Illustrated Taxonomy Manual of Weed Seeds,* Agronomy Publications, River Falls, Wis., 1970. (Photographs of seeds.)

Frankton, C.: *Weeds of Canada,* rev. ed., Canada Department of Agriculture Publication, no. 948, 1970.

Gilkey, H. M.: *Weeds of the Pacific Northwest,* Oregon State University Press, Corvallis, 1957.

Montgomery, F. H.: *Weeds of Northern United States and Canada,* Warne, New York, 1964.

Muenscher, W. C.: *Weeds,* 2d ed., Macmillan, New York, 1955.

*Nebraska Weeds,* Nebraska Department of Agriculture, Lincoln, 1975. (An excellent color picture book.)

Parker, K. F.: *An Illustrated Guide to Arizona Weeds,* University of Arizona Press, Tucson, 1972.

Reed, C. F.: *Selected Weeds of the United States,* U.S. Department of Agriculture, Agriculture Handbook, no. 366, 1970. (Illustrated; useful for the layperson.)

Wilkinson, R. E., and H. E. Jaques: *How to Know the Weeds,* 2d ed., Brown, Dubuque, Iowa, 1972.


## EDIBLE WILD PLANTS

Fernald, M. L., and A. C. Kinsey: *Edible Wild Plants of Eastern North America,* rev. ed. by R. C. Rollins, Harper and Row, New York, 1958. (The standard book for edible wild plants.)

Peterson, L.: *A Field Guide to Edible Wild Plants of Eastern and Central North America,* Houghton-Mifflin, Boston, 1978. (Peterson Field Guide Series.)

---

*Most states have illustrated publications identifying local weeds. These are available from county agents of the Cooperative Extension Service.

## POPULAR WILD FLOWER BOOKS

Armstrong, M. N.: *Field Book of Western Wild Flowers,* Putnam, New York, 1915. (A popular field guide.)

Batson, W. T.: *Wild Flowers in South Carolina,* The University of South Carolina Press, Columbia, 1964. (Color photos.)

Brown, C. A.: *Wildflowers of Louisiana and Adjoining States,* Louisiana State University Press, Baton Rouge, 1972. (Color photos.)

Budd, A. C.: *Wild Plants of the Canadian Prairies,* Canada Department of Agriculture Publication, no. 983, 1957. (Keys, descriptions, line drawings.)

Campbell, C. C., W. F. Hutson, and A. J. Sharp: *Great Smoky Mountains Wildflowers,* 3d ed., University of Tennessee Press, Knoxville, 1970. (Color photos.)

Clark, L. J., and T. G. Trelawny: *Wild Flowers of the Pacific Northwest from Alaska to Northern California,* Gray's Publishing Ltd., Sidney, B.C., Canada, 1976. (Color photos.)

Craighead, J. J., F. C. Craighead, and R. J. Davis: *A Field Guide to Rocky Mountain Wildflowers,* Houghton Mifflin, Boston, 1963. (Keys, descriptions, line drawings, and color photographs. Peterson Field Guide Series.)

Cuthbert, M. J.: *How to Know the Fall Flowers,* Brown, Dubuque, Iowa, 1948. (Keys, descriptions, line drawings.)

————: *How to Know the Spring Flowers,* Brown, Dubuque, Iowa, 1949. (Keys, descriptions, line drawings.)

Dean, B. E., A. Mason, and J. L. Thomas: *Wild Flowers of Alabama and Adjoining States,* University of Alabama Press, Tuscaloosa, 1973. (Color photos.)

Duncan, W. H., and L. E. Foote: *Wildflowers of the Southeastern United States,* University of Georgia Press, Athens, 1975. (Descriptions, color photos.)

Ferguson, M., and R. M. Saunders: *Wildflowers,* Van Nostrand, New York, 1976.

Fleming, G., P. Genelle, and R. W. Long: *Wild Flowers of Florida,* Banyan Books, Miami, 1976.

Grimm, W. C.: *Recognizing Wild Plants,* Stackpole, Harrisburg, Pa., 1968. (Keys, descriptions, line drawings.)

Haskin, L. L., and L. G. Haskin: *Wild Flowers of the Pacific Coast,* Binford and Mort, Portland, Ore., 1967. (Descriptions, black-and-white and color photos.)

Heller, C.: *Wild Flowers of Alaska,* Graphic Arts Center, Portland, Ore., 1966. (Descriptions, color photos.)

Hinds, H. R., and W. A. Hathaway: *Wildflowers of Cape Cod,* Chatham Press, Chatham, Mass., 1968.

Horn, E. L.: *The Sierra Nevada,* Touchstone Press, Beaverton, Ore., 1976. (Wildflowers in the Sierra Nevada Mountains.)

House, H. D.: *Wild Flowers,* Macmillan, New York, 1961. (Northeastern United States; black-and-white and color photos.)

Hylander, C. J., and E. F. Johnston: *Macmillan Wild Flower Book,* Macmillan, New York, 1954. (Eastern United States, watercolors.)

Jaeger, E. C.: *Desert Wild Flowers,* rev. ed., Stanford University Press, Stanford, Calif., 1969. (Desert plants arranged by family, many illustrations.)

Jaques, H. E.: *Plant Families: How to Know Them,* 2d ed., Brown, Dubuque, Iowa, 1949. (Keys, descriptions, line drawings.)

Justice, W. S., and C. R. Bell: *Wild Flowers of North Carolina,* University of North Carolina Press, Chapel Hill, 1968. (Descriptions, color photos.)

Klein, I. H.: *Wild Flowers of Ohio and Adjacent States,* Cleveland Museum of Natural History, Press of Case Western Reserve University, Cleveland, 1970.

Klimas, J. E., and J. A. Cunningham: *Wildflowers of Eastern North America,* Knopf, New York, 1974. (Color photos.)

Lemmon, R. S., and C. Johnson: *Wildflowers of North America,* Hanover House, Doubleday, Garden City, N.Y., 1961. (Descriptions, color photos.)

Lommasson, R. C.: *Nebraska Wildflowers,* University of Nebraska Press, Lincoln, 1973. (Color photos.)

Monserud, W., and G. B. Ownbey: *Common Wild Flowers of Minnesota,* The University of Minnesota Press, Minneapolis, 1971. (Line drawings.)

Montgomery, F. H.: *Plants from Sea to Sea,* Ryerson Press, Toronto, 1966. (Canada.)

Munz, P. A.: *California Spring Wildflowers,* University of California Press, Berkeley, 1961. (Descriptions, line drawings, color photos.)

————: *California Desert Wildflowers,* University of California Press, Berkeley, 1962. (Descriptions, line drawings, color photos.)

————: *California Mountain Wildflowers,* University of California Press, Berkeley, 1963. (Descriptions, line drawings, color photos.)

————: *Shore Wildflowers of California, Oregon and Washington,* University of California Press, Berkeley, 1964. (Descriptions, line drawings, color photos.)

Nelson, R. A.: *Handbook of Rocky Mountain Plants,* Dale Stuart King, Publisher, Tucson, Ariz., 1969. (Color photos.)

Niehaus, T. F., and C. L. Ripper: *A Field Guide to Pacific States Wildflowers,* Houghton Mifflin, Boston, 1976. (Peterson Field Guide Series.)

Orr, R. T., and M. C. Orr: *Wild Flowers of Western America,* Knopf, New York, 1974. (Descriptions, color photos.)

Page, N. M., and R. E. Weaver: *Wild Plants in the City,* Quadrangle, New York Times Book Company, New York, 1975. (Interesting color photos.)

Peterson, R. T., and M. McKenny: *A Field Guide to Wildflowers of Northeastern and North-central North America,* Houghton Mifflin, Boston, 1968. (Descriptions, line drawings, color photos; Peterson Field Guide Series.)

Petry, L. C., and M. G. Norman: *A Beachcomber's Botany,* Chatham Conservancy Foundation, Chatham, Mass., 1968. (Plants grouped by habitat, black-and-white drawings.)

Porsild, A. E.: *Rocky Mountain Wild Flowers,* National Museums of Canada, Ottawa, 1974.

Rickett, H. W. *The New Field Book of American Wild Flowers,* Putnam, New York, 1963. (Keys, descriptions, line drawings, and color photos.)

————: *Wild Flowers of the United States,* 6 vols., McGraw-Hill, New York, 1966–1975. (A series of volumes with color photographs and descriptions of selected wild flowers; undoubtedly the best of the wild flower photograph books; not suitable for use in the field because of size of the volumes.)

Smith, H. V.: *Michigan Wildflowers,* rev. ed., Cranbrook Institute of Science Bulletin, no. 42, 1966. (Color photos.)

Stevens, W. C.: *Kansas Wild Flowers,* 2d ed., University of Kansas Press, Lawrence, 1961. (Color photos.)

Stupka, A.: *Wildflowers in Color,* Harper and Row, New York, 1965. (Southern Appalachians; color photos.)

Taylor, R. J., and G. W. Douglas: *Moutain Wild Flowers of the Pacific Northwest,* Binford and Mort, Portland, Ore., 1975.

Thompson, E. R.: *Wildflower Portraits,* University of Oklahoma Press, Norman, 1964. (Central United States, especially Texas; watercolors.)

Welsh, S. L.: *Flowers of the Canyon Country,* rev. ed., Brigham Young University Press, Provo, Utah, 1977.

Wharton, M. E., and R. W. Barbour: *A Guide to the Wildflowers and Ferns of Kentucky,* University Press of Kentucky, Lexington, 1971.

Wills, M. M., and H. S. Irwin: *Roadside Flowers of Texas,* University of Texas Press, Austin, 1961. (Keys, notes, descriptions, watercolor drawings.)

## WOODY PLANTS

Baerg, H. J.: *How to Know the Western Trees,* 2d ed., Brown, Dubuque, Iowa, 1973.

Benson, L., and R. A. Darrow: *The Trees and Shrubs of the Southwestern Deserts,* 2d ed., University of Arizona Press, Tucson, and The University of New Mexico Press, Albuquerque, 1954.

Blackburn, B. C.: *Trees and Shrubs in Eastern North America,* Oxford University Press, New York, 1952. (Keys to native and horticultural species, excludes conifers.)

Blackwell, W. H.: *Guide to the Woody Plants of the Tri-State Area, Southwestern Ohio, Southern Indiana, and Northern Kentucky,* Kendall-Hunt, Dubuque, Iowa, 1976.

Braun, E. L.: *The Woody Plants of Ohio,* Ohio State University Press, Columbus, 1961. (Good line drawings, maps, useful keys.)

Brockman, C. F.: *Trees of North America,* Golden Press, New York, 1968. (An inexpensive and popular field guide to the common trees north of Mexico; colored drawings, distribution maps.)

Brown, R. G., and M. L. Brown: *Woody Plants of Maryland,* Student Supply Store, University of Maryland, College Park, 1972. (Illustrated.)

Clark, R. C.: "The woody plants of Alabama," *Ann. Mo. Bot. Gard.,* **58:**99–242, 1971. (Keys, notes, and distribution maps.)

Coker, W. C., and H. R. Totten: *Trees of the Southeastern States,* 3d ed., University of North Carolina Press, Chapel Hill, 1945. (Comments, illustrations, keys, descriptions.)

Core, E. L., and N. P. Ammons: *Woody Plants in Winter: A Manual of Common Trees and Shrubs in Winter in the Northeastern United States and Southeastern Canada,* Boxwood Press, Pittsburgh, 1958. (Keys, line drawings.)

Deam, C. C., and T. E. Shaw: *Trees of Indiana,* 3d rev. ed., Indiana Department of Conservation Publication, no. 13a, 1953.

Dirr, M. A.: *Photographic Manual of Woody Landscape Plants,* Stipes, Champaign, Ill., 1978. (Illustrating form and function of woody landscape plants.)

Duncan, W. H.: *Woody Vines of the Southeastern States,* University of Georgia Press, Athens, 1975. (Distribution maps, keys.)

Ferris, R. S.: *Native Shrubs of the San Francisco Bay Region,* University of California Press, Berkeley, 1968. (A popular illustrated guide useful in much of northern California.)

Grimm, W. C.: *The Book of Trees,* 2d ed., Stackpole, Harrisburg, Pa., 1962. (Keys, black-and-white illustrations.)

————: *Recognizing Native Shrubs,* Stackpole, Harrisburg, Pa., 1966. (Keys, line drawings.)

Harlow, W. M.: *Trees of the Eastern and Central United States and Canada,* corrected ed., Dover Publications, New York, 1957. (Keys, descriptions, black-and-white photos.)

————: *Fruit Key and Twig Key to Trees and Shrubs,* Dover Publications, New York, 1959. (Keys, illustrations.)

Harrar, E. S., and J. G. Harrar: *Guide to Southern Trees,* 2d ed., Dover Publications, New York, 1962. (Illustrated.)

Hosie, R. C.: *Native Trees of Canada,* 7th ed., Queen's Printer, Ottawa, 1969. (Excellent illustrated treatment of the trees of Canada.)

Jepson, W. L.: *The Trees of California,* 2d ed., Sather Gate Bookshop, Berkeley, 1923.

Little, E. L., and B. H. Honkala: *Trees and Shrubs of the United States—A Bibliography for Identification,* U.S. Department of Agriculture Miscellaneous Publication, no. 1336, 1976. (A bibliography for identification.)

Kurz, H., and R. K. Godfrey: *Trees of Northern Florida,* University of Florida Press, Gainesville, 1962. (Illustrations, keys.)

McMinn, H. E.: *An Illustrated Manual of California Shrubs,* University of California Press, Berkeley, 1964. (Line drawings.)

———— and E. Maino: *An Illustrated Manual of Pacific Coast Trees,* 2d ed., University of California Press, Berkeley, 1937. (A manual of Pacific Coast trees, both native and introduced.)

Meijer, W.: *Tree Flora of Kentucky,* Thomas Hunt Morgan School Biological Sciences, University of Kentucky, Lexington, 1971. (Keys.)

Muenscher, W. C.: *Keys to Woody Plants,* 6th ed., Comstock, Ithaca, N.Y., 1950.

Otis, C. H.: *Michigan Trees,* 9th ed., The Regents, Ann Arbor, Mich., 1931.

Petrides, G. A.: *A Field Guide to Trees and Shrubs,* 2d ed., Houghton Mifflin, Boston, 1972. (Peterson Field Guide Series.)

Preston, R. J.: *North American Trees,* 3d ed., Iowa State University Press, Ames, 1976.

Randall, W. R.: *Manual of Oregon Trees and Shrubs,* Oregon State University Bookstore, Corvallis, 1958. (Revised by R. F. Keniston, 1968.)

Raven, P. H.: *Native Shrubs of Southern California,* University of California Press, Berkeley, 1966. (Keys, illustrations.)

Rosendahl, C. O.: *Trees and Shrubs of the Upper Mid-west,* University of Minnesota Press, Minneapolis, 1955.

Sargent, C. S.: *Manual of the Trees of North America,* 2d corrected ed., 2 vols., Dover, New York, reprinted 1965. (Technical descriptions, line drawings.)

Stephens, H. A.: *Woody Plants of the North Central States,* University Press of Kansas, Lawrence, 1973. (Contains excellent illustrations, descriptions, distribution maps.)

Sudword, G. B.: *Forest Trees of the Pacific Slope,* Dover, New York, 1967. (Reprint of 1908 edition; illustrations, descriptions, no keys.)

Symonds, G. W. D.: *The Shrub Identification Book,* Barrows, New York, 1963. (Photographs of leaves, flowers, twigs, and so on.)

Thomas, J. H., and D. R. Parnell: *Native Shrubs of the Sierra Nevada,* University of California Press, Berkeley, 1974. (An illustrated guide.)

Trelease, W.: *Winter Botany,* 3d ed., Dover, New York, 1967. (Reprint of 1931 edition; keys to trees in the winter condition.)

Treshow, M., S. L. Welsh, and G. Moore: *Guide to the Woody Plants of the Mountain States,* Brigham Young University Press, Provo, Utah, 1970. (Includes both summer and winter key to genera.)

Vance, F. R.: *Wildflowers Across the Prairies,* Western Producer Prairie Books, Saskatoon, Sask., 1977.

Viereck, L. A., and E. L. Little: *Alaska Trees and Shrubs,* U.S. Department of Agriculture, Agriculture Handbook, no. 410, 1972. (Illustrations, maps.)

———— and ————: *Guide to Alaska Trees,* U.S. Department of Agriculture, Agriculture Handbook, no. 472, 1974.

Vines, R. A.: *Trees, Shrubs and Woody Vines of the Southwest,* University of Texas Press, Austin, 1960.

————: *Trees of East Texas,* University of Texas Press, Austin, 1977.

Wharton, M. E., and R. W. Barbour: *Trees and Shrubs of Kentucky,* University Press of Kentucky, Lexington, 1973.

## SELECTED PERIODICALS OF INTEREST TO PLANT TAXONOMISTS IN NORTH AMERICA

Much research and many taxonomic revisions are published in periodicals. Scientific publications are frequently published by a learned society, such as the American Society of Plant Taxonomists, or by herbaria or botanical gardens, such as the Field Museum or Missouri Botanical Garden. Some of the more important periodicals containing systematic research on plants from North America are listed below. These periodical citations are in the form customarily used in libraries.

*Aliso.* Journal of the Rancho Santa Ana Botanic Garden. 1948–
*American Fern Journal.* American Fern Society. 1910–
*American Journal of Botany.* Botanical Society of America. 1914–
*American Midland Naturalist.* 1909–
*Brittonia.* New York Botanical Garden. 1931–
*California. University. University of California Publications in Botany.* 1902–
*Canadian Journal of Botany.* 1951–
*Castanea.* Southern Appalachian Botanical Club. 1936–
*Evolution.* Society of the Study of Evolution. 1947–
*Field Museum of Natural History, Chicago. Fieldiana: Botany.* 1895–
*Harvard University. Arnold Arboretum. Journal of the Arnold Arboretum.* 1920–
*Harvard University. Gray Herbarium. Contributions.* 1891–
*Madroño.* California Botanical Society. 1916–
*New York. Botanical Garden. Memoirs.* 1900–
*Rhodora.* New England Botanical Club. 1899–
*St. Louis. Missouri Botanical Garden. Annals.* 1914–
*Sida.* 1962–
*Systematic Botany.* American Society of Plant Taxonomists. 1976–
*Taxon.* International Association for Plant Taxonomy. 1951–
*Torrey Botanical Club. Bulletin.* 1870–
*Torrey Botanical Club. Memoirs.* 1889–
*U.S. National Herbarium. Contributions.* 1891–

# A List of the Classes, Subclasses, Orders, and Families of the Angiosperms According to the Cronquist System of Classification*

Division MAGNOLIOPHYTA

Class MAGNOLIOPSIDA (Dicots)

Subclass I. Magnoliidae
  Order 1. Magnoliales
    Family  1. Winteraceae
           2. Degeneriaceae
           3. Himantandraceae
           4. Magnoliaceae
           5. Lactoridaceae
           6. Austrobaileyaceae
           7. Eupomatiaceae
           8. Annonaceae
           9. Myristicaceae
         10. Canellaceae

Order 2. Laurales
    Family  1. Amborellaceae
           2. Trimeniaceae
           3. Monimiaceae
           4. Gomortegaceae
           5. Calycanthaceae
           6. Idiospermaceae
           7. Lauraceae
           8. Hernandiaceae
Order 3. Piperales
    Family  1. Chloranthaceae
           2. Saururaceae
           3. Piperaceae
Order 4. Aristolochiales
    Family  1. Aristolochiaceae

*This treatment has been very kindly furnished by Dr. Arthur Cronquist and is used with his permission.

Order 5.  Illiciales
Family   1.  Illiciaceae
            2.  Schisandraceae
Order 6.  Nymphaeales
Family   1.  Nelumbonaceae
            2.  Nymphaeaceae
            3.  Cabombaceae
            4.  Ceratophyllaceae
Order 7.  Ranunculales
Family   1.  Ranunculaceae
            2.  Circaeasteraceae
            3.  Berberidaceae
            4.  Sargentodoxaceae
            5.  Lardizabalaceae
            6.  Menispermaceae
            7.  Coriariaceae
            8.  Corynocarpaceae
            9.  Sabiaceae
Order 8.  Papaverales
Family   1.  Papaveraceae
            2.  Fumariaceae

Subclass II.  Hamamelidae
Order 1.  Trochodendrales
Family   1.  Tetracentraceae
            2.  Trochodendraceae
Order 2.  Hamamelidales
Family   1.  Cercidiphyllaceae
            2.  Eupteliaceae
            3.  Platanaceae
            4.  Hamamelidaceae
Order 3.  Daphniphyllales
Family   1.  Daphniphyllaceae
Order 4.  Didymelales
Family   1.  Didymelaceae
Order 5.  Eucommiales
Family   1.  Eucommiaceae
Order 6.  Urticales
Family   1.  Barbeyaceae
            2.  Ulmaceae
            3.  Moraceae
            4.  Cannabaceae
            5.  Urticaceae
Order 7.  Leitneriales
Family   1.  Leitneriaceae
Order 8.  Juglandales
Family   1.  Rhoipteleaceae
            2.  Juglandaceae
Order 9.  Myricales
Family   1.  Myricaceae

Order 10.  Fagales
Family   1.  Balanopaceae
            2.  Fagaceae
            3.  Betulaceae
Order 11.  Casuarinales
Family   1.  Casuarinaceae

Subclass III.  Caryophyllidae
Order 1.  Caryophyllales
Family   1.  Phytolaccaceae
            2.  Achatocarpaceae
            3.  Nyctaginaceae
            4.  Aizoaceae
            5.  Didiereaceae
            6.  Cactaceae
            7.  Chenopodiaceae
            8.  Amaranthaceae
            9.  Portulacaceae
           10.  Basellaceae
           11.  Molluginaceae
           12.  Caryophyllaceae
Order 2.  Polygonales
Family   1.  Polygonaceae
Order 3.  Plumbaginales
Family   1.  Plumbaginaceae

Subclass IV.  Dilleniidae
Order 1.  Dilleniales
Family   1.  Dilleniaceae
            2.  Paeoniaceae
            3.  Crossosomataceae
Order 2.  Theales
Family   1.  Ochnaceae
            2.  Sphaerosepalaceae
            3.  Sarcolaenaceae
            4.  Dipterocarpaceae
            5.  Stachyuraceae
            6.  Caryocaraceae
            7.  Theaceae
            8.  Actinidiaceae
            9.  Pentaphylacaceae
           10.  Oncothecaceae
           11.  Tetrameristaceae
           12.  Pellicieraceae
           13.  Marcgraviaceae
           14.  Quiinaceae
           15.  Elatinaceae
           16.  Medusagynaceae
           17.  Clusiaceae (Guttiferae)
Order 3.  Malvales
Family   1.  Elaeocarpaceae

2. Scytopetalaceae
3. Huaceae
4. Tiliaceae
5. Sterculiaceae
6. Bombacaceae
7. Malvaceae
Order 4. Lecythidales
Family   1. Lecythidaceae
Order 5. Nepenthales
Family   1. Sarraceniaceae
2. Nepenthaceae
3. Droseraceae
Order 6. Violales
Family   1. Flacourtiaceae
2. Peridiscaceae
3. Bixaceae
4. Cistaceae
5. Lacistemataceae
6. Scyphostegiaceae
7. Violaceae
8. Tamaricaceae
9. Frankeniaceae
10. Dioncophyllaceae
11. Ancistrocladaceae
12. Turneraceae
13. Malesherbiaceae
14. Passifloraceae
15. Fouquieriaceae
16. Hoplestigmataceae
17. Achariaceae
18. Caricaceae
19. Cucurbitaceae
20. Datisaceae
21. Begoniaceae
22. Loasaceae
Order 7. Salicales
Family 1. Salicaceae
Order 8. Capparales
Family   1. Tovariaceae
2. Capparaceae
3. Brassicaceae (Cruciferae)
4. Moringaceae
5. Resedaceae
Order 9. Batales
Family   1. Gyrostemonaceae
2. Bataceae
Order 10. Ericales
Family   1. Cyrillaceae
2. Clethraceae
3. Grubbiaceae

4. Empetraceae
5. Epacridaceae
6. Ericaceae
7. Pyrolaceae
8. Monotropaceae
Order 11. Diapensiales
Family   1. Diapensiaceae
Order 12. Ebenales
Family   1. Sapotaceae
2. Ebenaceae
3. Styracaceae
4. Lissocarpaceae
5. Symplocaceae
Order 13. Primulales
Family   1. Theophrastaceae
2. Myrsinaceae
3. Primulaceae

Subclass V. Rosidae
Order 1. Rosales
Family   1. Brunelliaceae
2. Connaraceae
3. Eucryphiaceae
4. Cunoniaceae
5. Davidsoniaceae
6. Dialypetalanthaceae
7. Pittosporaceae
8. Byblidaceae
9. Hydrangeaceae
10. Columelliaceae
11. Grossulariaceae
12. Greyiaceae
13. Bruniaceae
14. Anisophylleaceae
15. Alseuosmiaceae
16. Crassulaceae
17. Cephalotaceae
18. Saxifragaceae
19. Donatiaceae
20. Rosaceae
21. Neuradaceae
22. Chrysobalanaceae
23. Surianaceae
24. Rhabdodendraceae
Order 2. Fabales
Family   1. Mimosaceae
2. Caesalpiniaceae
3. Fabaceae
Order 3. Podostemales
Family   1. Podostemaceae

Order 4. Haloragales
Family  1. Haloragaceae
          2. Gunneraceae
Order 5. Myrtales
Family  1. Sonneratiaceae
          2. Lythraceae
          3. Penaeaceae
          4. Crypteroniaceae
          5. Thymelaeaceae
          6. Trapaceae
          7. Myrtaceae
          8. Punicaceae
          9. Onagraceae
         10. Oliniaceae
         11. Melastomataceae
         12. Combretaceae
Order 6. Proteales
Family  1. Elaeagnaceae
          2. Proteaceae
Order 7. Rhizophorales
Family  1. Rhizophoraceae
Order 8. Cornales
Family  1. Alangiaceae
          2. Nyssaceae
          3. Cornaceae
          4. Garryaceae
Order 9. Santalales
Family  1. Medusandraceae
          2. Dipentodontaceae
          3. Olacaceae
          4. Opiliaceae
          5. Santalaceae
          6. Misodendraceae
          7. Loranthaceae
          8. Viscaceae
          9. Eremolepidaceae
         10. Balanophoraceae
Order 10. Rafflesiales
Family  1. Hydnoraceae
          2. Mitrastemonaceae
          3. Rafflesiaceae
Order 11. Celastrales
Family  1. Geissolomataceae
          2. Celastraceae
          3. Hippocrateaceae
          4. Stackhousiaceae
          5. Salvadoraceae
          6. Aquifoliaceae
          7. Icacinaceae
          8. Aextoxicaceae

          9. Cardiopteridaceae
         10. Dichapetalaceae
Order 12. Euphorbiales
Family  1. Buxaceae
          2. Simmondsiaceae
          3. Pandaceae
          4. Euphorbiaceae
Order 13. Rhamnales
Family  1. Rhamnaceae
          2. Leeaceae
          3. Vitaceae
Order 14. Linales
Family  1. Erythroxylaceae
          2. Humiriaceae
          3. Ixonanthaceae
          4. Hugoniaceae
          5. Linaceae
Order 15. Polygalales
Family  1. Malpighiaceae
          2. Vochysiaceae
          3. Trigoniaceae
          4. Tremandraceae
          5. Polygalaceae
          6. Xanthophyllaceae
          7. Krameriaceae
Order 16. Sapindales
Family  1. Staphyleaceae
          2. Melianthaceae
          3. Bretschneideraceae
          4. Akaniaceae
          5. Sapindaceae
          6. Hippocastanaceae
          7. Aceraceae
          8. Burseraceae
          9. Anacardiaceae
         10. Julianiaceae
         11. Simaroubaceae
         12. Cneoraceae
         13. Meliaceae
         14. Rutaceae
         15. Zygophyllaceae
Order 17. Geraniales
Family  1. Oxalidaceae
          2. Geraniaceae
          3. Limnanthaceae
          4. Tropaeolaceae
          5. Balsaminaceae
Order 18. Apiales
Family  1. Araliaceae
          2. Apiaceae (Umbelliferae)

Subclass VI. Asteridae
  Order 1. Gentianales
    Family 1. Loganiaceae
         2. Gentianaceae
         3. Saccifoliaceae Maguire ined.
         4. Apocynaceae
         5. Asclepiadaceae
  Order 2. Solanales (Polemoniales)
    Family 1. Duckeodendraceae
         2. Nolanaceae
         3. Solanaceae
         4. Convolvulaceae
         5. Cuscutaceae
         6. Menyanthaceae
         7. Retziaceae
         8. Polemoniaceae
         9. Hydrophyllaceae
  Order 3. Lamiales
    Family 1. Lennoaceae
         2. Boraginaceae
         3. Verbenaceae
         4. Lamiaceae (Labiatae)
  Order 4. Callitrichales
    Family 1. Hippuridaceae
         2. Callitrichaceae
         3. Hydrostachyaceae
  Order 5. Plantaginales
    Family 1. Plantaginaceae
  Order 6. Scrophulariales
    Family 1. Buddlejaceae
         2. Oleaceae
         3. Scrophulariaceae
         4. Globulariaceae
         5. Myoporaceae
         6. Orobanchaceae
         7. Gesneriaceae
         8. Acanthaceae
         9. Pedaliaceae
        10. Bignoniaceae
        11. Mendonciaceae
        12. Lentibulariaceae
  Order 7. Campanulales
    Family 1. Pentaphragmataceae
         2. Sphenocleaceae
         3. Campanulaceae
         4. Stylidiaceae
         5. Brunoniaceae
         6. Goodeniaceae
  Order 8. Rubiales
    Family 1. Rubiaceae

         2. Theligonaceae
  Order 9. Dipsacales
    Family 1. Caprifoliaceae
         2. Adoxaceae
         3. Valerianaceae
         4. Dipsacaceae
         5. Calyceraceae
  Order 10. Asterales
    Family 1. Asteraceae (Compositae)

Class LILIOPSIDA (Monocots)

Subclass I. Alismatidae
  Order 1. Alismatales
    Family 1. Butomaceae
         2. Limnocharitaceae
         3. Alismataceae
  Order 2. Hydrocharitales
    Family 1. Hydrocharitaceae
  Order 3. Najadales
    Family 1. Aponogetonaceae
         2. Scheuchzeriaceae
         3. Juncaginaceae
         4. Potamogetonaceae
         5. Ruppiaceae
         6. Zosteraceae
         7. Najadaceae
         8. Zannichelliaceae
         9. Cymodoceaceae
  Order 4. Triuridales
    Family 1. Petrosaviaceae
         2. Triuridaceae

Subclass II. Arecidae
  Order 1. Arecales
    Family 1. Arecaceae (Palmae)
  Order 2. Cyclanthales
    Family 1. Cyclanthaceae
  Order 3. Pandanales
    Family 1. Pandanaceae
  Order 4. Arales
    Family 1. Araceae
         2. Lemnaceae

Subclass III. Commelinidae
  Order 1. Commelinales
    Family 1. Rapataeaceae
         2. Xyridaceae
         3. Mayacaceae
         4. Commelinaceae
  Order 2. Eriocaulales
    Family 1. Eriocaulaceae

Order 3. Restionales
  Family  1. Flagellariaceae
            2. Joinvilleaceae
            3. Restionaceae
            4. Centrolepidaceae
            5. Hydatellaceae
Order 4. Juncales
  Family  1. Juncaceae
            2. Thurniaceae
Order 5. Cyperales
  Family  1. Cyperaceae
            2. Poaceae (Gramineae)
Order 6. Typhales
  Family  1. Sparganiaceae
            2. Typhaceae

Subclass IV. Zingiberidae
Order 1. Bromeliales
  Family  1. Bromeliaceae
Order 2. Zingiberales
  Family  1. Strelitziaceae
            2. Heliconiaceae
            3. Musaceae
            4. Lowiaceae
            5. Zingiberaceae
            6. Costaceae
            7. Cannaceae
            8. Marantaceae

Subclass V. Liliidae
Order 1. Liliales
  Family  1. Philydraceae
            2. Pontederiaceae
            3. Haemodoraceae
            4. Cyanastraceae
            5. Liliaceae
            6. Iridaceae
            7. Velloziaceae
            8. Aloeaceae
            9. Agavaceae
          10. Xanthorrhoeaceae
          11. Hanguanaceae
          12. Taccaceae
          13. Stemonaceae
          14. Smilacaceae
          15. Dioscoreaceae
Order 2. Orchidales
  Family  1. Geosiridaceae
            2. Burmanniaceae
            3. Corsiaceae
            4. Orchidaceae

# Classification of the Angiosperms According to the Thorne System

The diagram on the following page shows R. F. Thorne's classification of the flowering plants.* Thorne first divides the angiosperms into the dicots and the monocots; these in turn are subdivided into superorders (heavy black lines in the diagram) and orders (groups within the heavy black lines). Thorne's classification—which is more complicated than the diagrams of Cronquist and Takhtajan shown in Chapter 6 (Figures 6-11 and 6-13)—is considered by some taxonomists to be an excellent graphic representation of angiosperm phylogeny.

*This diagram was furnished by R. F. Thorne and is used with his permission.

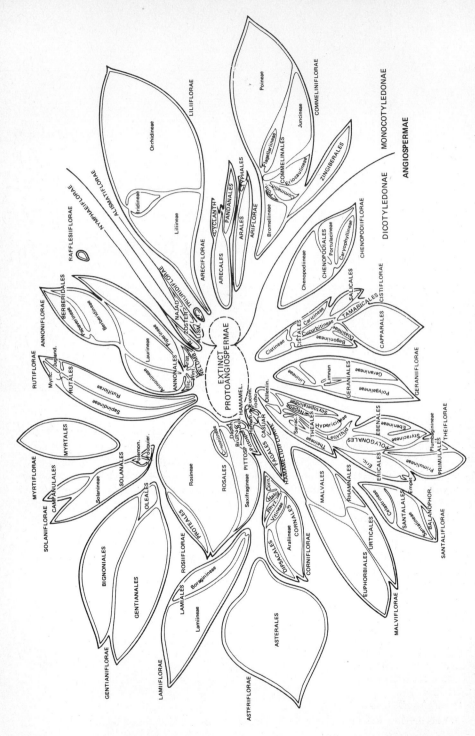

**Figure Appendix 4**  Classification of the angiosperms according to Thorne's system.

362

# Index of Family Descriptions

# Index to Taxa

Page numbers in *italic* indicate illustrations.

Abelia, 310
Abies, 227, 228
Abrus, 275, 276
Abutilon, 256, 257
Acacia, 273
Acalypha, 280
Acer, 285
Aceraceae, 50
  described, 284–285
Achillea, 119, 120
Acnida, 251
Acokanthera, 293
Aconitum, 240
Acorus, 318
Acrostichum, 215
Adiantaceae, described, 215
Adiantum, 215
Aeluropodeae, 325
Aesculus, 284
Agalinis, 306
Agavaceae, 66
Agave, 168
Aglaonema, 318

Agropyron, 325
Agrostis, 325
Ajuga, 303, 304
Albizia, 273
Aleurites, 280
Alisma, 315
Alismataceae:
  described, 314–315
  table, 234
Alismatales, 314–315
Alismatidae, 107, 314–316
Allamanda, 293
Allium, 328–329
Allotropa, 264
Alnus, 248
Alocasia, 318
Aloe, 327
Aloeaceae, 327
Alternanthera, 251
Althaea, 256, 257
Alyssum, 260
Amaranthaceae, described, 250–251
Amaranthus, 251

365

# Subject Index

Page numbers in *italic* indicate illustrations.